Linear and Nonlinear Inverse Problems with Practical Applications

Computational Science & Engineering

The SIAM series on Computational Science and Engineering publishes research monographs, advanced undergraduate- or graduate-level textbooks, and other volumes of interest to an interdisciplinary CS&E community of computational mathematicians, computer scientists, scientists, and engineers. The series includes both introductory volumes aimed at a broad audience of mathematically motivated readers interested in understanding methods and applications within computational science and engineering and monographs reporting on the most recent developments in the field. The series also includes volumes addressed to specific groups of professionals whose work relies extensively on computational science and engineering.

SIAM created the CS&E series to support access to the rapid and far-ranging advances in computer modeling and simulation of complex problems in science and engineering, to promote the interdisciplinary culture required to meet these large-scale challenges, and to provide the means to the next generation of computational scientists and engineers.

Editor-in-Chief
Donald Estep
Colorado State University

Editorial Board

Omar Ghattas
University of Texas at Austin

Max Gunzburger
Florida State University

Des Higham
University of Strathclyde

Michael Holst
University of California, San Diego

David Keyes
Columbia University and KAUST

Max D. Morris
Iowa State University

Alex Pothen
Purdue University

Padma Raghavan
Pennsylvania State University

Karen Willcox
Massachusetts Institute of Technology

Series Volumes

Mueller, Jennifer L. and Siltanen, Samuli, *Linear and Nonlinear Inverse Problems with Practical Applications*

Shapira, Yair, *Solving PDEs in C++: Numerical Methods in a Unified Object-Oriented Approach, Second Edition*

Borzì, Alfio and Schulz, Volker, *Computational Optimization of Systems Governed by Partial Differential Equations*

Ascher, Uri M. and Greif, Chen, *A First Course in Numerical Methods*

Layton, William, *Introduction to the Numerical Analysis of Incompressible Viscous Flows*

Ascher, Uri M., *Numerical Methods for Evolutionary Differential Equations*

Zohdi, T. I., *An Introduction to Modeling and Simulation of Particulate Flows*

Biegler, Lorenz T., Ghattas, Omar, Heinkenschloss, Matthias, Keyes, David, and van Bloemen Waanders, Bart, Editors, *Real-Time PDE-Constrained Optimization*

Chen, Zhangxin, Huan, Guanren, and Ma, Yuanle, *Computational Methods for Multiphase Flows in Porous Media*

Shapira, Yair, *Solving PDEs in C++: Numerical Methods in a Unified Object-Oriented Approach*

JENNIFER L. MUELLER

Colorado State University
Fort Collins, Colorado

SAMULI SILTANEN

University of Helsinki
Helsinki, Finland

Linear and Nonlinear Inverse Problems with Practical Applications

Society for Industrial and Applied Mathematics
Philadelphia

Copyright © 2012 by the Society for Industrial and Applied Mathematics

10 9 8 7 6 5 4 3 2 1

All rights reserved. Printed in the United States of America. No part of this book may be reproduced, stored, or transmitted in any manner without the written permission of the publisher. For information, write to the Society for Industrial and Applied Mathematics, 3600 Market Street, 6th Floor, Philadelphia, PA 19104-2688 USA.

Trademarked names may be used in this book without the inclusion of a trademark symbol. These names are used in an editorial context only; no infringement of trademark is intended.

MATLAB is a registered trademark of The MathWorks, Inc. For MATLAB product information, please contact The MathWorks, Inc., 3 Apple Hill Drive, Natick, MA 01760-2098 USA, 508-647-7000, Fax: 508-647-7001, info@mathworks.com, www.mathworks.com.

Orthopantomograph OP200 D and Volumetric Tomography device are registered trademarks of Instrumentarium Dental.

T-scan is a registered trademark of Tekscan, Inc.

Figures 9.14-15, 9.18-19 reprinted with permission from Palodex Group.
Figure 14.1 reprinted with permission from Alexandra Bellow.
Figure 16.3 © 2009 IEEE. Reprinted with permission from IEEE Transactions on Medical Imaging.
Figures 16.4, 16.6, 16.11, and 16.12 reprinted with permission from IOP Science.
Figures 16.13-15 reprinted with permission from American Institute of Mathematical Sciences.

Library of Congress Cataloging-in-Publication Data

Mueller, Jennifer (Jennifer L.) author.
  Linear and nonlinear inverse problems with practical applications / Jennifer Mueller Colorado State University, Fort Collins, Colorado [and] Samuli Siltanen, University of Helsinki, Helsinki, Finland.
     pages cm. – (Computational science and engineering series)
  Includes bibliographical references and index.
  ISBN 978-1-611972-33-7
  1. Inverse problems (Differential equations) I. Siltanen, Samuli, author. II. Title.
  QA378.5.M84 2012
  515'.357–dc23
                              2012020877

 is a registered trademark.

For Marcus, Katelyn, Melissa, and Carsten
-JM

For Airi
-SS

Contents

Preface		xi
I	**Linear Inverse Problems**	**1**
1	**Introduction**	**3**
2	**Naïve reconstructions and inverse crimes**	**7**
	2.1 Convolution	7
	2.2 Heat propagation	14
	2.3 Tomographic X-ray projection data	21
3	**Ill-posedness in inverse problems**	**35**
	3.1 Forward map and Hadamard's conditions	35
	3.2 Ill-posedness of the backward heat equation	36
	3.3 Ill-posedness in the continuous case	40
	3.4 Regularized inversion	47
	3.5 The SVD for matrices	49
	3.6 SVD for the guiding examples	51
4	**Truncated singular value decomposition**	**53**
	4.1 Minimum norm solution	53
	4.2 Truncated SVD	55
	4.3 Measuring the quality of reconstructions	56
	4.4 TSVD for the guiding examples	57
5	**Tikhonov regularization**	**63**
	5.1 Classical Tikhonov regularization	63
	5.2 Normal equations and stacked form	67
	5.3 Generalized Tikhonov regularization	69
	5.4 Choosing the regularization parameter	72
	5.5 Large-scale implementation	78
6	**Total variation regularization**	**83**
	6.1 What is total variation?	83
	6.2 Quadratic programming	86

		6.3	Sparsity-based parameter choice	88
		6.4	Large-scale implementation	90

7 Besov space regularization using wavelets — 95
- 7.1 An introduction to wavelets 95
- 7.2 Besov spaces and wavelets 98
- 7.3 Using B^1_{11} regularization to promote sparsity 99

8 Discretization-invariance — 103
- 8.1 Tikhonov regularization and discretizations 104
- 8.2 Total variation regularization and discretizations 107
- 8.3 Besov norm regularization and discretizations 108

9 Practical X-ray tomography with limited data — 111
- 9.1 Sparse full-angle tomography 114
- 9.2 Limited-angle tomography 119
- 9.3 Low-dose three-dimensional dental X-ray imaging 122

10 Projects — 131
- 10.1 Image deblurring 132
- 10.2 Inversion of the Laplace transform 133
- 10.3 Backward parabolic problem 133

II Nonlinear Inverse Problems — 137

11 Nonlinear inversion — 139
- 11.1 Analysis of nonlinear ill-posedness 140
- 11.2 Nonlinear regularization 143
- 11.3 Computational inversion 144
- 11.4 Examples of nonlinear inverse problems 145

12 Electrical impedance tomography — 159
- 12.1 Applications of EIT 160
- 12.2 Derivation from Maxwell's equations 162
- 12.3 Continuum model boundary measurements 163
- 12.4 Nonlinearity of EIT 165
- 12.5 Ill-posedness of EIT 165
- 12.6 Electrode models 170
- 12.7 Current patterns and distinguishability 173
- 12.8 Further reading 180

13 Simulation of noisy EIT data — 185
- 13.1 Eigenvalue data for symmetric σ 185
- 13.2 Continuum model data and FEM 187
- 13.3 Complete electrode model and FEM 191
- 13.4 Adding noise to EIT data matrices 197

14	**Complex geometrical optics solutions**	**199**
	14.1 Calderón's pioneering work	199
	14.2 The $\bar{\partial}$ operator and its kin	203
	14.3 CGO solutions for the Schrödinger equation	205
	14.4 CGO solutions for the Beltrami equation	215
15	**A regularized D-bar method for direct EIT**	**223**
	15.1 Reconstruction with infinite-precision data	224
	15.2 Regularization via nonlinear low-pass filtering	231
	15.3 Numerical solution of the boundary integral equation	234
	15.4 Numerical solution of the D-bar equation	237
	15.5 Regularized reconstructions	242
16	**Other direct solution methods for EIT**	**249**
	16.1 D-bar methods with approximate scattering transforms	249
	16.2 Calderón's method	259
	16.3 The Astala–Päivärinta method	266
	16.4 The enclosure method of Ikehata	277
17	**Projects**	**281**
	17.1 Enclosure method for EIT	281
	17.2 The D-bar method with Born approximation	283
	17.3 Calderón's method	285
	17.4 Inverse obstacle scattering	286
A	**Banach spaces and Hilbert spaces**	**291**
B	**Mappings and compact operators**	**293**
C	**Fourier transform and Sobolev spaces**	**297**
	C.1 Sobolev spaces on domains $\Omega \subset \mathbb{R}^n$	297
	C.2 Fourier series and spaces $H^s(\partial\Omega)$	301
	C.3 Traces of functions in $H^m(\Omega)$	306
D	**Iterative solution of linear equations**	**307**
Bibliography		**311**
Index		**349**

Preface

Inverse problems arise from the need to interpret indirect and incomplete measurements. As an area of contemporary mathematics, the field of inverse problems is strongly driven by applications and has been growing steadily in the past 30 years. This growth has been fostered both by advances in computation and by theoretical breakthroughs. Modern digital sensors provide vast amounts of data related to diverse areas including engineering, geophysics, medicine, biology, physics, chemistry, and finance. As a result, the need for computational inversion can be expected to increase in the future.

The main goal of this book is to provide a practical introduction to inverse problems from both a computational and theoretical perspective. A solid theoretical framework is mandatory for understanding why *ill-posed inverse problems* require a different set of solution methods than well-posed problems. Ill-posedness is related to interpretation tasks that are extremely sensitive to measurement and modeling errors. On the other hand, solving an inverse problem involves the implementation of a computational algorithm that recovers useful information from measured data (the word "useful" can best be understood in the context of a particular application). A successful inversion algorithm is robust against measurement noise, computationally effective, and mathematically justified by appropriate analysis and theorems.

Much of the literature on computational inversion considers tailored methods for linear problems (such as filtered back-projection for X-ray tomography) and generic iterative methods for nonlinear problems (such as Tikhonov regularization with nonlinear objective function). However, in this book we do exactly the opposite: we discuss a unified solution framework for linear problems and tailored direct methods for nonlinear problems. Our rationale is the following:

- Linear inverse problems are all essentially alike since they are completely described by the singular value expansion of the forward map. Thus it makes sense to apply a general methodology designed for complementing measurement data with a priori information, for example, by enforcing nonnegativity or by promoting sparsity in a basis.

- Nonlinear inverse problems are all different and need dedicated solution methods. One way to proceed is to use the results of the analytic-geometric inverse problems research tradition to construct regularized algorithms.

The book is organized into two parts. The first part, "Linear Inverse Problems," is suitable for a one-semester undergraduate course or for a part of a graduate course. We present both continuous and discrete inverse problems to instruct how the ill-posedness is inherent in the idealized inverse problem and how it shows up in the real-life problem and in its discretization. With this approach we hope the reader will develop a deeper understanding of the connection between the mathematical theory, the computational model, and the practical problem arising from the application.

The guiding examples in Part I are the problem of image deblurring, X-ray tomography, and backward heat propagation. We discuss how to realistically simulate measurement data for all three. A dangerous pitfall in algorithm development and testing is the act of committing an *inverse crime*, that is, obtaining a great reconstruction due to the fact that the simulated data resonates in some helpful way with the reconstruction algorithm. We explain how this can occur and how to avoid it.

The use of Besov spaces and wavelets as a means of regularization is included in addition to the classical methods of truncated singular value decomposition, Tikhonov regularization, and total variation regularization. Also, this is the first book to discuss the recently introduced sparsity-based parameter choice rules. Many practical problems demand the use of very large data sets, and appropriate large-scale variants of the above reconstruction methods are addressed as well.

Part I requires knowledge of basic linear algebra and matrix computations, some knowledge of PDEs, basic analysis, and some programming skills. Some material for Part I is provided in the appendices.

Part II addresses nonlinear inversion, and it is a suitable text for a graduate course in applied mathematics. Also, we have received many requests for a text on the D-bar method, and Part II is designed to fulfill this need. We hope that researchers in electrical impedance tomography (EIT) will welcome this exposition.

The guiding example for Part II is EIT, although several other examples are also discussed briefly. Taking one guiding example is in accordance with the above rationale of treating nonlinear inverse problems as unique cases needing tailored solution strategies. We hope that the detailed discussion of the regularized D-bar method for EIT serves as a model for further research regarding other nonlinear inverse problems.

Actually, we see the regularized D-bar method as a topic that combines and unifies several schools of thought. Namely, there appear to be rather separated research traditions in the field of inverse problems, including the following three traditions: (1) The *analytic-geometric tradition* treats inverse problems as coefficient recovery tasks for PDEs. The main questions studied are uniqueness and stability proofs for recovering coefficient functions from limited but infinite-precision information. (2) The *regularization tradition* studies the construction of continuous maps from the data space to the model space providing approximate reconstructions from indirect measurement data containing errors. The main questions studied are convergence rates of the reconstructions at the asymptotic limit of zero measurement noise. (3) The *engineering tradition* is involved with writing robust computational algorithms that recover useful information from practical data.

In the regularized D-bar method, the reconstruction technique is defined using complex geometric optics solutions and a nonlinear Fourier transform introduced by the analytic-geometric tradition, the regularization strategy is provided by a nonlinear low-pass filtering step that can be analyzed according to the standards of the regularization tradition, and the final result is a robust imaging algorithm applicable to *in vivo* medical EIT.

Part II requires some background in PDEs, complex analysis, functional analysis, the finite element method (FEM), and numerical solution methods for linear systems. However, we do include short introductions to these topics in the text and in the appendices.

Throughout the book we provide exercises and project works involving MATLAB programming. Selected pieces of software can be downloaded from the website www.siam.org/books/cs10.

We also wish to take this opportunity to thank the many people who have provided support, encouragement, and help with this text. We thank David Isaacson for his guidance and encouragement early in our careers. We are grateful to Kari Astala, Jutta Bikowski, Michael DeAngelo, Raul Gonzalez-Lima, Sarah Hamilton, Alemdar Hasanov, Claudia Natalia Lara Herrera, Masaru Ikehata, Kim Knudsen, Ville Kolehmainen, Matti Lassas, Miguel Fernando Montoya Vallejo, Ethan Murphy, Jon Newell, Allan Perämäki, Lassi Päivärinta, and Alan Von Herrmann for their help in reporting our joint works in this book. We thank Per Christian Hansen, Andreas Hauptmann, Olli Koskela, and Jussi Määttä for their helpful comments regarding the manuscript, and Aki Kallonen, Keijo Hämäläinen, Martti Kalke, Jyrki Saarinen, and the PaloDEx Group for providing measurement data and image material concerning X-ray imaging. We thank Kati Niinimäki, Esa Niemi, and Miguel Fernando Montoya Vallejo for creating some of the figures, and Ethan Murphy for providing some of the figures from his thesis. Furthermore, we thank Simon Arridge, Guillaume Bal, Martin Burger, Fioralba Cakoni, Martin Hanke, Tapio Helin, Slava Kurylev, Peter Maass, Adrian Nachman, Hanna Pikkarainen, Ronny Ramlau, Otmar Scherzer, Carola Schönlieb, John Schotland, Mikko Sillanpää, and Erkki Somersalo for valuable discussions. We thank Alexandra Bellow for providing the two photos of Alberto Calderón included in Chapter 14. Thanks to Elizabeth Greenspan from SIAM for her help, patience, and support during the preparation of the manuscript, and to the staff of SIAM for putting the book in its final form and catching more than a few typos along the way. Part of the time writing the book was spent at Mathematical Sciences Research Institute in Berkeley, California, and at Isaac Newton Institute in Cambridge, UK, whose hospitalities are greatly appreciated. Finally, we thank all our families for their love and support.

Jennifer Mueller
Samuli Siltanen

Fort Collins, Colorado
Helsinki, Finland
January, 2012

Part I

Linear Inverse Problems

Chapter 1
Introduction

Inverse problems are the opposites of direct problems. Informally, in a direct problem one finds an effect from a cause, and in an inverse problem one is given the effect and wants to recover the cause. The most usual situation giving rise to an inverse problem is the need to interpret indirect physical measurements of an unknown object of interest.

For example, in X-ray tomography the direct problem is to determine the X-ray projection images we would get from a physical body whose internal structure we know precisely. The corresponding inverse problem is to reconstruct the inner structure of an unknown physical body from the knowledge of X-ray images taken from different directions. Here is a two-dimensional example:

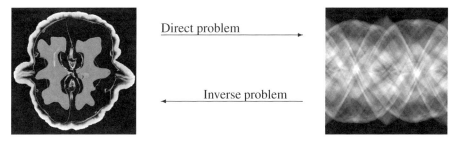

Here the slice through a walnut (left) is the cause and the collection of X-ray data (right) is the effect. The tomographic data is shown in the traditional *sinogram* form, which will be discussed in detail in Section 2.3.5. The slice image on the left is courtesy of Keijo Hämäläinen and Aki Kallonen from University of Helsinki, Finland.

Variants of the above tomographic problem appear also in monitoring ozone profiles in the upper atmosphere using spaceborne star occultation measurements [296], identifying molecules based on electron microscope imaging [139], interpretation of Doppler weather radar measurements [299], and measuring the temperature distribution of hot gases flowing through the window of a burning house using metal wires [30]. This demonstrates the general nature of mathematics: the same underlying problem may be found in very different application areas.

Another example comes from image processing. Define the direct problem as finding out how a given sharp photograph would look like if the camera were incorrectly focused.

The inverse problem known as *deblurring* is finding the sharp photograph from a given blurry image.

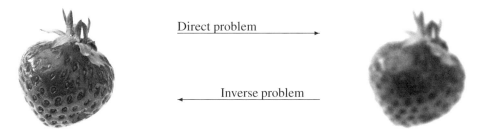

Here the cause is the sharp image and the effect is the blurred image. A famous example of deblurring is the correction algorithm used for the Hubble Space Telescope images after finding out a flaw in the construction of its main lens.

There is an apparent symmetry in the above explanation: without further restriction of the definitions, the direct problem and inverse problem would be in identical relation with each other. For example, we might take as the direct problem the determination of a positive photograph from the knowledge of the negative photograph.

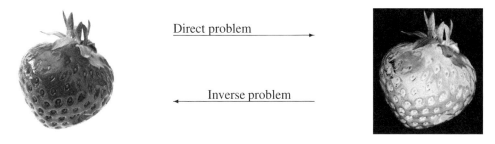

In this case the corresponding "inverse problem" would be inverting a given photograph to arrive at the negative. Here both problems are easy and stable, and one can move between them repeatedly.

However, we concentrate on *ill-posed inverse problems*, where the inverse problem is more difficult to solve than the direct problem. To explain this we need the notion of a *well-posed problem* introduced by Jacques Hadamard (1865–1963):

H_1: **Existence.** There should be at least one solution.

H_2: **Uniqueness.** There should be at most one solution.

H_3: **Stability.** The solution must depend continuously on data.

Now denote by $\mathcal{A}$ the *forward map*, defined conceptually by $\mathcal{A}(\text{cause}) = \text{effect}$. The direct problem must be well-posed, in other words $\mathcal{A}$ should be a well-defined, single-valued, and continuous function. The inverse problem is ill-posed if $\mathcal{A}^{-1}$ does not exist or is not continuous; then at least one of the conditions H_1–H_3 fails for $\mathcal{A}^{-1}$. In the positive-negative photograph example above both $\mathcal{A}$ and $\mathcal{A}^{-1}$ are well-defined and continuous, so it is not an ill-posed inverse problem.

Our general mathematical model is constructed as follows. We consider indirect linear measurements of the form
$$\mathbf{m} = \mathcal{A}f + \varepsilon, \tag{1.1}$$
where f is a piecewise continuous function defined on a subset of $\mathbb{R}^d$ and $\mathbf{m} \in \mathbb{R}^k$ is a vector of numbers given by a measurement device. Models of the form (1.1) arise from various situations in technology or physics; the linear operator $\mathcal{A}$ may be related, for example, to a PDE or to an integral equation. We will discuss examples of practically relevant operators $\mathcal{A}$ in Chapter 2.

The vector $\varepsilon \in \mathbb{R}^k$ in (1.1) models errors coming from measurement noise, which is inevitable in practical situations. Sometimes ε is modeled as a random variable with certain statistics. However, in this book we think of ε as a deterministic but unknown error; the information we have on ε is an inequality $\|\varepsilon\| \leq \delta$ with a known constant $\delta > 0$. Such a number δ can often be found by calibration of the measurement device; higher-quality device typically gives smaller δ. Our deterministic approach does include the possibility that ε is a fixed realization of a random process; this is actually a quite accurate model of many realistic measurements.

The reason for the term "indirect measurement" is the following. We are interested in the function f but cannot measure its values directly. However, f is connected to another physical quantity $\mathbf{m}$, which is available for measurement. The connection is modeled by the linear operator $\mathcal{A}$. Now the direct problem is "Given f, determine $\mathbf{m} = \mathcal{A}f$." The corresponding inverse problem is

$$\begin{array}{l}\text{Given noisy measurement } \mathbf{m} = \mathcal{A}f + \varepsilon \text{ and } \delta > 0 \\ \text{with } \|\varepsilon\| \leq \delta, \text{ extract information about } f.\end{array} \tag{1.2}$$

Part I of this book is about practical extraction of information from indirect linear measurements using computational methods. Consequently, we need to introduce a finite-dimensional approximation $\mathbf{f} \in \mathbb{R}^n$ to the function f and to build a matrix model for the linear operator $\mathcal{A}$.

In case of discrete linear inverse problems we consider measurements of the form
$$\mathbf{m} = A\mathbf{f} + \varepsilon, \tag{1.3}$$
where $\mathbf{m} \in \mathbb{R}^k$ and $\mathbf{f} \in \mathbb{R}^n$. Moreover, A is a matrix of size $k \times n$ (k rows and n columns). Strictly speaking, we abuse notation by using $\mathbf{m}$ in both (1.1) and (1.3) although they are different models. Whenever there is the possibility of confusion, we denote the measurement from the finite model (1.3) by $\mathbf{m}^{(n)}$.

Once the computational model (1.3) has been constructed, it is tempting to try to solve the inverse problem (1.2) by the naïve reconstruction
$$\mathbf{f} \approx A^{-1}\mathbf{m}. \tag{1.4}$$
However, in the case of ill-posed inverse problems the approach (1.4) will fail. In Sections 2.1–2.3 we describe some important indirect measurements and demonstrate the failure of the naïve reconstruction (1.4) numerically.

Regularization is what really needs to be done for successful and noise-robust solution of linear inverse problems. We discuss the theory and implementation of various regularization methods in Chapters 3–7. We demonstrate the properties of the various methods using the example problems developed in Chapter 2. Large-scale computational methods are emphasized throughout the text because practical applications often lead to very high-dimensional problems.

Discretization of the continuum model (1.1) using discrete models of the form (1.3) involves choosing the dimension n of the discrete vector **f**. It is desirable to design computational inversion methods that give consistent results at different resolutions n. This so-called *discretization invariance* is discussed in Chapter 8.

Practical examples of linear inversion are described in Chapter 9 in the case of X-ray tomography.

It is impossible to cover all useful and important material related to computational inversion in this book. We list here some further reading that complements our approach.

Regularization theory is discussed in the classical texts [439, 340] and in the more recent books [132, 254, 468]. The mathematical foundations of inverse problems are explained more generally in the books [236, 269].

Some of the computational inversion methods presented in this book (such as truncated singular value decomposition, Tikhonov, regularization, and total variation regularization) are discussed also in [461, 194, 196, 195, 354]. Useful methodologies that are not covered here due to restrictions of space include truncated iterative solvers [193, 191, 65], approximate inverses [407], statistical inversion [247, 434, 67], and variational methods [370, 405].

Application-oriented texts are available for inverse problems in medical imaging [249, 353, 133], geophysical inversion [331, 434], and signal processing [403].

Chapter 2
Naïve reconstructions and inverse crimes

This book is about developing computational solution methods for real-life inverse problems. The design of reconstruction algorithms is best done by first testing the code extensively with simulated data because every new aspect of the code can be systematically tested. Working directly with measured data may lead to very hard debugging problems, as the source of difficulties can be hard to track.

What happens if proper simulation of errors is neglected? For example, using the same computational grid for the data simulation and reconstruction sometimes results in perfect reconstructions from noise-free data. Such a situation is not realistic and is referred to as an *inverse crime*. Excellent inversion results may be obtained, but these are not representative of any realistic inverse problem, since noise is present in any experimental setting. Such studies are inconclusive at best since robustness against modeling and measurement errors is not tested.

In this chapter, we will introduce these concepts in the context of the three guiding examples in Part I: deconvolution, the backward heat equation, and X-ray tomography.

2.1 Convolution

Linear convolution is a useful process for modeling a variety of practical measurements. *Deconvolution*, the corresponding inverse problem, is related to many engineering problems such as removing unwanted echoes from sound recordings or sharpening a misfocused photograph.

One-dimensional deconvolution will serve as a basic example throughout Part I of the book. Two-dimensional deconvolution is a project topic in Section 10.

2.1.1 Continuum model for one-dimensional convolution

We build a computational model for one-dimensional convolution with periodic boundary conditions. We consider 1-periodic functions $f : \mathbb{R} \to \mathbb{R}$ satisfying $f(x) = f(x+n)$ with any integer $n \in \mathbb{Z}$. Essentially the function f is defined on an interval of length 1 such as $[0,1]$ or $[-\frac{1}{2}, \frac{1}{2}]$ with the endpoints identified; another way of thinking about this is to consider $f(x)$ defined on a circle with radius $(2\pi)^{-1}$ and x being the arc length variable.

The reason for considering periodic functions is that we can avoid some technicalities related to boundary conditions that would obscure the main message about ill-posedness. Also, the Fourier transform and the wavelet transform are easily defined and implemented in the periodic setting.

The continuum measurement model concerns a 1-periodic signal $f : \mathbb{R} \to \mathbb{R}$ blurred by a 1-periodic *point spread function* ψ. Other common names for the point spread function include *device function, impulse response, blurring kernel, convolution kernel*, and *transfer function*.

Let us first construct the point spread function using a building block ψ_0 defined in the interval $[-a,a] \subset \mathbb{R}$ with some constant $0 < a < 1/2$:

$$\psi_0(x) = C_a(x+a)^2(x-a)^2 \quad \text{for } -a \leq x \leq a, \tag{2.1}$$

where the constant $C_a := (\int_{-a}^{a}(x+a)^2(x-a)^2 dx)^{-1}$ is chosen to enforce the following normalization:

$$\int_{-a}^{a} \psi_0(x)\,dx = 1. \tag{2.2}$$

The periodic point spread function is defined by copying $\psi_0(x)$ to every interval $[n-a, n+a]$ with $n \in \mathbb{Z}$ and setting $\psi(x)$ to zero outside those intervals. The resulting ψ is a non-negative and even function:

$$\psi(x) \geq 0 \quad \text{and} \quad \psi(x) = \psi(-x) \quad \text{for all } x \in \mathbb{R}. \tag{2.3}$$

See Figure 2.1 for a plot of the point spread function with $a = 0.04$.

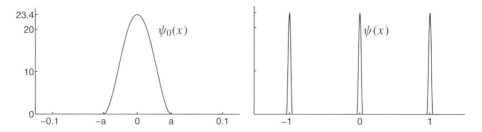

Figure 2.1. Point spread function according to (2.4) with $a = 0.04$ for one-dimensional convolution. Left: the continuously differentiable building block $\psi_0(x)$ used for constructing the periodic point spread function. Right: the periodic point spread function $\psi(x)$.

We remark that instead of (2.2) one often requires $\int_{-a}^{a} \psi_0(x)^2 dx = 1$. However, we prefer (2.2) since then constant functions remain unchanged in convolution with ψ; this will be convenient below when we compare plots of reconstructions to the plot of the true signal by showing them in the same figure.

Definition 2.1.1. *The continuum model of convolution, or blurring, is given by the following integral:*

$$(\psi * f)(x) = \int_{-a}^{a} \psi(x')f(x-x')\,dx'. \tag{2.4}$$

2.1. Convolution

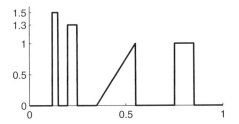 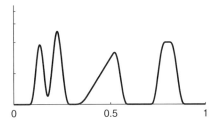

Figure 2.2. Effect of convolution on a piecewise continuous function. Left: target function $f(x)$. Right: the function $(\psi * f)(x)$.

An example of the effect of convolution with the point spread function is found in Figure 2.2. The smoothing effect of the convolution is evident, and motivates the terminology blurring kernel for the point spread function.

Note that formula (2.4) is not of the form (1.1) since the left-hand side is not a k-dimensional vector. However, suppose the function f is defined on an interval $[b, b+1]$, and assume that we have a device that measures the values of the convolution function $(\psi * f)(x)$ at a collection of k equally spaced points $\tilde{x}_1 = b, \tilde{x}_2 = b + \frac{1}{k}, \tilde{x}_3 = b + \frac{2}{k}, \ldots, \tilde{x}_k = b + \frac{k-1}{k}$ and define

$$\mathbf{m} := [(\psi * f)(\tilde{x}_1), (\psi * f)(\tilde{x}_2), \ldots, (\psi * f)(\tilde{x}_k)]^T \in \mathbb{R}^k. \tag{2.5}$$

Then $\mathcal{A}f = \mathbf{m}$ is of the form (1.1).

2.1.2 Discrete convolution model

Next we need to discretize the continuum model to arrive at a finite-dimensional measurement model of the form (1.3). Define

$$x_j = b + \frac{j-1}{n} \quad \text{for } j = 1, 2 \ldots, n; \tag{2.6}$$

then the 1-periodic real-valued function $f(x)$ is represented by a vector $\mathbf{f}$ containing values at the grid points:

$$\mathbf{f} = [\mathbf{f}_1, \mathbf{f}_2, \ldots, \mathbf{f}_n]^T = [f(x_1), f(x_2), \ldots, f(x_n)]^T \in \mathbb{R}^n. \tag{2.7}$$

Furthermore, denote $\Delta x := x_2 - x_1 = 1/n$.

We can approximate the integral appearing in (2.4) by numerical quadrature. For any reasonably well-behaved function $g : [b, b+1] \to \mathbb{R}$ we have

$$\int_b^{b+1} g(x)\,dx \approx \Delta x \sum_{j=1}^n g(x_j), \tag{2.8}$$

the approximation becoming better as n increases.

For convenience, let us take $k = n$ and measure the convolution at the same points (2.6) as where the unknown function f is sampled. This is not necessary in general, but it will lead to a square-shaped matrix A, making it easy to illustrate naïve reconstructions and inverse crimes.

Let us construct an $n \times n$ matrix A so that $A\mathbf{f} \in \mathbb{R}^k$ approximates $\mathcal{A}f$ defined by (2.4). We define a discrete point spread function denoted by

$$\mathbf{p} = [\mathbf{p}_{-\nu}, \mathbf{p}_{-\nu+1}, \ldots, \mathbf{p}_{-1}, \mathbf{p}_0, \mathbf{p}_1, \ldots, \mathbf{p}_{\nu-1}, \mathbf{p}_\nu]^T$$

as follows. Recall that $\psi_0(x) \equiv 0$ for $|x| > a > 0$. Take $\nu > 0$ to be the smallest integer satisfying the inequality $(\nu + 1)\Delta x > a$ and set

$$\widetilde{\mathbf{p}}_j = \psi_0(j\Delta x) \quad \text{for } j = -\nu, \ldots, \nu.$$

For example, with $a = 0.04$ as in Figure 2.1 and $n = 64$, we get $\nu = 2$. By (2.8) the normalization condition (2.2) almost holds: $\Delta x \sum_{j=-\nu}^{\nu} \widetilde{p}_j \approx 1$. However, in practice it is a good idea to normalize the discrete point spread function explicitly by the formula

$$\mathbf{p} = \left(\Delta x \sum_{j=-\nu}^{\nu} \widetilde{\mathbf{p}}_j \right)^{-1} \widetilde{\mathbf{p}}; \tag{2.9}$$

then it follows that

$$\Delta x \sum_{j=-\nu}^{\nu} \mathbf{p}_j = 1. \tag{2.10}$$

Now

$$\int_{-a}^{a} \psi(x')f(x_j - x')dx' \approx \Delta x \sum_{\ell=-\nu}^{\nu} \psi(x_\ell)f(x_j - x_\ell)$$

$$\approx \Delta x \sum_{\ell=-\nu}^{\nu} \mathbf{p}_\ell \mathbf{f}_{j-\ell}.$$

Hence discrete convolution is defined by the formula

$$(\mathbf{p} * \mathbf{f})_j = \sum_{\ell=-\nu}^{\nu} \mathbf{p}_\ell \mathbf{f}_{j-\ell}, \tag{2.11}$$

where $\mathbf{f}_{j-\ell}$ is defined using periodic boundary conditions for the cases $j - \ell < 1$ and $j - \ell > n$. Then

$$\Delta x(\mathbf{p} * \mathbf{f}) \approx \mathcal{A}f, \tag{2.12}$$

and we define the measurement vector $\mathbf{m} = [\mathbf{m}_1, \ldots, \mathbf{m}_k]^T$ by

$$\mathbf{m}_j = \Delta x(\mathbf{p} * \mathbf{f})_j + \varepsilon_j. \tag{2.13}$$

We would like to write formula (2.13) using a matrix A so that we would arrive at the desired model (1.3). To this end, set

$$\begin{bmatrix} \mathbf{m}_1 \\ \vdots \\ \mathbf{m}_k \end{bmatrix} = \begin{bmatrix} a_{11} & \cdots & a_{1n} \\ \vdots & \ddots & \vdots \\ a_{k1} & \cdots & a_{kn} \end{bmatrix} \begin{bmatrix} \mathbf{f}_1 \\ \vdots \\ \mathbf{f}_n \end{bmatrix} + \begin{bmatrix} \varepsilon_1 \\ \vdots \\ \varepsilon_k \end{bmatrix}.$$

2.1. Convolution

The answer is to build a circulant matrix having the elements of **p** appearing systematically on every row of A.

Let us illustrate the structure of the convolution matrix A by an example in the case $n = 64$. As observed above, if $a = 0.04$, then $\nu = 2$, and the point spread function takes the form $p = [p_{-2}\ p_{-1}\ p_0\ p_1\ p_2]^T$. According to (2.11) we have

$$(\mathbf{p} * \mathbf{f})_1 = \mathbf{p}_0 \mathbf{f}_1 + \mathbf{p}_{-1} \mathbf{f}_2 + \mathbf{p}_{-2} \mathbf{f}_3 + \mathbf{p}_2 \mathbf{f}_{n-1} + \mathbf{p}_1 \mathbf{f}_n,$$
$$(\mathbf{p} * \mathbf{f})_2 = \mathbf{p}_1 \mathbf{f}_1 + \mathbf{p}_0 \mathbf{f}_2 + \mathbf{p}_{-1} \mathbf{f}_3 + \mathbf{p}_{-2} \mathbf{f}_4 + \mathbf{p}_2 \mathbf{f}_n,$$
$$(\mathbf{p} * \mathbf{f})_3 = \mathbf{p}_2 \mathbf{f}_1 + \mathbf{p}_1 \mathbf{f}_2 + \mathbf{p}_0 \mathbf{f}_3 + \mathbf{p}_{-1} \mathbf{f}_4 + \mathbf{p}_{-2} \mathbf{f}_5,$$
$$\vdots$$
$$(\mathbf{p} * \mathbf{f})_n = \mathbf{p}_{-1} \mathbf{f}_1 + \mathbf{p}_{-2} \mathbf{f}_2 + \mathbf{p}_2 \mathbf{f}_{n-2} + \mathbf{p}_1 \mathbf{f}_{n-1} + \mathbf{p}_0 \mathbf{f}_n.$$

Consequently the matrix A looks like this:

$$A = \Delta x \begin{bmatrix} \mathbf{p}_0 & \mathbf{p}_{-1} & \mathbf{p}_{-2} & 0 & 0 & 0 & \cdots & \mathbf{p}_2 & \mathbf{p}_1 \\ \mathbf{p}_1 & \mathbf{p}_0 & \mathbf{p}_{-1} & \mathbf{p}_{-2} & 0 & 0 & \cdots & 0 & \mathbf{p}_2 \\ \mathbf{p}_2 & \mathbf{p}_1 & \mathbf{p}_0 & \mathbf{p}_{-1} & \mathbf{p}_{-2} & 0 & \cdots & 0 & 0 \\ 0 & \mathbf{p}_2 & \mathbf{p}_1 & \mathbf{p}_0 & \mathbf{p}_{-1} & \mathbf{p}_{-2} & \cdots & 0 & 0 \\ \vdots & & & & \ddots & & & & \\ \vdots & & & & & \ddots & & & \\ 0 & 0 & \cdots & \mathbf{p}_2 & \mathbf{p}_1 & \mathbf{p}_0 & \mathbf{p}_{-1} & \mathbf{p}_{-2} \\ \mathbf{p}_{-2} & 0 & \cdots & 0 & \mathbf{p}_2 & \mathbf{p}_1 & \mathbf{p}_0 & \mathbf{p}_{-1} \\ \mathbf{p}_{-1} & \mathbf{p}_{-2} & \cdots & 0 & 0 & \mathbf{p}_2 & \mathbf{p}_1 & \mathbf{p}_0 \end{bmatrix}; \quad (2.14)$$

note the systematic band-diagonal structure, which characterizes A as a circulant matrix. Linear systems involving circulant matrices can be quickly solved using fast Fourier transforms, a topic we will return to later.

Returning to the general case of **p** defined by (2.9), the approximation formula (2.12) can be written in the form

$$A\mathbf{f} \approx \mathcal{A}f. \quad (2.15)$$

Figure 2.3 shows data computed by the discrete model $A\mathbf{f}$ and compares the result to the continuous data $(\psi * f)(x)$ defined by (2.4).

Now let's add a little noise to the data. For example, we might take $k = 64 = n$ and construct the measurement noise in a probabilistic manner by taking a realization of a random vector with 64 independently distributed Gaussian elements having standard deviation $\sigma = 0.01 \cdot \max|f(x)|$. This corresponds to a relative noise level of 1%.

2.1.3 Naïve deconvolution and inverse crimes

We illustrate numerically the failure of the following naïve reconstruction attempt:

$$\mathbf{f} \approx A^{-1}\mathbf{m} \approx A^{-1}(A\mathbf{f} + \varepsilon) = \mathbf{f} + A^{-1}(\varepsilon). \quad (2.16)$$

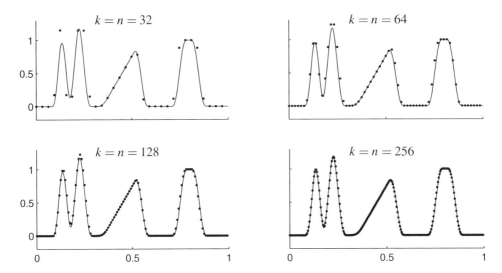

Figure 2.3. Illustration of the approximation $A\mathbf{f} \approx \mathcal{A}f$ of formula (2.15) for different choices of $k = n$. The actual function $(\psi * f)(x)$ defined by (2.4) is shown with a thin solid line, and the data points are indicated as dots. Note how the discrete approximation becomes better as the discretization is refined.

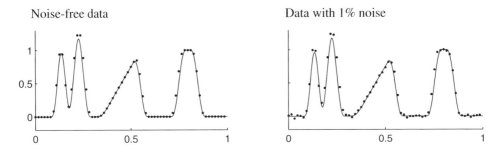

Figure 2.4. Illustration of simulated measurement noise. The actual function $(\psi * f)(x)$ defined by (2.4) is shown with a thin solid line, and the data points are indicated as dots. Left: noise-free discrete data $A\mathbf{f}$ with $n = 64 = k$. Right: the same data corrupted with 1% white noise.

In the case of no added noise ($\varepsilon = 0$) we use the data shown in the left plot of Figure 2.4 and get the left plot in Figure 2.5. The naïve reconstruction seems perfect! However, there is a catch. This apparently accurate reconstruction is not to be trusted; it is an example of an *inverse crime*. We will show how to avoid inverse crimes in Section 2.1.4.

If we apply naïve reconstruction (2.16) to the slightly noisy data shown in the right plot of Figure 2.4, we get the result shown in the right plot in Figure 2.5. It is completely useless. This example shows how sensitive inverse problems are to the smallest errors in measurement. We need to introduce *regularization* to overcome extreme sensitivity to measurement errors.

2.1. Convolution

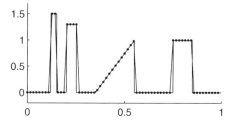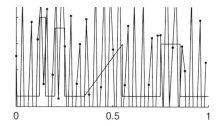

Figure 2.5. Two naïve deconvolutions by applying the inverse matrix A^{-1} to data. The original target function $f(x)$ is shown with a thin solid line, and the reconstruction is shown as dots. Left: naïve reconstruction (involving inverse crime) from the noise-free discrete data $A\mathbf{f}$ with $n = k = 64$ shown in the left plot in Figure 2.4. Right: naïve reconstruction from the noisy data shown in the right plot of Figure 2.4.

2.1.4 Naïve reconstruction without inverse crime

In the case of the deconvolution problem, we first simulate the measurements by convolving our known function f with a known discretized point spread function. In reality, when a blurred signal or image is encountered, the point spread function that "caused" the blurring is both unknown and can unlikely be expressed in simple terms. Thus, using the same point spread function for simulating a blurred signal and deconvolving the signal constitutes a serious inverse crime. Using the same point spread function *and* the same discretization mesh is an inverse felony!

We show one simple way to avoid inverse crime. We use a modified point spread function by taking $a = 0.041$ in (2.1) when simulating data. We compute the function $(\psi * f)(x)$ defined in (2.4) approximately at 1000 uniformly spaced points in the interval $[0, 1]$ using trapezoidal rule with 400 quadrature points for the evaluation of the integral. Finally, we interpolate the values of $\psi * f$ at the 64 grid points using splines.

Now the data has been simulated completely differently than using the 64×64 model matrix A as was (criminally) done in Section 2.1.3.

We apply naïve inversion (2.16) to the crime-free data and show the results in Figure 2.6. Compare the left plots in Figures 2.5 and 2.6. Proper simulation of crime-free data

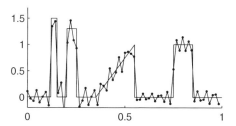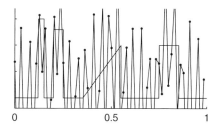

Figure 2.6. Two naïve deconvolutions by applying the inverse matrix A^{-1} to data generated avoiding inverse crime. The original target function $f(x)$ is shown with a thin solid line, and the reconstruction is shown as dots. Left: naïve reconstruction from noise-free discrete data with $n = k = 64$. Right: naïve reconstruction from noisy data. Compare to Figure 2.5.

reveals the ill-posedness of the deconvolution problem: The slightest perturbations in the data are amplified in naïve reconstruction using (2.16).

Exercise 2.1.1. *Determine whether the point spread function ψ is a $C^\infty(\mathbb{R})$ function.*

Exercise 2.1.2. *What is the effect of increasing the support of ψ_0 on v? Use the MATLAB programs* `DCcontdatacomp.m` *and* `DCcontdataplot.m` *to study the effect of increasing a on the convolved function. What do you observe?*

Exercise 2.1.3. *Plot a constant function of height 2 on $[0,1]$ before and after convolution with ψ. Use the MATLAB program* `DC2discretedatacomp.m` *to add noise to the convolved function and* `DC2naiveplot.m` *to compute a naïve reconstruction. Plot your results.*

2.2 Heat propagation

A classic ill-posed problem is that of determining the temperature distribution in a region from knowledge of the temperature distribution at the present time. This problem is known as the backward heat equation. We will begin with a discussion of the governing PDEs and their origins and then move to a simple discrete model.

2.2.1 Diffusion processes

The heat equation is the prototypical equation for modeling processes governed by pure diffusion. Following a probabilistic description as in, for example, [181], it can be derived by modeling the Brownian motion of the individual molecules in what we will assume to be a homogeneous material.

Suppose we have a material, such as depicted in Figure 2.7, containing n molecules, each of mass m, and suppose each molecule in this small volume is continually in motion. We will derive a model for one-dimensional spatial motion for simplicity, and so assume each molecule can only move to the left or to the right a distance Δx, representing an average displacement in time period Δt. To extend to higher dimensions, discrete motion in each of the three Cartesian coordinates would be permissible. Let p be the probability that the molecule moves to the right, and let q be the probability that the molecule moves to the left. Note that $p + q = 1$. Let $u(x,t)$ be the probability per unit length that a molecule

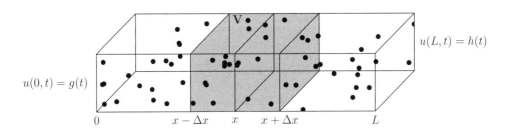

Figure 2.7. An illustration of the molecules in Brownian motion.

2.2. Heat propagation

is located in the interval $[x - \frac{1}{2}\Delta x, x + \frac{1}{2}\Delta x]$ at time t. The integral $\int_a^b u(x,t)dx$ is the probability that a molecule is located in $[a,b]$ at time t and $u(x,t)$ is a *probability density*.

The mass of molecules between $x - \Delta x$ and x is approximately

$$mnu\left(x - \frac{1}{2}\Delta x, t\right)\Delta x,$$

so the mass of the molecules crossing the plane at x from left to right at time t is approximately $pmnu(x - \frac{1}{2}\Delta x, t)\Delta x$. Similarly, the mass of the molecules crossing the plane at x from right to left at time t is approximately $qmnu(x + \frac{1}{2}\Delta x, t)\Delta x$. Thus, the net mass flux ϕ across the plane at x in the positive x-direction over a time interval Δt is approximately

$$\phi(x,t) \approx mn\frac{\Delta x}{\Delta t}\left(pu\left(x - \frac{1}{2}\Delta x, t\right) - qu\left(x + \frac{1}{2}\Delta x, t\right)\right).$$

Taking a Taylor series expansion for $u(x \pm \frac{1}{2}\Delta x, t)$ about (x,t),

$$u\left(x \pm \frac{1}{2}\Delta x, t\right) = u(x,t) + u_x(x,t)\left(\pm\frac{1}{2}\Delta x\right) + \frac{1}{2!}u_{xx}(x,t)\left(\pm\frac{1}{2}\Delta x\right)^2 + \cdots,$$

gives the linear approximations

$$u\left(x + \frac{1}{2}\Delta x, t\right) \approx u(x,t) + \frac{\Delta x}{2}u_x(x,t)$$

$$u\left(x - \frac{1}{2}\Delta x, t\right) \approx u(x,t) - \frac{\Delta x}{2}u_x(x,t).$$

Now the net mass flux across the plane at x is approximately

$$\phi(x,t) \approx pmn\frac{\Delta x}{\Delta t}\left(u(x,t) + \frac{\Delta x}{2}u_x(x,t)\right) - qmn\frac{\Delta x}{\Delta t}\left(u(x,t) - \frac{\Delta x}{2}u_x(x,t)\right)$$

$$= (p-q)mn\frac{\Delta x}{\Delta t}u(x,t) - \frac{mn(\Delta x)^2}{2\Delta t}u_x(x,t)(p+q).$$

The concentration $c(x,t)$ of molecules between $x - \frac{1}{2}\Delta x$ and $x + \frac{1}{2}\Delta x$ is approximately

$$c(x,t) \approx \frac{1}{\Delta x}(mnu(x,t)\Delta x).$$

So in terms of concentration, our expression for net flux becomes

$$\phi(x,t) \approx (p-q)\frac{\Delta x}{\Delta t}c(x,t) - \frac{1}{2}\frac{(\Delta x)^2}{\Delta t}c_x(x,t).$$

As $\Delta t \to 0$ assume that

$$\lim_{\Delta t \to 0}(p-q)\frac{\Delta x}{\Delta t} = \lambda,$$

where λ is a constant known as the drift constant, and

$$\lim_{\Delta t \to 0}\frac{1}{2}\frac{(\Delta x)^2}{\Delta t} = \frac{1}{2}D > 0,$$

where D is a constant known as the diffusion coefficient. Then in the limit as $\Delta t \to 0$ the net flux becomes

$$\phi(x,t) = \lambda c(x,t) - \frac{1}{2}Dc_x(x,t). \tag{2.17}$$

For a volume unit V, the quantity $\int_V c(x,t)dx$ represents the total mass in V, and conservation of mass implies that the time rate of change of the total mass in the volume V equals the flux across the boundary plus any mass created by sources f inside the volume V:

$$\frac{d}{dt}\int_V c(x,t)dx = -\int_{\partial V} \phi \cdot n\,ds + \int_V f\,dx. \tag{2.18}$$

By the divergence theorem

$$\int_{\partial V} \phi \cdot n\,ds = \int_V \nabla \cdot \phi\,dx,$$

and here in our one-dimensional model, equation (2.18) takes the form

$$\frac{d}{dt}\int_{x-\Delta x}^{x+\Delta x} c(x,t)dx = -\int_{x-\Delta x}^{x+\Delta x} \frac{\partial}{\partial x}\phi(x,t)dx + \int_{x-\Delta x}^{x+\Delta x} f\,dx. \tag{2.19}$$

Since the interval $[x - \Delta x, x + \Delta x]$ is arbitrary, we have

$$c_t(x,t) + \phi_x(x,t) = f(x,t)$$

or, from (2.17),

$$c_t(x,t) + \lambda c_x(x,t) - \frac{1}{2}Dc_{xx}(x,t) = f(x,t).$$

Since $c(x,t) = mnu(x,t)$, under the assumption of no sources or sinks ($f = 0$) and the assumption that $\lambda = 0$ (which can also be thought of as $p = q$), we have the familiar heat equation with initial and boundary conditions:

$$\begin{aligned}
u_t - Du_{xx} &= 0, \quad 0 < x < L, \quad t > 0, \\
u(0,t) &= g(t), \quad t > 0, \\
u(L,t) &= h(t), \quad t > 0, \\
u(x,0) &= f(x), \quad 0 < x < L.
\end{aligned} \tag{2.20}$$

The forward problem is to determine the temperature distribution $u(x,t)$ throughout the domain at time t from knowledge of $u(x,0)$. If the endpoints of the bar are kept at zero temperature, we have $g(t) = h(t) = 0$, which we will henceforth take for simplicity. Problems with nonzero boundary conditions can be transformed to zero boundary conditions through a change of variables.

The backward problem is to determine the temperature distribution $u(x,t)$ at some prior time $t < T$ from knowledge of $u(x,T)$,

$$\begin{aligned}
u_t - Du_{xx} &= 0, \quad 0 < x < L, \quad t > 0, \\
u(0,t) &= 0, \quad t > 0, \\
u(L,t) &= 0, \quad t > 0, \\
u(x,T) &= m(x), \quad 0 < x < L.
\end{aligned} \tag{2.21}$$
$$\tag{2.22}$$

2.2.2 A finite difference discrete model

In the following section, we will look at the fundamental solution for the heat equation and discretize the integral equation for the solution of the backward problem. Here, we consider another elementary approach: a finite difference discretization with explicit time stepping.

Define a mesh on the spatial domain by

$$x_0 = 0, \quad x_1 = \frac{L}{M+1}, \quad x_2 = \frac{2L}{M+1}, \ldots, x_{M+1} = L.$$

Then $\Delta x = 1/(M+1)$. Define a sequence of uniform time steps up to time T by

$$t_0 = 0, \quad t_1 = \frac{T}{N+1}, \quad t_2 = \frac{2T}{N+1}, \ldots, t_{N+1} = T.$$

Then $\Delta t = 1/(N+1)$. Denote $u(x_i, t_j)$ by u_{ij}. The Taylor series expansion for $u(x+\Delta x, t)$ about (x,t) is

$$u(x+\Delta x, t) = u(x,t) + (\Delta x)u_x(x,t) + \frac{(\Delta x)^2}{2!}u_{xx}(x,t) + O((\Delta x)^3). \tag{2.23}$$

For a linear approximation to $u(x+\Delta x, t)$, the terms of order $(\Delta x)^2$ and higher are dropped. Solving for $u_x(x,t)$ in the linear approximation results in the *forward difference* formula

$$u_x(x,t) = \frac{1}{\Delta x}(u(x+\Delta x, t) - u(x,t)) + O(\Delta x),$$

or in subscript notation

$$(u_x)_{ij} \approx \frac{1}{\Delta x}(u_{i+1,j} - u_{ij}).$$

Similarly, in the time variable

$$(u_t)_{ij} \approx \frac{1}{\Delta t}(u_{i,j+1} - u_{ij}). \tag{2.24}$$

Alternatively, by evaluating the Taylor expansion with respect to t about (x,t) at $(x, t - \Delta t)$ one obtains the *backward difference formula*

$$(u_t)_{ij} \approx \frac{1}{\Delta t}(u_{ij} - u_{i,j-1}). \tag{2.25}$$

An approximation to the second spatial derivative of u can be obtained by adding (2.23) and the Taylor expansion in (2.23) evaluated at $(x - \Delta x, t)$ to obtain

$$(u_{xx})_{ij} \approx \frac{1}{(\Delta x)^2}(u_{i+1,j} - 2u_{ij} + u_{i-1,j}). \tag{2.26}$$

Combining the approximations (2.24) and (2.26), the finite difference discretization of the PDE (2.20) is

$$\frac{u_{i,j+1} - u_{i,j}}{\Delta t} = \frac{D}{\Delta x^2}(u_{i+1,j} - 2u_{i,j} + u_{i-1,j}). \tag{2.27}$$

In the solution of the forward model, we step forward in time solving for $u_{i,j}$ for $i = 1, \ldots, M$ and $j = 1, 2, \ldots$ until we reach the desired time at which we want to compute the solution. An analysis of the error in making these approximations to the derivatives shows that the finite difference method will converge to the solution of the forward problem, provided

$$0 < \frac{\Delta t}{(\Delta x)^2} < \frac{1}{2}. \tag{2.28}$$

Most texts on the numerical solution of PDEs contain a proof of this result. See, for example, [9]. The boundary conditions are assumed to be known, and they are included in the solution as follows:

$$u_{0,j} = g(j\Delta t), \tag{2.29}$$

$$u_{M+1,j} = h(j\Delta t). \tag{2.30}$$

The initial condition is included by setting

$$u_{i,0} = f(i\Delta x).$$

Now we can write the solution at time step t_{i+1} as a linear system as follows. For simplicity, let $d = \frac{D\Delta t}{(\Delta x)^2}$ and denote by A the banded matrix

$$A = \begin{bmatrix} 1-2d & d & 0 & \cdots & \cdots & 0 \\ d & 1-2d & d & 0 & \cdots & 0 \\ 0 & d & 1-2d & d & \cdots & 0 \\ \vdots & & \ddots & \ddots & \ddots & \\ 0 & \cdots & 0 & d & 1-2d & d \\ 0 & \cdots & \cdots & 0 & d & 1-2d \end{bmatrix}.$$

Let $\mathbf{u^j}$ denote the vector at the jth time step $\mathbf{u^j} = [u_{1,j}, \ldots, u_{M,j}]^T$ and $\mathbf{v_j}$ the vector at the jth time step $\mathbf{v^j} = [du_{0,j}, 0, \ldots, 0, du_{M+1,j}]^T$. Provided the stability criterion (2.28) holds, the solution at the $(j+1)$st time step is approximated by computing

$$\mathbf{u^{j+1}} = A\mathbf{u^j} + \mathbf{v^j}. \tag{2.31}$$

Example. Consider the forward problem

$$u_t - u_{xx} = 0, \quad 0 < x < \pi, \quad t > 0, \tag{2.32}$$

$$u(0,t) = 0, \quad t > 0, \tag{2.33}$$

$$u(L,t) = 0, \quad t > 0, \tag{2.34}$$

$$u(x,0) = 10\sin 2x, \quad 0 < x < \pi. \tag{2.35}$$

One can show that the actual solution to the forward problem is $u(x,t) = 10e^{-4t}\sin 2x$. The relative errors for the solution computed out to times $T = 0.1, 0.2, 0.3, 0.4$ are given in Table 2.1. A plot of the evolution of the solution in time is found in Figure 2.8. Notice that the solution has decayed to nearly zero by time $T = 1$.

Data at time $T = 0.4$ was simulated using (2.31) with $\Delta x = 0.1366$ and $\Delta t = 0.0037$, which corresponds to $M + 1 = 24$ in our spatial discretization and $K = 108$ time steps. The noise-free data and data with 2% noise is found in Figure 2.8.

2.2. Heat propagation

Table 2.1. Accuracy of the finite difference forward solver on the problem (2.32)–(2.35).

Time T	Δx	Δt	$\|u(x,T)\|_\infty$	Relative sup-norm error
0.1	0.0668	4.65e-4	6.67	2.24e-4
0.2	0.0806	0.002	4.49	0.0015
0.3	0.0668	0.0014	3.01	0.0016
0.4	0.1366	0.0037	2.01	0.0020

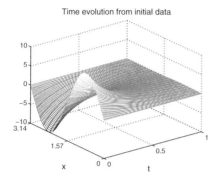
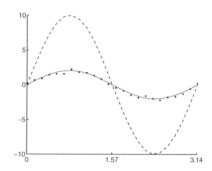

Figure 2.8. Left: illustration of the time evolution of the forward solution of the heat equation with initial condition $u_0(x) = 10\sin 2x$ computed to time $T = 1$. Right: illustration of simulated measurement noise for the heat equation. The dashed line is the initial profile. The noise-free data at time $T = 0.4$ is the solid line. The computed solution corrupted with 2% white noise is plotted with large dots.

2.2.3 Naïve reconstruction of the initial temperature

A naïve approach to the inverse problem of determining u at the previous time step would be to solve

$$\mathbf{u}^j = A^{-1}(\mathbf{u}^{j+1} - \mathbf{v}^j). \qquad (2.36)$$

This casts the problem in a discrete form. Since the stepsize in time is limited by the stability criterion, it will take numerous time steps to reach the initial condition, but this can be achieved by iterating the method (2.36). The results of applying (2.36) and iterating backward in time with and without noisy data at $T = 0.4$ are shown in Figure 2.9. The results are displayed at the iterate at which the solution to the backward problem begins to become unstable. In the case of noise-free data, this occurs at approximately 22 backward steps, or at time $t = 0.3215$. However, the method is very sensitive to noise in the data, and with just 0.01% noise, it is only stable for approximately four backward steps, or $t = 0.3888$. In either case, a serious inverse crime is being committed here. The same method on the same mesh is being used to both generate and reconstruct the data, and the results are therefore better than they should be!

The approach (2.36) is not equivalent to using backward differences and time-stepping backward in that manner since A^{-1} is not equal to the matrix that arises from that approach. Let us next investigate that approach. Stepping backward in time from $j = K - 1$, we solve

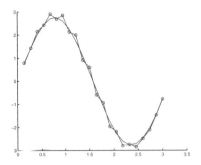 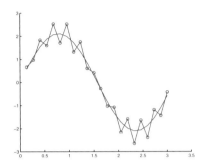

Figure 2.9. Left: reconstruction of the heat profile at time $t = 0.3215$ from noise-free data measured at $T = 0.4$. The solid line is the actual solution and the line with dots is the reconstruction computed by iterating (2.36) 22 steps. Right: reconstruction of the heat profile at time $t = 0.3888$ from data with 0.01% random noise measured at $T = 0.4$. The solid line is the actual solution and the line with dots is the reconstruction computed by iterating (2.36) 4 steps.

for each $u_{i,j}$ from

$$u_{i,j} = u_{i,j+1} - \frac{D\Delta t}{\Delta x^2}(u_{i+1,j+1} - 2u_{i,j+1} + u_{i-1,j+1}). \tag{2.37}$$

Plots of the results from four time steps backward with and without noisy data are found in Figures 2.10 and 2.11. We see that this method is somewhat more stable than (2.36), but it is still not useful for long times. In the computations resulting in Figure 2.10 an inverse crime is still being committed since the same mesh is used for the solution of the inverse

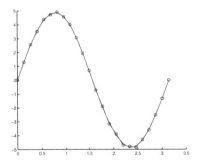 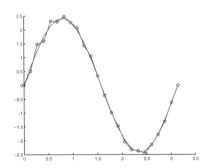

Figure 2.10. Left: reconstruction of the heat profile at time $t = 0.1794$ from noise-free data measured at $T = 0.4$. The solid line is the actual solution and the line with dots is the reconstruction computed by iterating (2.37) 60 steps. Here, an inverse crime was still committed since the same mesh and time steps were used to construct the data and compute the solution. The results are undeservingly good. Right: reconstruction of the heat profile at time $t = 0.3551$ from data with 0.01% random noise measured at $T = 0.4$. The solid line is the actual solution and the line with dots is the reconstruction computed by iterating (2.37) 13 steps. The same inverse crime is committed here.

2.3. Tomographic X-ray projection data

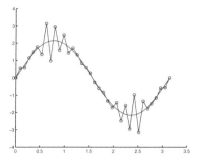 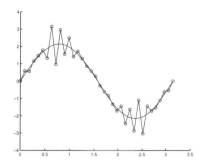

Figure 2.11. Left: reconstruction of the heat profile at time $t = 0.3844$ from noise-free data measured at $T = 0.4$. The solid line is the actual solution and the line with dots is the reconstruction computed by iterating (2.37) 8 steps. Equation (2.31) was used to compute the data on a different mesh from (2.37). Right: reconstruction of the heat profile at time $t = 0.3844$ from data with 0.01% random noise measured at $T = 0.4$. The solid line is the actual solution and the line with dots is the reconstruction computed by iterating (2.37) 8 steps.

problem as was used for the construction of the data by (2.31). By using a different mesh and time steps for the solution of (2.31) and (2.37) and interpolating the data to the mesh of (2.37), an inverse crime is avoided. This was the approach used in computing the results in Figure 2.11.

Exercise 2.2.1. *Complete Table 2.1 for later times $T = 0.5, 0.6, \ldots, 1.0$. Observe what happens to the solution when the stability criterion is violated.*

Exercise 2.2.2. *Compute reconstructions by the methods in this section for the same example but use data at time $T = 0.1$. Do not commit any inverse crimes. Is the method more stable than from the final time data at $T = 0.4$?*

Exercise 2.2.3. *Study the numerical forward solution of the heat equation by adding noise to the initial data $u_0(x) = 10\sin 2x$ in the MATLAB program* InverseHeatCondData-Simulator.m *and computing the solution at time $T = 0.4$. Plot the difference between the computed $u(x,T)$ from a noisy and noise-free initial condition. How does this differ from what we see in Figures 2.9 and 2.11?*

Exercise 2.2.4. *Use the MATLAB program* InvHeatCondNaiveSolver.m *and modify it to use the initial temperature distribution $f(x) = 10\chi_{[\pi/4, 3\pi/4]}(x)$, where χ is the characteristic function. Compute noise-free and noisy simulated data, and compute naive reconstructions at four prior time steps using methods (2.36) and (2.37). Plot your results. Give the reason that the discontinuity in the initial condition can never be reconstructed.*

2.3 Tomographic X-ray projection data

In tomographic X-ray imaging one takes X-ray projection images of an object from several different directions and attempts to recover the inner structure of the object from the data.

We show how such a measurement can be written in the form $\mathbf{m} = A\mathbf{f} + \varepsilon$ and illustrate numerically how the naïve reconstruction approach (1.4) fails.

2.3.1 A simple example: Probing two aluminum slabs

Let us first demonstrate the exponential attenuation law of X-rays using a very simple example, where two aluminum slabs are probed as shown in Figure 2.12. Typically, X-rays emanate from a roughly point-like location inside an X-ray tube. That point is called the *X-ray source* and shown as a black dot in Figures 2.12 and 2.13. Three X-rays are sent traveling towards a detector, each consisting initially of 1000 photons. The detector is capable of counting how many photons arrive at each point. One of the rays arrives at the detector through empty space, delivering all 1000 photons. Another ray travels through an aluminum slab whose width is chosen to be the *half-thickness* of the X-radiation used here. This means that half of the photons entering the slab will be absorbed inside the slab. The third ray encounters two such aluminum slabs. We call these three X-rays the empty-space ray, the one-slab ray, and the two-slab ray, respectively.

The photon count data can now be transformed into line integral data via two simple steps. First, take the logarithm of each photon count. Then, realizing that the integral of the empty-space ray must be zero, subtract each logarithm from the logarithm corresponding to the empty-space ray. As seen from the actual numbers shown in Figure 2.12, the resulting attenuation data is zero for the empty-space ray, a positive number (0.693) for the one-slab ray, and twice that number (1.386) for the two-slab ray.

We have described the basic calibration process for ideal photon count data based on the exponential attenuation law. However, we ignored at least a couple of properties of real-world measurements. First, practical detectors (for instance, charge-coupled devices or CCDs do not provide the actual photon count but rather an integer that is proportional to the photon count. However, this is not a serious problem, as you can find out in Exercise 2.3.2 below. Second, the photon count is not a deterministic number; it is better modeled as a random variable with Poisson distribution. This results in random measurement noise in the data; we will discuss this below.

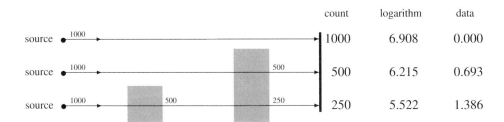

Figure 2.12. Simple experiment illustrating the attenuation of X-rays and interpretation of measurements. The three black dots show the positions of the X-ray source, and the horizontal lines depict X-rays. The gray boxes are slabs of attenuating material, and their width has been chosen to be the half-thickness of the X-radiation. The vertical thick line is the detector counting how many photons arrive at each point.

2.3.2 From photon count data to line integral data

The two-slab example in Section 2.3.1 is quite simple as it concerns only homogeneous material. Consider now an X-ray traveling through a phantom[1] representing a two-dimensional cross-section of a patient's head along a straight line, as shown in the left panel of Figure 2.13. We place the target slice inside the unit square defined by $0 \leq x_1 \leq 1$ and $0 \leq x_2 \leq 1$. For the sake of argument, assume that the X-ray travels along the horizontal path defined by $0 \leq x_1 \leq 1$ and $x_2 = \frac{1}{2}$.

Interaction between radiation and matter lowers the intensity of the ray. We think of the X-ray having initial intensity $I_0 := I(0)$ when entering the patient's head and smaller intensity $I_1 := I(1)$ when exiting. Also, we denote by $I(x_1)$ the intensity of the X-ray at the point $(x_1, \frac{1}{2})$ while traveling from the source to the detector.

In contrast to the simple homogeneous slab example above, the cross-section of a head contains various tissues with different X-ray attenuation properties. We model this situation using a nonnegative attenuation coefficient function $f(x_1, x_2)$, whose value gives the relative intensity loss of the X-ray within a small distance dx:

$$\frac{dI(x_1)}{I(x_1)} = -f\left(x_1, \frac{1}{2}\right) dx_1.$$

For example, bone has higher attenuation coefficient than brain tissue, and cerebrospinal fluid (white ovals in the left panel of Figure 2.13) provides practically zero attenuation. See the right panel in Figure 2.13 for a plot of the profile $f(x_1, \frac{1}{2})$.

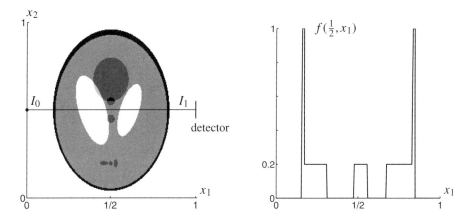

Figure 2.13. X-ray measurement. Left: an X-ray traveling through a simulated cross-section of a human head (a low-contrast version of the infamous Shepp–Logan phantom). Note that high attenuation is shown here as darker shades of gray and low attenuation as lighter shades. Right: plot of the attenuation coefficient along the path of the X-ray.

---

[1] A *phantom* can be either a physical calibration device or a mathematical model. Here it is a mathematical model to simulate an idealized cross-section of a human head. The use of "phantom" will be clear from the context.

Integration along the X-ray from source to detector gives

$$\int_0^1 f\left(x_1, \frac{1}{2}\right)dx_1 = -\int_0^1 \frac{I'(x_1)}{I(x_1)}dx_1 = \log I_0 - \log I_1. \tag{2.38}$$

Now the right-hand side of (2.38) is known: I_0 by calibration and I_1 from the measurement. The left-hand side of (2.38) consists of an integral of the unknown function f over a straight line, as wished.

Regarding noise, the quantity I_1 is a constant multiple of a Poisson-distributed random variable. It is typically sampled in practice using an analog-to-digital converter that produces integer output containing truncation errors and additional electronic noise. Taking logarithm of I_1 leads to a random variable with remarkably complicated statistics. However, it is usually quite plausible to model the measurement as

$$\log I_0 - \log I_1 = \int_0^1 f\left(x_1, \frac{1}{2}\right)dx_1 + \varepsilon, \tag{2.39}$$

where $\varepsilon \sim \mathcal{N}(0, \sigma^2)$ is a normally distributed random variable. The standard deviation σ of the noise can be estimated, for example, by measuring the same target repeatedly and calculating the standard deviation of the samples. This procedure is a reasonably accurate model when the photon count is large enough; see [414, Appendix].

We remark that in the above model we neglect the energy dependence of the attenuation function. Namely, most X-ray sources produce a multispectral beam, and an energy-dependent f may result in different measured line integrals depending on the propagation direction of the X-ray along the line. This is called *beam hardening*.

2.3.3 Continuous tomographic data: The Radon transform

In the previous section we described how to turn attenuation data from one single X-ray into line integral data concerning a nonnegative, compactly supported attenuation coefficient $f : \mathbb{R}^2 \to \mathbb{R}$. The aim of tomographic imaging is to collect information about f using different angles of view.

Let us define the *Radon transform*, denoted by $\mathfrak{R}$, as follows. We interpret $\theta \in \mathbb{R}$ as an angle measured in radians, and denote by

$$\vec{\theta} := \begin{bmatrix} \cos\theta \\ \sin\theta \end{bmatrix} \in \mathbb{R}^2$$

the unit vector with angle θ with respect to the x_1-axis. The Radon transform of the function f depends on the angular parameter θ and on a linear parameter $s \in \mathbb{R}$ in the following way:

$$\mathfrak{R}f(s,\theta) = \int_{x \cdot \vec{\theta} = s} f(x)dx^\perp, \tag{2.40}$$

where $dx^\perp$ denotes the one-dimensional Lebesgue measure along the line defined by $\{x \in \mathbb{R}^2 : x \cdot \vec{\theta} = s\}$. We remark that the parametrization of tomographic data provided by formula (2.40) is related to the so-called parallel-beam geometry used in the first-generation computed tomography (CT) scanners in the 1970s.

2.3. Tomographic X-ray projection data

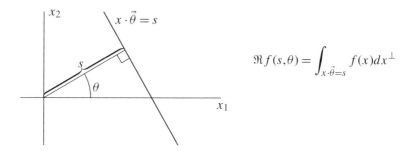

Figure 2.14. Illustration of the definition (2.40) of the Radon transform. The line $\{x \in \mathbb{R}^2 : x \cdot \vec{\theta} = s\}$ is drawn with a thick line.

Many variations in the data geometry are possible, such as limited angle data, local tomography data, exterior tomography, and combinations thereof. We refer the reader to the classical texts by Natterer [353] and Kak and Slaney [249]. See [353] for analytic inversion formulas, a thorough analysis of the mapping properties of $\mathfrak{R}$, and its generalizations to higher dimensions. See [354] for another perspective on image reconstruction.

The Fourier transform and Radon transform are connected in a simple way. This result is known as the central slice theorem. First, define the Fourier transform in one dimension as follows. See Figure 2.14 for an illustration.

Definition 2.3.1. *The Fourier transform of a function defined on $\mathbb{R}$ is given by*

$$\mathcal{F}(x)(\xi) = \widehat{f}(\xi) = \frac{1}{(2\pi)^n} \int_\mathbb{R} f(x) e^{-ix\xi} dx.$$

Theorem 2.1. *Let f be an absolutely integrable function defined on the whole real line. For any real number r and unit vector $\vec{\theta}$, we have the identity*

$$\int_{-\infty}^\infty \mathfrak{R} f(s, \vec{\theta}) e^{-isr} ds = \widehat{f}(r\vec{\theta}). \tag{2.41}$$

Proof. By the definition of the Radon transform

$$\int_{-\infty}^\infty \mathfrak{R} f(s, \vec{\theta}) e^{-isr} ds = \int_{-\infty}^\infty \int_{x \cdot \vec{\theta} = s} f(x) e^{-isr} dx^\perp ds$$
$$= \int_{-\infty}^\infty \int_{-\infty}^\infty f(x) e^{-ix \cdot (r\vec{\theta})} dx_1 dx_2$$
$$= \widehat{f}(r\vec{\theta}). \qquad \square$$

It will prove convenient to have a notation for the one-dimensional Fourier transform of a function in the scalar parameter as appears in the central slice theorem. Let $\tilde{h}(s, \vec{\theta})$ denote such a Fourier transform:

$$\tilde{h}(s, \vec{\theta}) = \int_{-\infty}^\infty h(t, \vec{\theta}) e^{-its} dt. \tag{2.42}$$

Then the central slice theorem says
$$\widetilde{\Re f}(r,\vec{\theta}) = \hat{f}(r\vec{\theta}).$$

The Radon inversion formula provides a way to obtain f from its Radon transform in the ideal case.

Theorem 2.2. *If f is an absolutely integrable function defined on the real line and $\hat{f}$ is absolutely integrable, then*
$$f(x) = \frac{1}{(2\pi)^2} \int_0^\pi \int_{-\infty}^\infty e^{isx\cdot\vec{\theta}} \widetilde{\Re f}(s,\vec{\theta}) |s| ds d\theta. \tag{2.43}$$

Proof. First note that since the Radon transform satisfies $\Re f(-s,-\vec{\theta}) = \Re f(s,\vec{\theta})$,
$$\begin{aligned}
\widetilde{\Re f}(-s,-\vec{\theta}) &= \int_{-\infty}^\infty \Re f(t,-\vec{\theta}) e^{-it(-s)} dt \\
&= \int_{-\infty}^\infty \Re f(t,-\vec{\theta}) e^{-i(-t)s} dt \\
&= \int_{-\infty}^\infty \Re f(-t,-\vec{\theta}) e^{-its} dt \\
&= \int_{-\infty}^\infty \Re f(t,\vec{\theta}) e^{-its} dt \\
&= \widetilde{\Re f}(s,\vec{\theta}).
\end{aligned}$$

Now by the Fourier inversion formula, with $\xi = (r\cos\theta, r\sin\theta)$,
$$\begin{aligned}
f(x) &= \frac{1}{(2\pi)^2} \int_{\mathbb{R}^2} \hat{f}(\xi) e^{ix\cdot\xi} d\xi \\
&= \frac{1}{(2\pi)^2} \int_0^{2\pi} \int_0^\infty \hat{f}(r\vec{\theta}) e^{irx\cdot\vec{\theta}} r dr d\theta \\
&= \frac{1}{(2\pi)^2} \int_0^{2\pi} \int_0^\infty \widetilde{\Re f}(r,\vec{\theta}) e^{irx\cdot\vec{\theta}} r dr d\theta \\
&= \frac{1}{(2\pi)^2} \int_0^\pi \int_{-\infty}^\infty \widetilde{\Re f}(r,\vec{\theta}) e^{irx\cdot\vec{\theta}} |r| dr d\theta,
\end{aligned}$$

where the last equality follows from the fact that $\widetilde{\Re f}(-s,-\vec{\theta}) = \widetilde{\Re f}(s,\vec{\theta})$. □

To summarize, this results in the following idealized reconstruction algorithm for X-ray CT imaging:

- Let f be the attenuation coefficient of a two-dimensional slice of a three-dimensional object. Then the intensity $I_{(s,\vec{\theta})}$ of the beam satisfies the differential equation
$$\frac{dI_{(s,\vec{\theta})}}{I_{(s,\vec{\theta})}} = -f(s,\vec{\theta}) ds.$$

2.3. Tomographic X-ray projection data

- We measure the Radon transform of f,

$$\Re f(s,\vec{\theta}) = \log\left(\frac{I_0}{I_d}\right),$$

where I_0 is the intensity of the beam at the source, and I_d is the intensity of the beam at the detector.

- Reconstruct f from the Radon inversion formula (2.43).

For the filtered back-projection algorithm, we regard the radial integral in the Radon Inversion Formula as a filter. We denote the output of the filter by $\mathcal{G}\Re f(t,\vec{\theta})$, where

$$\mathcal{G}\Re f(t,\vec{\theta}) = \frac{1}{2\pi}\int_{-\infty}^{\infty} \widetilde{\Re f}(r,\vec{\theta})e^{irt}|r|dr.$$

Then, with $t = x \cdot \vec{\theta}$,

$$f(x) = \frac{1}{2\pi}\int_0^{\pi} \mathcal{G}\Re f(x \cdot \vec{\theta},\vec{\theta})d\theta.$$

Note that one sees from this formula that low-frequency components are suppressed by $|r|$ and high-frequency components are amplified. Let's look at the filter a little more carefully. Recall that the Fourier transform of $g'(t)$ is

$$\mathcal{F}(\partial_t g)(\xi) = i\xi\hat{g}(\xi).$$

Thus if we had r instead of $|r|$ in the Radon inversion formula, we would have had the

$$\text{"inversion formula"} = \frac{1}{2\pi i}\int_0^{\pi} \partial_r \Re f(r,\theta)d\theta.$$

If f is real-valued, this quantity is purely imaginary! Thus, the $|r|$ is very important!

The MATLAB function `iradon.m` in the Image Processing Toolbox implements filtered back-projection. In the subsequent sections, we will be comparing the results of filtered back-projection implemented with `iradon.m` to other inversion techniques.

2.3.4 Discrete tomographic data

We model practical tomographic X-ray data by a bounded set $\Omega \subset \mathbb{R}^2$, a nonnegative attenuation coefficient f supported in $\overline{\Omega}$, and some finite collection $\{L_j\}_{j=1}^k$ of lines $L_j \subset \mathbb{R}^2$ intersecting Ω.

As a first example, we will use the following data set. It is an example of *parallel-beam geometry* illustrated in Figure 2.15. The angular variable is sampled with equidistant steps over the half circle:

$$\theta_j = \theta_1 + \left(\frac{j-1}{J}\right)\pi, \quad 1 \leq j \leq J, \quad (2.44)$$

Figure 2.15. Parallel beam X-ray measurement geometry. Here $J = 5$ and $N = 11$. Black dots show the locations of the X-ray source at different times of measurement. The thick line represents the detector measuring the intensity of the X-rays after passing through the target. High attenuation is shown here as darker shades of gray and low attenuation as lighter shades.

where $\theta_1 \in \mathbb{R}$ is an appropriate constant, a reference angle. The linear parameter s is also sampled uniformly over a suitable interval:

$$s_\nu = -S + 2\left(\frac{\nu - 1}{N}\right)S, \quad 1 \leq \nu \leq N, \tag{2.45}$$

where $S > 0$.

Defining $k = JN$, the measurement (1.1) then takes the form

$$\mathbf{m} = \mathcal{A}f + \varepsilon = \begin{bmatrix} \int_{L_1} f(x_1, x_2) ds_1 \\ \vdots \\ \int_{L_k} f(x_1, x_2) ds_k \end{bmatrix} + \varepsilon, \tag{2.46}$$

where ds_j denotes the one-dimensional Lebesgue measure along the line L_j. Each integral in (2.46) can be understood as a suitable rotation and scaling of formula (2.39).

For the computational solution we need to build a finite-dimensional measurement model of the form (1.3). We discretize the tomographic problem by dividing the unknown area into n pixels and assume that attenuation values are constant within each pixel. We number the pixels from 1 to n and call the corresponding attenuation values $\mathbf{f}_j \geq 0$ for $j = 1, \ldots, n$.

The measurement $\mathbf{m}_i$ of the line integral of f over line L_i is approximated by

$$\mathbf{m}_i = \int_{L_i} f(x_1, x_2) ds \approx \sum_{j=1}^{n} a_{ij} \mathbf{f}_j, \tag{2.47}$$

where a_{ij} is the distance that L_i travels in the jth pixel. Note that only pixels that intersect the beam L_i are included in this sum. Further, if we have k measurements in the vector $\mathbf{m} \in \mathbb{R}^k$, then (2.47) yields a matrix equation $\mathbf{m} = A\mathbf{f}$, where the matrix is defined by $A = (a_{ij})$.

Consider the following discretization and measurements, where $J = 2, k = 6, N = 3$ and the total number of pixels is $N^2 = 9$:

2.3. Tomographic X-ray projection data

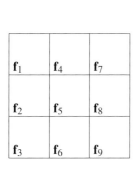
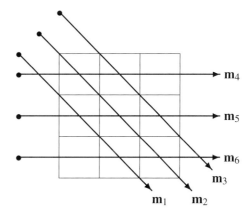

Here we have divided the square-shaped domain $\Omega \subset \mathbb{R}^2$ into 9 pixels, denoted by thin lines. The length of the side of each pixel is 1. Inside the pixels there is a constant value $\mathbf{f}_j$ of attenuation. The six arrows are X-rays used for probing the inner structure of Ω. Measurement data is the vector $\mathbf{m} = [\mathbf{m}_1, \ldots, \mathbf{m}_6]^T$ modeled by (2.47). The resulting measurement model is

$$\begin{bmatrix} 0 & \sqrt{2} & 0 & 0 & 0 & \sqrt{2} & 0 & 0 & 0 \\ \sqrt{2} & 0 & 0 & 0 & \sqrt{2} & 0 & 0 & 0 & \sqrt{2} \\ 0 & 0 & 0 & \sqrt{2} & 0 & 0 & 0 & \sqrt{2} & 0 \\ 1 & 0 & 0 & 1 & 0 & 0 & 1 & 0 & 0 \\ 0 & 1 & 0 & 0 & 1 & 0 & 0 & 1 & 0 \\ 0 & 0 & 1 & 0 & 0 & 1 & 0 & 0 & 1 \end{bmatrix} \begin{bmatrix} f_1 \\ f_2 \\ f_3 \\ f_4 \\ f_5 \\ f_6 \\ f_7 \\ f_8 \\ f_9 \end{bmatrix} = \begin{bmatrix} m_1 \\ m_2 \\ m_3 \\ m_4 \\ m_5 \\ m_6 \end{bmatrix}. \quad (2.48)$$

The model (2.48) is low-dimensional and simple. However, it already demonstrates one feature typical for inverse problems: nonuniqueness of solution. Namely, as can be seen in Exercise 2.3.5, there are several targets that produce exactly the same data. Thus the inverse problem cannot be uniquely solved using the measurement information alone.

Let us build a more realistic (higher-dimensional) data simulation model. We work with the so-called Shepp–Logan phantom, which is a piecewise constant model of a cross-section of a human head. The phantom is defined using ellipses and can be realized at any desired discrete resolution. See Figure 2.16 for pictures of the Shepp–Logan phantom at discretizations with $16 \times 16 = 256$ pixels, $50 \times 50 = 2500$ pixels, and $512 \times 512 = 262144$ pixels.

Let us construct the measurement matrix A corresponding to the low-resolution case with 16×16 pixels and projection directions specified by taking $J = 16$ in (2.44). In this case the size of A is not too large and we can show a picture of the nonzero elements of A for observing its structure.

We use MATLAB's command `radon.m` to simulate parallel-beam X-ray projection data from 16×16 pixel images with zero entries except one pixel with value 1. The pixel value 1 is first located in pixel 1 in the numeration shown in Figure 2.17(b), then in pixel 2, and so on. This way, column by column, we construct a measurement matrix A for a computational tomography model of the form (1.3).

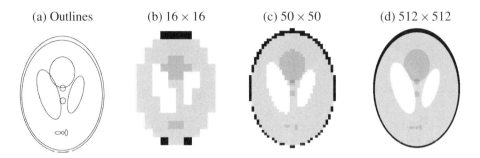

Figure 2.16. (a) Schematic illustration of the Shepp–Logan phantom. Areas of different attenuation values are bounded by various ellipses. (b)–(d) Plots of grayscale images of the Shepp–Logan phantom at different resolutions. High attenuation is shown here as darker shades of gray and low attenuation as lighter shades.

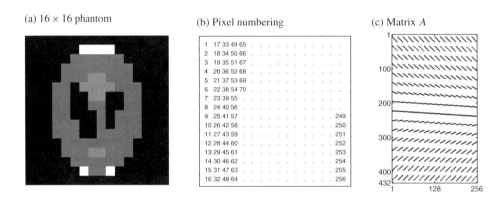

Figure 2.17. (a) Low-resolution Shepp–Logan phantom (16×16). Here black denotes zero attenuation, and white denotes maximum attenuation. (b) Numbering of the pixels in a 16×16 image when interpreted as a vector in $\mathbb{R}^{256}$. (c) Nonzero elements of the 432×256 tomographic measurement matrix.

What is the size of A? Obviously, the number of columns must be 256, the number of pixels in a 16×16 image. The `radon.m` algorithm picked automatically the value $N = 27$ in formula (2.45), so A has $JN = 16 \cdot 27 = 432$ rows. We observe that there are 11086 nonzero elements out of the total 110592; this means that roughly 90% of the elements in A are zero.

Three-dimensional X-ray tomography problems can be approached similarly to the above explanation using voxelization instead of pixelization and by tracing the paths of X-rays through the voxels in a three-dimensional manner. However, for illustration and simplicity purposes we stick to the two-dimensional case in this book. In principle there is no essential difference between the two- and three-dimensional cases, only the computations will be significantly more demanding in three dimensions.

2.3.5 Naïve reconstruction

The 16×16 Shepp–Logan phantom used in Section 2.3.4 has too low a resolution to really show the intended anatomic features properly. In the rest of the book we work with the 50×50 Shepp–Logan phantom when we need to construct the matrix A explicitly, and with the 512×512 phantom when we illustrate matrix-free large-scale methods in Section 9. See Figure 2.16.

Next we wish to experiment with naïve reconstructions of the 50×50 Shepp–Logan phantom shown in Figure 2.16(c). We choose the number of projection directions to be $J = 50$ in formula (2.44). We construct the measurement matrix A column by column as explained in Section 2.3.4. MATLAB's `radon.m` algorithm picked automatically the value $N = 75$ in formula (2.45), so A has $JN = 50 \cdot 75 = 3750$ rows. We arrive at the following measurement model:

$$A \begin{bmatrix} \mathbf{f}_1 \\ \vdots \\ \mathbf{f}_{2500} \end{bmatrix} = \begin{bmatrix} \mathbf{m}_1 \\ \vdots \\ \mathbf{m}_{3750} \end{bmatrix}, \qquad (2.49)$$

where the elements of the 50×50 pixel image $\mathbf{f}$ and the elements of the 75×50 sinogram $\mathbf{m}$ are numbered similarly to Figure 2.17(b). The phantom and sinogram are shown in Figure 2.18.

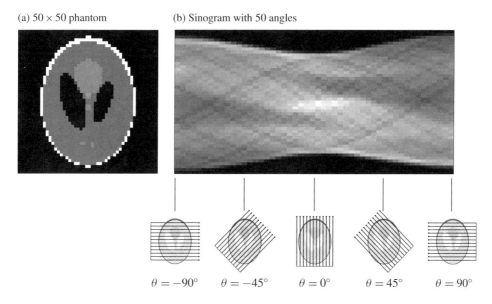

Figure 2.18. (a) Shepp–Logan phantom at resolution 50×50. Here black denotes zero attenuation, and white denotes maximum attenuation. (b) Measured data (involving inverse crime) in sinogram form, where the horizontal axis is the angle θ and the vertical axis the variable s in (2.40). We have removed some purely zero rows from the top and bottom of the sinogram for clarity. Underneath the sinogram we show some of the projection directions to illustrate the structure of the sinogram.

Once A is in place, we can try out naïve inversion, but not in the sense of (1.4) since A is not a square matrix. Instead we use least-squares naïve inversion defined as follows:

$$\mathbf{f} \approx (A^T A)^{-1} A^T \mathbf{m}. \tag{2.50}$$

Derivation and interpretation of formula (2.50) is postponed to Section 5.2. The result of applying (2.50) to ideal (noise-free) tomographic data is shown in Figure 2.19(b), and it looks very good indeed. The relative error of this reconstruction is very small. Perhaps we can conclude that we succeeded in reconstructing the 50×50 phantom from indirect tomographic measurements?

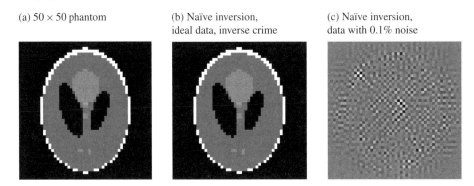

(a) 50×50 phantom   (b) Naïve inversion, ideal data, inverse crime   (c) Naïve inversion, data with 0.1% noise

Figure 2.19. (a) Shepp–Logan phantom at resolution 50×50. (b) Result of naïve inversion (2.50) from noise-free data. The seemingly successful result is not to be trusted because an inverse crime was committed. (c) Result of naïve inversion (2.50) from data contaminated by 0.1% noise. The much worse performance of (c) compared to (b) indicates that naïve inversion is not stable with respect to noise.

Before jumping to such a (wrong) conclusion, let us see what a small amount of measurement noise does to the naïve reconstruction. We add white noise of relative amplitude 0.1% to the sinogram and try formula (2.50) again. The result is shown in Figure 2.19(c), and it consists merely of numerical garbage. This shows that formula (2.50) is not practically useful since real measurements always contain noise.

2.3.6 Naïve reconstruction without inverse crime

We wish to avoid the inverse crime evident in Figure 2.19. To this end, we interpolate our data from tomographic data simulated using the Shepp–Logan phantom on a twice finer grid (100×100) than the grid used in the naïve reconstruction. This can be done conveniently as the phantom is defined analytically using ellipses, so it can be evaluated with arbitrary resolution. The measurement angles are the same. See Figure 2.20 for plots of the data.

Figure 2.21 shows the result of applying naïve reconstruction to the crime-free data. Now the result has unacceptable quality even when there is no added noise.

Exercise 2.3.1. *Let $\mathbf{f} \in \mathbb{R}^8$ be a signal and $\mathbf{p} = [\mathbf{p}_{-1} \quad \mathbf{p}_0 \quad \mathbf{p}_1]^T$ a point spread function. Write down the 8×8 matrix A modeling the one-dimensional convolution (2.11) with the assumption that $\mathbf{f}_{j-\ell} = 0$ for the cases $j - \ell < 1$ and $j - \ell > 8$.*

2.3. Tomographic X-ray projection data

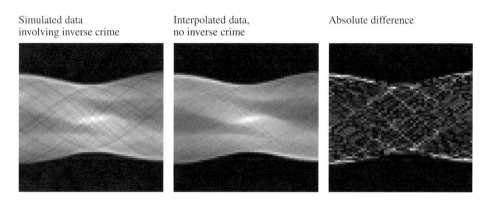

Figure 2.20. Tomographic data with and without inverse crime. Left: ideal data obtained by applying the measurement model matrix to the Shepp–Logan phantom at the final reconstruction resolution of 50×50 pixels. Middle: tomographic data computed from 100×100 Shepp–Logan phantom (at same measurement angles but finer arrangement of X-rays) and interpolated to lower resolution. Right: absolute difference between the two data sets.

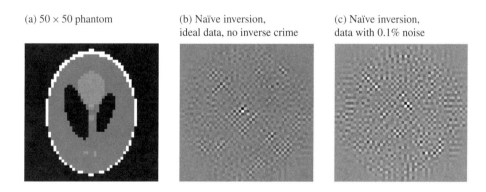

Figure 2.21. (a) Shepp–Logan phantom at resolution 50×50. (b) Result of naïve inversion (2.50) from noise-free data with no inverse crime. (c) Result of naïve inversion (2.50) from data contaminated by 0.1% noise. Compare to Figure 2.19.

Exercise 2.3.2. Assume that an X-ray detector provides the proportional number Mc instead of the actual photon count c. Here $M > 0$ is a positive constant. Show that the calibration procedure described in Figure 2.12 works fine even if M is unknown.

Exercise 2.3.3. Define f and I_0 and I_1 appropriately in the context of the simple example shown in Figure 2.12. Furthermore, describe the measurement data in Figure 2.12 in terms of formula (2.38).

Exercise 2.3.4. In Figure 2.22, thin lines depict pixels and thick lines X-rays. Give a numbering to the nine pixels ($\mathbf{f} \in \mathbb{R}^9$) and to the six X-rays ($\mathbf{m} \in \mathbb{R}^6$), and construct the matrix A for the measurement model $\mathbf{m} = A\mathbf{f}$. The length of the side of a pixel is one.

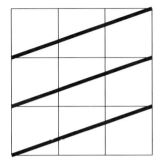 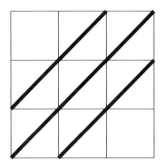

Figure 2.22. Tomographic X-ray measurement configuration related to Exercise 2.3.4. Thick lines depict X-rays, and the length of the side of a pixel is one.

Exercise 2.3.5. *Show that the following targets produce exactly the same data in the measurement model* (2.48):

4	4	5
1	3	4
1	0	2

5	6	2
1	5	2
4	0	−1

(a) *What's wrong with the negative value* −1 *above?* (b) *Can you find more examples that produce the same data but have only nonnegative entries?*

Chapter 3
Ill-posedness in inverse problems

Inverse problems are characterized by ill-posedness—in other words extreme sensitivity to measurement noise and modeling errors. In this chapter we will look at ill-posedness of the infinite-dimensional inverse problem, which requires some knowledge of operator theory, and ill-posedness of the finite-dimensional inverse problem, which typically arises as a discretization of an infinite-dimensional inverse problem. For the necessary background on Banach spaces and compact linear operators, see Appendix B and references therein. We introduce the concept of regularization and the singular value decomposition.

3.1 Forward map and Hadamard's conditions

The examples in Sections 2.1–2.3 should convince the reader that there is something suspicious going on with the inversion of those three simple and indirect measurements. The reason for the observed instability is *ill-posedness*, which we discuss next.

The core of any inverse problem is the *forward map* $\mathcal{A} : \mathcal{D}(A) \to Y$, a mathematical model of the corresponding direct problem. Here X and Y are suitable Hilbert spaces called *model space* and *data space*, respectively, and the subset $\mathcal{D}(A) \subset X$ is the domain of definition of the bounded linear operator $\mathcal{A}$. The forward map is used as a mathematical model of the indirect measurement

$$m = \mathcal{A}f + \varepsilon. \tag{3.1}$$

Here $f \in \mathcal{D}(A) \subset X$ is the quantity of interest, $m \in Y$ is measurement data, and ε is noise satisfying $\|\varepsilon\|_Y \leq \delta$ with some known $\delta > 0$. (Here we denote the measurement by m instead of the vector notation **m** because we want to cover more general data spaces than just $Y = \mathbb{R}^k$.)

Given a particular inverse problem, it is not necessarily straightforward to choose the spaces X and Y and the forward map $\mathcal{A}$. Constructing them must be considered as a nontrivial mathematical modeling task. The following aspects need to be modeled: physical processes involved, technical properties of the measurement device, geometry of the measurement, and possible limitations in data sets.

According to Hadamard, a solution method is called *well-posed* if the following three conditions are satisfied:

H_1: **Existence.** There should be at least one solution.

H_2: **Uniqueness.** There should be at most one solution.

H_3: **Stability.** The solution must depend continuously on data.

We need to study the well-posedness of the naïve inversion $\mathcal{A}^{-1}m$ as a solution to the inverse problem "Given m, find f."

If the forward map is bijective from X to Y and allows a continuous inverse $\mathcal{A}^{-1}$, then naïve inversion satisfies all conditions H_1–H_3 and we are dealing with a well-posed inverse problem.

This book is about *ill-posed inverse problems*, defined as the complement of well-posed problems: at least one of the conditions H_1–H_3 must fail for naïve inversion $\mathcal{A}^{-1}m$. Condition H_1 is violated if the measured noisy data does not belong to the range of the forward map: $\mathcal{A}f + \varepsilon \notin \mathcal{A}(\mathcal{D}(\mathcal{A}))$. Condition H_2 fails if two quantities $f, g \in \mathcal{D}(\mathcal{A})$ give the same measurement: $\mathcal{A}f = \mathcal{A}g$, leading to nonuniqueness. The forward map does not always allow a continuous inverse, not even when restricted to the range $\mathcal{A}(\mathcal{D}(\mathcal{A}))$, so condition H_3 does not hold. See Figure 3.1 for a schematic illustration of the forward map and the related definitions.

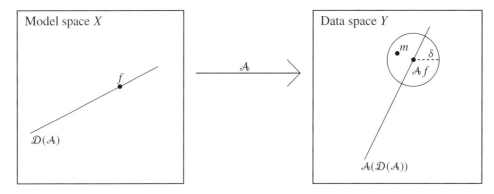

Figure 3.1. Schematic illustration of the linear forward map $\mathcal{A}$. The radius δ indicates the known maximum amplitude of measurement noise.

3.2 Ill-posedness of the backward heat equation

The first effort related to the numerical solution of the backward heat equation was by Fritz John [240] for the problem in one spatial dimension on the whole real line. In this work he also showed that the problem is ill-posed in the sense that the solution does not depend continuously on the data. In [240] he considered the problem on $(-\infty, \infty)$ and expressed the time-dependent solution in terms of the fundamental solution. Taking the same approach for (2.20) leads to an analytic expression for the solution that can be used to solve both the forward and inverse problem. The solution is useful both for understanding the ill-posedness of the problem and as another method of either simulating data or computing a solution. Since one way to completely avoid an inverse crime is to use unrelated methods for reconstruction and data simulation, this is particularly useful here.

3.2. Ill-posedness of the backward heat equation

The forward problem considered here can be solved by the method of eigenfunction expansions, also known as separation of variables, to obtain an integral equation for the solution of the backward heat equation. This will then be discretized by a Galerkin method to obtain another formulation of $\mathcal{A}f = m$.

Consider the forward problem (2.20) on $(0,\pi)$ repeated below for convenience:

$$
\begin{aligned}
u_t - u_{xx} &= 0, \quad 0 < x < \pi, \quad t > 0, \\
u(0,t) &= 0, \quad t > 0, \\
u(\pi,t) &= 0, \quad t > 0, \\
u(x,0) &= f(x), \quad 0 < x < \pi.
\end{aligned}
\tag{3.2}
$$

The technique of separation of variables can be found in nearly every introductory text on PDEs, and we will just include a brief outline of the method as applied to this problem, allowing the reader to go to a PDE's text for a more thorough treatment. Suppose $u(x,t)$ can be written as the product of a function depending on x and a function depending on t. That is, $u(x,t) = X(x)\mathcal{T}(t)$. Substituting this into the PDE results in $X(x)\mathcal{T}'(t) = X''(x)\mathcal{T}(t)$ or

$$
\frac{\mathcal{T}'(t)}{\mathcal{T}(t)} = \frac{X''(x)}{X(x)} = -\lambda^2,
\tag{3.3}
$$

where $-\lambda^2$ is the separation constant. The equation in the spatial variable is a Sturm–Liouville problem with boundary conditions $X(0) = X(\pi) = 0$, which results in eigenvalues n and eigenfunctions $\sin nx$. Thus, $-\lambda^2 = -n^2$ and $\mathcal{T}(t) = A_n e^{-n^2 t}$. By the superposition principle, we then have the general solution

$$
u(x,t) = \sum_{n=1}^{\infty} c_n e^{-n^2 t} \sin nx, \quad c_n = \frac{2}{\pi} \int_0^{\pi} f(y) \sin ny\, dy.
\tag{3.4}
$$

The solution at time t is then given by

$$
\begin{aligned}
u(x,t) &= \sum_{n=1}^{\infty} \frac{2}{\pi} \left(\int_0^{\pi} f(y) \sin ny\, dy \right) e^{-n^2 t} \sin nx \\
&= \int_0^{\pi} k(x,y,t) f(s)\, ds
\end{aligned}
$$

where $k(x,y,t) = \frac{2}{\pi} \sum_{n=1}^{\infty} e^{-n^2 t} \sin nx \sin ny$. Knowledge of the solution at a final time T results in a Fredholm integral equation of the first kind for the initial temperature distribution $f(x)$:

$$
u(x,T) = \int_0^{\pi} k(x,y,t) f(y)\, dy.
\tag{3.5}
$$

It is clear from (3.5) that a unique solution to the backward problem exists, but a straightforward example illustrates that continuous dependence of the solution on the data is violated. Suppose $f(x) = M \sin mx$, where $M > 0$ and m is fixed but arbitrary. Then $u(x,T) = M e^{-m^2 T} \sin mx$. Now $|u(x,T)|$ can be made arbitrarily small by choosing m sufficiently large, but $\|f(x)\|_\infty = M$. Exercise 3.2.1 is to write this in proof format.

To discretize equation (3.5), we will use a Galerkin method and define a partition of $[0,\pi]$ by
$$x_0 = 0, x_1 = \pi/K, \ldots, x_j = j\pi/K, \ldots, x_K = \pi.$$

Let ϕ_i, $i = 1, \ldots, K$, be the orthonormal basis functions
$$\phi_i(x) = \begin{cases} 1, & x_{i-1} \leq x \leq x_i, \\ 0, & \text{otherwise.} \end{cases}$$

Denote the final time data $u(x,T)$ by $g(x)$. Expand f and g in terms of these basis functions as follows:
$$f(x) = \sum_{j=1}^{K} f_j \phi_j(x), \quad g(x) = \sum_{j=1}^{K} b_j \phi_j(x),$$

where
$$f_j = \int_0^{\pi} f(x)\phi_j(x)dx, \quad b_j = \int_0^{\pi} g(x)\phi_j(x)dx.$$

Then
$$\sum_{j=1}^{K} b_j \phi_j(x) = \int_0^{\pi} k(x,y,T) \sum_{j=1}^{K} f_j \phi_j(y) dy.$$

Multiply by $\phi_i(x)$ and integrate from 0 to π:
$$\int_0^{\pi} \sum_{j=1}^{K} b_j \phi_j(x) \phi_i(x) dx = \int_0^{\pi} \int_0^{\pi} k(x,y,T) \sum_{j=1}^{K} f_j \phi_j(y) \phi_i(x) dy\, dx.$$

By the orthogonality of the ϕ_i this becomes
$$b_i \int_{x_{i-1}}^{x_i} 1\, dx = \Delta x\, b_i = \sum_{j=1}^{K} f_j \int_{x_{i-1}}^{x_i} \int_{x_{j-1}}^{x_j} k(x,y,T) \phi_j(y) \phi_i(x) dy\, dx.$$

Letting $i = 1, \ldots, K$, we will write this as a linear system,
$$\left[\begin{array}{c} \\ A_{ij} \\ \\ \end{array}\right] \left[\begin{array}{c} f_1 \\ \vdots \\ f_K \end{array}\right] = \left[\begin{array}{c} b_1 \\ \vdots \\ b_K \end{array}\right],$$

where the elements for the A matrix are given by
$$A_{ij} = \frac{1}{\Delta x} \int_{x_{i-1}}^{x_i} \int_{x_{j-1}}^{x_j} k(x,y,T) \phi_j(y) \phi_i(x) dy\, dx. \tag{3.6}$$

3.2. Ill-posedness of the backward heat equation

Approximating the series for $k(x, y, T)$ by a finite sum with N terms results in

$$A_{ij} \approx \frac{2}{\pi} \sum_{n=1}^{N} e^{-n^2 T} \int_{x_{i-1}}^{x_i} \int_{x_{j-1}}^{x_j} \sin(nx) \sin(ny) \, dy \, dx$$

$$= \frac{2}{\pi} \sum_{n=1}^{N} e^{-n^2 T} \int_{x_{i-1}}^{x_i} \sin(nx) \, dx \int_{x_{j-1}}^{x_j} \sin(ny) \, dy$$

$$= \frac{2}{\pi} \sum_{n=1}^{N} \frac{e^{-n^2 T}}{n^2} (\cos(nx_i) - \cos(nx_{i-1}))((\cos(nx_j) - \cos(nx_{j-1})).$$

Note that no time-stepping is necessary with this method since we have a closed form of the solution. The accuracy of the method on the example of Section 2.2.3 is demonstrated in Table 3.1. One sees that the accuracy improves linearly with the discretization, and it requires quite a fine discretization of $[0, \pi]$ (512 discretization points) to attain an accuracy on the nearly same order as that of the finite difference method of Section 2.2.3.

Table 3.1. Accuracy of the Galerkin method forward solver on the backward heat equation (2.32)–(2.35).

Time T	Δx	$\|u(x,T)\|_\infty$	Relative sup-norm error
0.1	0.0245	6.67	0.0242
0.1	0.0123	6.67	0.0122
0.1	0.0061	6.67	0.0061
0.2	0.0245	4.49	0.0245
0.2	0.0123	4.49	0.0123
0.2	0.0061	4.49	0.0061
0.3	0.0245	3.01	0.0255
0.3	0.0123	3.01	0.0127
0.3	0.0061	3.01	0.0061
0.4	0.0245	2.01	0.0286
0.4	0.0123	2.01	0.0143
0.4	0.0061	2.01	0.0071

Exercise 3.2.1. *Give an ϵ, δ proof that the choice $u(x,0) = M \sin mx$ and $u(x,T) = M e^{-m^2 T} \sin mx$ for $M > 0$ and $m \in \mathbb{N}$ fixed but arbitrary demonstrates that the solution to the backward heat equation does not depend continuously on the data.*

Exercise 3.2.2. *Add a column to Table 3.1 displaying the condition number of A for each time T and Δx.*

Exercise 3.2.3. *Study the numerical forward solution of the heat equation by adding noise to the initial data $u_0(x) = 10 \sin 2x$ in the MATLAB program* `InverseHeatCondData-Simulator.m` *and computing the solution at time $T = 0.4$. Plot the difference between* the computed $u(x,T)$ from a noisy and noise-free initial condition. How does this differ from what we see in Figure 2.8?

3.3 Ill-posedness in the continuous case

This section relies on some knowledge of operator theory, and the reader may need to first cover some or all of the material in the appendices before covering this section. Recall from Appendix B that $\mathcal{L}(X,Y)$ denotes the space of bounded linear mappings from a normed linear space X to a normed linear space Y.

In terms of the linear operator $\mathcal{A} : U \to V$ with U and V being subsets of normed spaces X and Y, the conditions of well-posedness imply

- $\mathcal{A}$ is surjective (onto),
- $\mathcal{A}$ is injective (one-to-one),
- $\mathcal{A}^{-1}$ is continuous.

These three properties are not independent. The following theorem is a consequence of the closed graph theorem, which implies that the range of a continuous operator from a Banach space to a Banach space is closed.

Theorem 3.1 (see [396, Theorem 7.72]). *Let X and Y be Banach spaces. If $\mathcal{A} \in \mathcal{L}(X,Y)$ is bijective, then $\mathcal{A}^{-1} \in \mathcal{L}(Y,X)$.*

By Theorem B.1 of Appendix B, boundedness implies continuity, so the problem $\mathcal{A}f = m$ is well-posed. From the following theorem, we see that an ill-posed problem is guaranteed to result when a compact linear operator acts on an infinite-dimensional Banach space.

Theorem 3.2. *Let X and Y be Banach spaces. Suppose $\mathcal{A} : U \subset X \to Y$ is a compact linear operator, and $\dim U$ is not finite. Then the problem $\mathcal{A}f = m$ is ill-posed.*

Proof. Suppose $\dim U = \infty$ and $\mathcal{A}^{-1} \in \mathcal{L}(Y,U)$, that is, the inverse of $\mathcal{A}$ exists and is continuous. Then $I_U = \mathcal{A}^{-1}\mathcal{A}$ is the composition of a compact and a continuous operator and thus compact. However, the identity map I_U on an infinite-dimensional Banach space is not compact. Thus, $\dim(U)$ is finite. □

Two straightforward consequences of this theorem are given in the following two corollaries. Their proofs are Exercises 3.3.1 and 3.3.2.

Corollary 3.3.1. *If $\mathcal{A}$ is a compact, linear operator from H to H, where H is an infinite-dimensional Hilbert space, and $\mathcal{A}^{-1}$ exists, then $\mathcal{A}^{-1}$ is unbounded.*

Corollary 3.3.2. *If $\mathcal{A} : H \to H$ is a compact, linear operator and $\mathcal{A}^{-1}$ exists, then $\dim(H)$ is finite.*

The following theorem provides one way to prove that an operator is compact. Theorems 3.3 and 3.4 are very important because they explain why the ill-posed behavior of a

3.3. Ill-posedness in the continuous case

large class of linear systems, $\mathcal{A}f = m$, cannot be rigorously detected by examining a single approximation matrix A_k to the operator $\mathcal{A}$, and why one needs to examine the sequence of approximations $\{A_k\}$ for large k. This leads to the test defined by condition (3.18).

Theorem 3.3. *Let* $\mathcal{A} : H \to H$ *be a linear operator and* $K_n \in \mathcal{L}(H,H)$ *a sequence of compact operators. If* $K_n \to \mathcal{A}$ *in the operator norm, then* $\mathcal{A}$ *is a compact operator.*

The proof can be found in [396]; it uses a diagonal sequence argument.

If K is a Hilbert–Schmidt kernel of an integral operator, the resulting integral equation, a Fredholm integral equation of the first kind when $\mathcal{A}f = m$, is an ill-posed problem, as we see from the following theorem.

Theorem 3.4. *Let*

$$(\mathcal{A}f)(x) = \int_\Omega K(x,y) f(y) \, dy$$

with kernel $K \in L^2(\Omega \times \Omega)$ *(i.e., K is a Hilbert–Schmidt kernel from $\Omega \times \Omega \to \mathbb{R}$). Then* $\mathcal{A} \in \mathcal{L}(L^2(\Omega), L^2(\Omega))$ *is compact.*

Proof. First note that if Φ_i is an orthonormal basis for $L^2(\Omega)$, then $\Phi_i(x)\Phi_j(y)$ is an orthonormal basis for $L^2(\Omega \times \Omega)$. Express K in this basis:

$$K(x,y) = \sum_{i,j=1}^\infty k_{i,j} \, \Phi_i(x) \, \Phi_j(y)$$

with

$$k_{i,j} = \int_\Omega \int_\Omega K(x,y) \, \Phi_i(x) \, \Phi_j(y) \, dx \, dy.$$

The convergence is in L^2, and we find

$$\|K\|_{L^2}^2 = \sum_{i,j=1}^\infty k_{i,j}^2.$$

Define

$$K_n(x,y) = \sum_{i,j=1}^n k_{i,j} \, \Phi_i(x) \, \Phi_j(y).$$

Then

$$(\mathcal{A}_n f)(x) = \int_\Omega K_n(x,y) f(y) \, dy = \sum_{i,j=1}^n k_{i,j} \int_\Omega \Phi_i(x) \, \Phi_j(y) f(y) \, dy.$$

Thus, $\mathcal{A}_n$ maps from $L^2(\Omega)$ to a finite-dimensional subspace of $L^2(\Omega)$, which we will denote by $\tilde{L}^2(\Omega)$. By the converse of Corollary 3.3.1, due to the fact that the Range($\mathcal{A}_n$) is finite-dimensional, $\mathcal{A}_n$ is compact. Now we see that

$$\|(\mathcal{A}-\mathcal{A}_n)f\|^2_{\tilde{L}^2(\Omega)} = \left\|\int_\Omega (K(x,y)-K_n(x,y))\,f(y)\,dy\right\|^2_{\tilde{L}^2(\Omega)}$$

$$= \int_\Omega \left|\int_\Omega (K(x,y)-K_n(x,y))\,f(y)\,dy\right|^2 dx$$

$$\le \int_\Omega \left(\int_\Omega |K(x,y)-K_n(x,y)|\,|f(y)|\,dy\right)^2 dx$$

by Cauchy–Schwarz

$$\le \int_\Omega \left(\int_\Omega |K(x,y)-K_n(x,y)|^2\,dx \int_\Omega |f(y)|^2\,dy\right) dx$$

$$= \left(\int_\Omega \int_\Omega |K(x,y)-K_n(x,y)|^2\,dx\,dy\right) \int_\Omega |f(y)|^2\,dy.$$

Therefore,

$$\|\mathcal{A}-\mathcal{A}_n\|^2 \le \int_\Omega \int_\Omega |K(x,y)-K_n(x,y)|^2\,dx\,dy$$

$$= \sum_{i,j=n+1}^\infty |k_{i,j}|^2 \longrightarrow 0 \text{ as } n\to\infty.$$

Thus, $\mathcal{A}_n$ converges to $\mathcal{A}$ in the operator norm, and $\mathcal{A}_n$ is compact. We conclude that $\mathcal{A}$ is compact. $\square$

As we saw in Section 3.2, the one-dimensional heat equation (3.2) on $[0,\pi]$ is solved by

$$u(x,t) = \int_0^\pi k(x,y;t)\,f(y)\,dy,$$

where

$$k(x,y;t) = \frac{2}{\pi} \sum_{n=1}^\infty e^{-n^2 t} \sin(nx)\sin(ny). \tag{3.7}$$

Using an appropriate approximation k_n to the kernel k, one can show that the operator is compact. (See Exercise 3.3.4.) Thus, the inverse problem of determining the initial temperature $u(x,0)$ given $u(x,T)$ is an ill-posed problem. It is also a Fredholm integral equation of the first kind.

Definition 3.3.1. *Let $G \subset \mathbb{R}^n$ be a bounded, open set and $x,y \in G$. A kernel K is* weakly singular *if and only if K is smooth for $x \ne y$ and there exist constants $b > 0$ and $\nu < n$ such that*

$$|K(x,y)| \le b\,|x-y|^{-\nu}.$$

3.3. Ill-posedness in the continuous case

Under certain conditions (see Appendix B or [448]) a weakly singular kernel K results in a compact operator $\mathcal{A}$ defined by

$$(\mathcal{A}f)(x) = \int_\Omega K(x,y) f(y) dy.$$

The condition $\nu < n$ for weakly singular kernels K guarantees that K is absolutely integrable (which is why the singularity is called "weak"). To see this, suppose that $|K(x,y)| \leq b|x-y|^{-\nu}$ and that τ_n is the area of the unit sphere in $\mathbb{R}^n$. Let G be an open bounded subset and $y \in G$ with $\mathbb{B}_y(R) \subset G$. Then

$$\int_G |K(x,y)|\, dx \leq C + \int_{\mathbb{B}_y(R)} |K(x,y)|\, dx \leq C + \int_{\mathbb{B}_y(R)} \frac{b}{|x-y|^\nu}\, dx$$

$$\leq C + \int_{|z| \leq R} \frac{b}{|z|^\nu}\, dz = C + \int_0^R \frac{b\sigma_n r^{n-1}}{r^\nu}\, dr$$

$$= C + \int_0^R b\sigma_n r^{n-1-\nu}\, dr = C + \frac{b\sigma_n R^{n-\nu}}{n-\nu},$$

which is finite.

Returning to the example of X-ray tomography, recall from (2.39) that the data can be modeled as

$$\log \frac{I(x)}{I_0} = -\int_{-y(x)}^{+y(x)} \beta(x,y)\, dy,$$

where $\beta(x,y)$ is the attenuation coefficient and x is in a circle of radius R. If β is radially symmetric, that is, $\beta(x,y) = \beta(\sqrt{x^2+y^2}) = \beta(r)$, a change of variables gives

$$p(x) = \log \frac{I(x)}{I_0} = -2 \int_0^{\sqrt{R^2-x^2}} \beta\left(\sqrt{x^2+y^2}\right) dy$$

$$r = \sqrt{x^2+y^2}, \qquad dr = \frac{y}{r} dy$$

$$= -2 \int_x^R \frac{r}{\sqrt{r^2-x^2}} \beta(r)\, dr$$

$$z = R^2 - r^2, \qquad dz = -2r\, dr$$

$$= \int_{R^2-x^2}^0 \frac{1}{\sqrt{R^2-x^2-z}} \beta(\sqrt{R^2-z})\, dz.$$

With the notations $\tau = R^2 - x^2$ and $\Phi(z) = \beta(\sqrt{R^2-z})$, we find that

$$P(\tau) = p(\pm\sqrt{R^2-\tau}) = -\int_0^\tau \frac{\Phi(z)}{\sqrt{\tau-z}}\, dz.$$

Since $k(\tau,z) = |\tau-z|^{-1/2}$, this kernel is weakly singular.

3.3.1 The singular value expansion

A very thorough treatment of the singular value expansion (SVE) of a kernel can be found in, for example, the text by Hansen [194]. We include some of that material here to provide a better understanding of the underlying continuous problem and also to motivate the SVD. In this section we will consider the Fredholm integral equation of the first kind, in which we wish to find $f(t)$ such that

$$\int_0^1 K(s,t)f(t)dt = g(s), \quad 0 \leq s \leq 1, \quad \|K\|^2_{L^2([0,1]\times[0,1])} \leq C. \tag{3.8}$$

The SVE theorem states that any kernel K with $\|K\|_{L_2} < \infty$ can be written in the form (where the upper limit of the sum is finite in the case of a degenerate kernel)

$$K(s,t) = \sum_{i=1}^\infty \mu_i u_i(s) v_i(t),$$

where u_i and v_i are the *singular functions* of K and μ_i are the *singular values* of K. The singular functions are orthonormal and the singular values satisfy

$$\mu_1 \geq \mu_2 \geq \cdots \geq 0 \quad \text{and} \quad \sum_{i=1}^\infty \mu_i^2 = \|K\|_{L_2}.$$

Also

$$\int_0^1 K(s,t)v_i(t) = \mu_i u_i(s).$$

Multiplying (3.8) by $\mu_i u_i(s)$, then integrating with respect to s and using the fact that the u_i are orthonormal, implies

$$\int_0^1 \mu_i u_i(s) g(s) \, ds = \int_0^1 \mu_i u_i(s) \int_0^1 \left(\sum_{j=1}^\infty \mu_j u_j(s) v_j(t) f(t) \right) dt \, ds$$

$$= \sum_{j=1}^\infty \int_0^1 v_j(t) f(t) \mu_i \mu_j \left(\int_0^1 u_i(s) u_j(s) \, ds \right) dt$$

$$= \int_0^1 v_j(t) f(t) \mu_i^2 \, dt.$$

Thus, $\langle v_i, f \rangle v_i(t) = \frac{1}{\mu_i} \langle u_i, g \rangle v_i(t)$, and we have the following expression for the solution of (3.8):

$$f(t) = \sum_{i=1}^\infty \langle v_i, f \rangle v_i(t) = \sum_{i=1}^\infty \frac{1}{\mu_i} \langle u_i, g \rangle v_i(t). \tag{3.9}$$

3.3. Ill-posedness in the continuous case

The behavior of the singular values and singular functions depends on the properties of the kernel K. As observed in [194], the smoother the kernel K, the faster the singular values μ_i decay. The proof is due to Smithies [421]. The smaller the μ_i, the more oscillatory the functions u_i and v_i will be. There is no known proof of this result in general, but it has been frequently observed. Also, it is noteworthy that the factor $\frac{1}{\mu_i}$ amplifies highly oscillatory contributions in g.

A useful characterization of ill-posedness in terms of the singular values was introduced in [213].

Definition 3.3.2. *If there exists a real number $\alpha > 0$ such that the singular values satisfy $\mu_n = O(n^{-\alpha})$, then α is the* degree of ill-posedness.

1. *If $0 < \alpha \leq 1$, the problem is* mildly ill-posed.

2. *If $\alpha > 1$, the problem is* moderately ill-posed.

3. *If $\mu_n = \mathcal{O}(e^{-\alpha n})$, the problem is* severely ill-posed.

For the problem (3.8) to have a solution f given by (3.9), g must satisfy

$$\sum_{i=1}^{\infty} \left(\frac{\langle u_i, g \rangle}{\mu_i} \right)^2 < \infty.$$

This is known as the *Picard condition*. This is a stronger condition than $g \in L^2(0,1)$ since $g \in L^2$ requires $|\langle u_n, g \rangle| \leq \frac{1}{\sqrt{n}}$ while Picard's condition requires $|\langle u_n, g \rangle| \leq \frac{\mu_n}{\sqrt{n}}$. The Picard condition is equivalent to $g \in \text{Range}(K)$ since

$$\sum_{i=1}^{\infty} \mu_i \langle v_i, f \rangle u_i(t) = \sum_{i=1}^{\infty} \langle u_i, g \rangle u_i(t)$$

implies the right-hand side is a projection of g onto $\text{span}\{u_i\}_{i \in \mathbb{N}}$. Let g_k be the approximation of g obtained by truncating the SVE after k terms:

$$g_k(t) = \sum_{i=1}^{k} \langle u_i, g \rangle u_i(t).$$

Clearly, the Picard condition is satisfied and the solution is given by

$$f_k(t) = \sum_{i=1}^{k} \frac{\langle u_i, g \rangle}{s_i} v_i(t).$$

If $g \notin \text{Range}(K)$, then $g_k \to g$ as $k \to \infty$, but $\|f_k\|_{L^2}^2 = \langle f_k, f_k \rangle \to \infty$. This shows the lack of stability.

In practice g will contain measurement errors and can be expressed as $g = g^{exact} + \eta$, where $g^{exact} \in \text{Range}(K)$ but $\eta \notin \text{Range}(K)$.

3.3.2 Discretization

We introduce two discretizations of (3.8).

• **Quadrature method**: A discretization of (3.8) using a quadrature method based on the points $0 \leq t_1 < t_2 < t_3 \cdots < t_n \leq 1$ in which

$$\int_0^1 \Phi(t)\,dt \approx \sum_{j=1}^n w_j\,\Phi(t_j)$$

can be applied at M discrete points $0 \leq s_1 < s_2 < t_3 < \cdots < s_M \leq 1$ leading to M linear equations

$$\sum_{j=1}^n w_j\,K(s_i,t_j)\,f(t_j) = g(s_i).$$

Defining a matrix A with entries $a_{ij} \equiv w_j\,K(s_i,t_j)$, a vector $\vec{b}$ with entries $b_i \equiv g(s_i)$, and a vector $\vec{f}$ with entries $f_j \equiv f(t_j)$, the system of equations becomes $A\mathbf{f} = \mathbf{g}$. The number of unknowns n does not have to coincide with the number of equations M.

• **Galerkin method**: Let ϕ_i and ψ_j be orthonormal basis functions in $L^2(0,1)$ and define vectors $\mathbf{b}, \xi$ and matrix A by

$$b_i = \int_0^1 g(s)\phi_i(s)\,ds,$$

$$\xi_j = \int_0^1 f(t)\psi_j(t)\,dt,$$

$$a_{ij} = \int_0^1 \int_0^1 K(s,t)\phi_i(s)\psi_j(t)\,ds\,dt,$$

and write

$$K(s,t) = \sum_i \sum_j a_{ij}\,\phi_i(s)\psi_j(t) \text{ and } g(s) = \sum_i b_i\,\phi_i(s).$$

Then the solution ξ of $A\xi = \mathbf{b}$ generates an approximate solution $f(t) = \sum_j \xi_j \psi_j(t)$. If $\psi_i = \phi_i$ and the kernel K is symmetric ($K(t,s) = K(s,t)$), then the resulting matrix A will be symmetric, and this is known as the *Rayleigh–Ritz method*.

Exercise 3.3.1. *Prove Corollary* 3.3.1.

Exercise 3.3.2. *Prove Corollary* 3.3.2.

Exercise 3.3.3. *Prove that if $K \in \mathcal{L}(H,H)$ is compact and $\{e_n\}$ is an orthonormal basis for H, then $K e_n \to 0$ as $n \to \infty$.*

Exercise 3.3.4. *Using an appropriate approximation k_n to the kernel k (3.7), show that the operator $\mathcal{A}$ defined by*

$$(\mathcal{A}f)(x,t) \equiv \int_0^\pi k(x,y;t)f(y)\,dy$$

is compact.

Exercise 3.3.5. *Let $\{e_n\}$ be an orthonormal sequence (not necessarily a basis) in a Hilbert space H. Let $\{\lambda_n\}_{n=1}^{\infty} \subset \mathbb{C} \setminus \{0\}$ with $\sum_{n=1}^{\infty} |\lambda_n|^2 < \infty$. Define $K : H \to H$ by*

$$Kx = \sum_{n=1}^{\infty} \lambda_n \langle x, e_n \rangle e_n.$$

Prove that K is compact and linear.

Exercise 3.3.6. *Let $H = L^2(0, \pi)$ and let*

$$k(x,y) = \sum_{n=1}^{\infty} \frac{1}{n^2} \sin((n+1)x) \sin(ny), \quad x, y \in (0, \pi).$$

Show that the operator $\mathcal{A}$ defined by

$$(\mathcal{A}f)(x,t) \equiv \int_0^{\pi} k(x,y;t) f(y) dy$$

is compact by showing that k is a Hilbert–Schmidt kernel.

3.4 Regularized inversion

Having established that many inverse problems are ill-posed, we now introduce an indispensable technique known as *regularization* or *regularized inversion* to deal with the ill-posedness.

The basic inverse problem related to the indirect measurement (3.1) is this:

(IP$_1$) **Inverse problem:** *Let $m = \mathcal{A}f + \varepsilon$ as in (3.1).*
Given m and $\delta > 0$ with $\|m - \mathcal{A}f\|_Y \leq \delta$, recover f approximately.

In ill-posed inverse problems there does not exist any continuous function from Y to X that would map $\mathcal{A}f \in Y$ to $f \in X$. This can be viewed as extreme sensitivity to perturbations in $\mathcal{A}f$, which is inevitable because of measurement noise. Consequently, it is not straightforward to design a computational method that would map $m = \mathcal{A}f + \varepsilon$ to some point in X near f.

The naïve way of approaching the inverse problem IP$_1$ would be to approximate f by $\mathcal{A}^{-1}m$. Because m and $\mathcal{A}f$ are close to each other, the point $f = \mathcal{A}^{-1}\mathcal{A}f$ must be close to $\mathcal{A}^{-1}m$. This approach is fine for well-posed problems but does not work for ill-posed inverse problems since $\mathcal{A}$ may fail to be either injective or surjective, and even if $\mathcal{A}$ is invertible, its inverse may not be continuous. So what can be done?

Sometimes it helps to consider a restricted problem, such as the following:

(IP$_2$) **Restricted inverse problem:** *Let $m = \mathcal{A}f + \varepsilon$ as in (3.1).*
Given m and $\delta > 0$ with $\|m - \mathcal{A}f\|_Y \leq \delta$, extract any information about f.

For example, one might look for the locations of inclusions in known background material. In any case, the most important property of an inversion method is robustness against noise.

Let us define the notions of *regularization strategy* and *admissible choice of regularization parameter*. We need to assume that $\text{Ker}(\mathcal{A}) = \{0\}$; however, this is not a serious lack of generality since we can always consider the restriction of $\mathcal{A}$ to $(\text{Ker}(\mathcal{A}))^{\perp}$ by working in the linear space of equivalence classes $[f + \text{Ker}(\mathcal{A})]$.

Definition 3.4.1. *Let X and Y be Hilbert spaces. Let $\mathcal{A}: X \to Y$ be an injective bounded linear operator. Consider the measurement $m = \mathcal{A}f + \varepsilon$. A family of linear maps $\mathcal{R}_\alpha : Y \to X$ parameterized by $0 < \alpha < \infty$ is called a* regularization strategy *if*

$$\lim_{\alpha \to 0} \mathcal{R}_\alpha \mathcal{A} f = f \qquad (3.10)$$

for every $f \in X$.

Further, assume we are given a noise level $\delta > 0$ so that $\|m - \mathcal{A}f\|_Y \leq \delta$. A choice of regularization parameter $\alpha = \alpha(\delta)$ as a function of δ is called admissible *if*

$$\alpha(\delta) \to 0 \text{ as } \delta \to 0, \text{ and} \qquad (3.11)$$

$$\sup_m \left\{ \|\mathcal{R}_{\alpha(\delta)} m - f\| : \|\mathcal{A}f - m\| \leq \delta \right\} \to 0 \quad \text{as } \delta \to 0 \text{ for every } f \in X.$$

Figure 3.2 shows a schematic illustration of regularized inversion. See [132, 269] for more details about regularization.

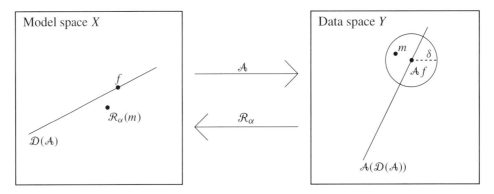

Figure 3.2. Schematic illustration of regularization for linear inverse problems. The linear forward map $\mathcal{A}$ does not necessarily have a continuous inverse. The regularized approximate inverses $\mathcal{R}_\alpha : Y \to X$ are continuous for any choice of regularization parameter $0 < \alpha < \infty$.

This book studies computational inversion methods that apply to practical inverse problems and allow precise mathematical analysis. Developing such methods typically involves considering the following series of questions:

- Q_1: **Uniqueness.** Is the forward map $\mathcal{A}$ injective on $\mathcal{D}(\mathcal{A})$? In other words, are there two different objects $f \neq \widetilde{f}$ producing exactly the same infinite-precision measurement data: $\mathcal{A}f = \mathcal{A}\widetilde{f}$?

- Q_2: **Reconstruction from ideal data.** Assume that $\mathcal{A}$ is injective on $\mathcal{D}(\mathcal{A})$. Given infinite-precision data $\mathcal{A}f$, how to calculate f?

- Q_3: **Conditional stability.** Assume that $\mathcal{A}$ is injective on $\mathcal{D}(\mathcal{A})$. Is it possible to derive a formula of the type $\|f - \widetilde{f}\|_X \leq g(\|\mathcal{A}f - \mathcal{A}\widetilde{f}\|_Y)$, where $g : \mathbb{R} \to \mathbb{R}$ is a continuous function satisfying $g(0) = 0$?

- Q_4: **Characterization of the range.** How to decide whether a given $m \in Y$ belongs to $\mathcal{A}(\mathcal{D}(\mathcal{A}))$, in other words, whether $m = \mathcal{A}f$ for some $f \in X$?

Q5: **Reconstruction from practical data.** Given the noisy data $m = \mathcal{A}f + \varepsilon$, how to approximate f in a noise-robust fashion?

3.5 The SVD for matrices

In practice the continuum measurement model (1.1) of the form $\mathbf{m} = \mathcal{A}f + \varepsilon$ needs to be approximated by a discrete model of the form $\mathbf{m} = A\mathbf{f} + \varepsilon$, where A is a matrix, $\mathbf{f} \in \mathbb{R}^n$, and $\mathbf{m} \in \mathbb{R}^k$. Let us now discuss a tool that allows explicit analysis of Hadamard's conditions in this finite-dimensional setting, namely, *singular value decomposition* (SVD) of A.

We know from matrix algebra that any matrix $A \in \mathbb{R}^{k \times n}$ can be written in the form

$$A = UDV^T, \tag{3.12}$$

where $U \in \mathbb{R}^{k \times k}$ and $V \in \mathbb{R}^{n \times n}$ are orthogonal matrices, that is,

$$U^T U = UU^T = I, \quad V^T V = VV^T = I,$$

and $D \in \mathbb{R}^{k \times n}$ is a diagonal matrix. The rigt-hand side of (3.12) is called the SVD of matrix A, and the diagonal elements d_j are the *singular values* of A. The properties of d_j, and the columns u_i of U, and the columns V_i of V correspond to those of the SVE.

In the case $k = n$ the matrix D is square-shaped: $D = \mathrm{diag}(d_1, \ldots, d_k)$. If $k > n$, then

$$D = \begin{bmatrix} \mathrm{diag}(d_1, \ldots, d_n) \\ \mathbf{0}_{(k-n) \times n} \end{bmatrix} = \begin{bmatrix} d_1 & 0 & \cdots & 0 \\ 0 & d_2 & & \vdots \\ \vdots & & \ddots & \\ 0 & \cdots & \cdots & d_n \\ 0 & \cdots & \cdots & 0 \\ \vdots & & & \vdots \\ 0 & \cdots & \cdots & 0 \end{bmatrix}, \tag{3.13}$$

and in the case $k < n$ the matrix D takes the form

$$\begin{aligned} D &= [\mathrm{diag}(d_1, \ldots, d_k), \mathbf{0}_{k \times (n-k)}] \\ &= \begin{bmatrix} d_1 & 0 & \cdots & 0 & 0 & \cdots & 0 \\ 0 & d_2 & & \vdots & \vdots & & \vdots \\ \vdots & & \ddots & \vdots & \vdots & & \vdots \\ 0 & \cdots & \cdots & d_k & 0 & \cdots & 0 \end{bmatrix}. \end{aligned} \tag{3.14}$$

The diagonal elements d_j are nonnegative and in decreasing order:

$$d_1 \geq d_2 \geq \cdots \geq d_{\min(k,n)} \geq 0. \tag{3.15}$$

Note that some or all of the d_j can be equal to zero.

Recall the definitions of the following linear subspaces related to the matrix A:

$$\begin{aligned} \mathrm{Ker}(A) &= \{\mathbf{f} \in \mathbb{R}^n : A\mathbf{f} = 0\}, \\ \mathrm{Range}(A) &= \{\mathbf{m} \in \mathbb{R}^k : \text{there exists } \mathbf{f} \in \mathbb{R}^n \text{ such that } A\mathbf{f} = \mathbf{m}\}, \\ \mathrm{Coker}(A) &= (\mathrm{Range}(A))^\perp \subset \mathbb{R}^k. \end{aligned}$$

See Figure 3.3 for a diagram illustrating these concepts.

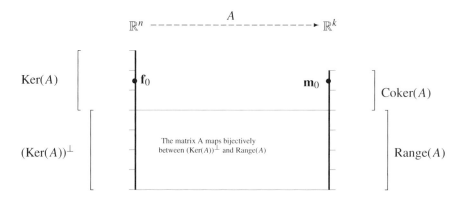

Figure 3.3. This diagram illustrates various linear subspaces related to a matrix mapping $\mathbb{R}^n$ to $\mathbb{R}^k$. The two thick vertical lines represent the linear spaces $\mathbb{R}^n$ and $\mathbb{R}^k$; in this schematic picture we have $n = 7$ and $k = 6$. Furthermore, $\dim(\text{Ker}(A)) = 3$ and $\dim(\text{Range}(A)) = 4$ and $\dim(\text{Coker}(A)) = 2$. Note that the four-dimensional orthogonal complement of $\text{Ker}(A)$ in R^n is mapped in a bijective manner to $\text{Range}(A)$. The points $\mathbf{f}_0 \in \text{Ker}(A)$ and $\mathbf{m}_0 \in \text{Coker}(A)$ are used in the text.

Failure of Hadamard's existence and uniqueness conditions can now be detected from the matrix D. If $k > n$, then $\dim(\text{Range}(A)) < k$ and we can choose a nonzero $\mathbf{m}_0 \in \text{Coker}(A)$ as shown in Figure 3.3. Even in the case $\varepsilon = 0$ we have problems since there does not exist any $\mathbf{f} \in \mathbb{R}^n$ satisfying $A\mathbf{f} = \mathbf{m}_0$, and consequently the existence condition H_1 fails since the output $A^{-1}\mathbf{m}_0$ is not defined for the input $\mathbf{m}_0$. In case of nonzero random noise the situation is even worse since even though $A\mathbf{f} \in \text{Range}(A)$, it might happen that $A\mathbf{f} + \varepsilon \notin \text{Range}(A)$. If $k < n$, then $\dim(\text{Ker}(A)) > 0$ and we can choose a nonzero $\mathbf{f}_0 \in \text{Ker}(A)$ as shown in Figure 3.3. Then even in the case $\varepsilon = 0$ we have a problem of defining $A^{-1}\mathbf{m}$ uniquely since both $A^{-1}\mathbf{m}$ and $A^{-1}\mathbf{m} + \mathbf{f}_0$ satisfy $A(A^{-1}\mathbf{m}) = \mathbf{m} = A(A^{-1}\mathbf{m} + \mathbf{f}_0)$. Thus the uniqueness condition H_2 fails unless we specify an explicit way of dealing with the nullspace of A. Note that if $d_{\min(k,n)} = 0$, then both conditions H_1 and H_2 fail.

The above problems with existence and uniqueness are quite clear since they are related to integer-valued dimensions. In contrast, ill-posedness related to the continuity condition H_3 is more tricky in our finite-dimensional context. Consider the case $n = k$ so A is a square matrix, and assume that A is invertible. In that case we can write

$$A^{-1}\mathbf{m} = A^{-1}(A\mathbf{f} + \varepsilon) = \mathbf{f} + A^{-1}\varepsilon, \tag{3.16}$$

where the error $A^{-1}\varepsilon$ can be bounded by

$$\|A^{-1}\varepsilon\| \leq \|A^{-1}\|\|\varepsilon\|.$$

Now if $\|\varepsilon\|$ is small and $\|A^{-1}\|$ has reasonable size, then the error $A^{-1}\varepsilon$ is small. However, if $\|A^{-1}\|$ is large, then the error $A^{-1}\varepsilon$ can be huge even when ε is small. This is the kind of amplification of noise we see in Figures 2.5, 2.9, and 2.19.

Note that if $\varepsilon = 0$ in (3.16), then we do have $A^{-1}\mathbf{m} = \mathbf{f}$ even if $\|A^{-1}\|$ is large. However, in practice the measurement data always has some noise, and even computer simulated data is corrupted with roundoff errors. Those inevitable perturbations prevent using $A^{-1}\mathbf{m}$ as a reconstruction method for an ill-posed problem.

3.6. SVD for the guiding examples

To define ill-posedness related to the continuity condition H$_3$ rigorously, we must consider the relative sizes of the singular values. Consider the case $n = k$ and $d_n > 0$, when we do not have the above problems with existence or uniqueness. It seems that nothing is wrong since we can invert the matrix A as

$$A^{-1} = VD^{-1}U^T, \qquad D^{-1} = \mathrm{diag}\left(\frac{1}{d_1}, \ldots, \frac{1}{d_k}\right),$$

and define $\mathcal{R}(\mathbf{m}) = A^{-1}\mathbf{m}$ for any $\mathbf{m} \in \mathbb{R}^k$. The problem comes from the *condition number*

$$\mathrm{Cond}(A) := \frac{d_1}{d_k} \qquad (3.17)$$

being large. Namely, if d_1 is several orders of magnitude greater than d_k, then numerical inversion of A becomes difficult since the diagonal inverse matrix D^{-1} contains floating point numbers of very different sizes. This in turn leads to uncontrollable amplification of truncation errors.

Strictly mathematically speaking, though, A is an invertible matrix even in the case of large condition number. For a rigorous definition, we must return to the continuum problem approximated by the matrix model. Suppose that we model the continuum measurement by a sequence of matrices A_k having size $k \times k$ for $k = k_0, k_0 + 1, k_0 + 2, \ldots$ so that the approximation to the forward problem becomes better as k grows. Then we say that condition H$_3$ fails if

$$\lim_{k \to \infty} \mathrm{Cond}(A_k) = \infty. \qquad (3.18)$$

Thus, the ill-posedness cannot be rigorously detected from one approximation matrix A_k but only from the sequence $\{A_k\}_{k=k_0}^{\infty}$. Theorem 3.3 tells us further that the ill-posedness of the problem is not evident from a single approximation matrix A_k to the operator $\mathcal{A}$, but only from a sequence of approximations.

3.6 SVD for the guiding examples

We start with the one-dimensional convolution example introduced in Section 2.1. We compute the singular values of measurement matrices for the two resolutions $k = n = 64$ and $k = n = 128$. See Figure 3.4 for a logarithmic plot of the singular values. The singular values decrease very quickly towards zero but nevertheless stay positive; this is a sign of ill-posedness.

Next we consider the heat propagation model discussed in Section 2.2. Singular values of the matrix A defined by (3.6) for $T = 0.1$ and $T = 0.4$ are shown in Figure 3.5. The distribution of the singular values does not change significantly with Δx, but does have some dependence on the final time T. It is an exercise to compute the condition number of A.

Finally, Figure 3.6 shows singular values of a measurement matrix related to the X-ray tomography problem. The matrix is the one constructed in Section 2.3.5 for the resolution 50×50 and with 50 projection directions.

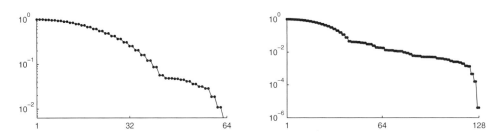

Figure 3.4. Left: plot of the singular values of the 64×64 convolution matrix in logarithmic scale. Right: plot of the singular values of the 128×128 convolution matrix in logarithmic scale.

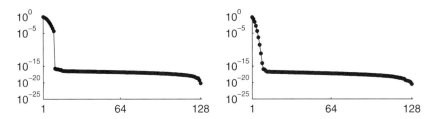

Figure 3.5. Singular values of the matrix A defined by (3.6) related to the backward heat problem. Left: final time $T = 0.1$. Right: final time $T = 0.4$.

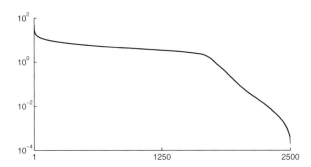

Figure 3.6. Singular values of the X-ray tomography measurement matrix corresponding to the 50×50 resolution and 50 uniformly distributed projection directions.

Exercise 3.6.1. *Use the singular values of the SVD of the guiding examples to classify the problems as mildly ill-posed, moderately ill-posed, or severely ill-posed.*

Exercise 3.6.2. *Calculate the condition number of the matrix A defined by (3.6) for an $n \times n$ matrix with $n = 8, 16, 32, 64,$ and 128. Make a table and compare your results when $T = 0.1$ and $T = 0.4$.*

Chapter 4

Truncated singular value decomposition

In this chapter we introduce the concept of least-squares inversion, arguably the most-used method for solving inverse problems and the method we recommend in general for linear problems. For ill-posed problems the minimization of the discrete least-squares cost functional will ultimately result in an ill-posed linear system that must be regularized. Truncated SVD is the first regularized inversion method discussed in this book. It also provides a way to get a quick estimate on the degree of ill-posedness of the inverse problem. It has the virtues of being straightforward to implement and is popular starting point for regularized inversion.

4.1 Minimum norm solution

Let us define the minimum norm solution of the matrix equation $A\mathbf{f} = \mathbf{m}$, where $\mathbf{f} \in \mathbb{R}^n$ and $\mathbf{m} \in \mathbb{R}^k$ and A has size $k \times n$.

Definition 4.1.1. *A vector* $\mathcal{L}(\mathbf{m}) \in \mathbb{R}^n$ *is called a* least-squares solution *of the equation* $A\mathbf{f} = \mathbf{m}$ *if*

$$\|A\mathcal{L}(\mathbf{m}) - \mathbf{m}\| = \min_{\mathbf{z} \in \mathbb{R}^n} \|A\mathbf{z} - \mathbf{m}\|. \tag{4.1}$$

Furthermore, $\mathcal{L}(\mathbf{m})$ *is called the* minimum norm solution *if*

$$\|\mathcal{L}(\mathbf{m})\| = \inf\{\|\mathbf{z}\| : \mathbf{z} \text{ is a least-squares solution of } A\mathbf{f} = \mathbf{m}\}. \tag{4.2}$$

We need a method for computing minimum norm solutions. For this, write A in the form of its SVD $A = UDV^T$ as explained in Section 3.5. Recall that the singular values are ordered from largest to smallest as shown in (3.15), and let r be the largest index for which the corresponding singular value is nonzero:

$$r = \max\{j \mid 1 \leq j \leq \min(k,n),\ d_j > 0\}. \tag{4.3}$$

The definition of index r is essential in the following analysis, so we will be extra specific:

$$d_1 > 0,\quad d_2 > 0,\quad \cdots\quad d_r > 0,\quad d_{r+1} = 0,\quad \cdots\quad d_{\min(k,n)} = 0.$$

Of course, it is also possible that all singular values are zero, in which case r is not defined and A is the zero matrix, or none of the singular values may be zero.

The next result gives a method to determine the minimum norm solution.

Theorem 4.1. *Let A be a $k \times n$ matrix and denote by $A = UDV^T$ the SVD of A. The minimum norm solution of the equation $A\mathbf{f} = \mathbf{m}$ is given by $A^+\mathbf{m}$, where*

$$A^+\mathbf{m} = VD^+U^T\mathbf{m}$$

and where

$$D^+ = \begin{bmatrix} 1/d_1 & 0 & \cdots & 0 & \cdots & 0 \\ 0 & 1/d_2 & & & & \vdots \\ \vdots & & \ddots & & & \\ & & & 1/d_r & & \\ & & & & 0 & \\ \vdots & & & & \ddots & \vdots \\ 0 & \cdots & & & \cdots & 0 \end{bmatrix} \in \mathbb{R}^{n \times k}.$$

Proof. Write the singular matrix V in the form $V = [V_1 \ V_2 \ \cdots \ V_n]$ and note that the column vectors $V_1, \ldots, V_n$ form an orthogonal basis for $\mathbb{R}^n$. We write $\mathbf{f} \in \mathbb{R}^n$ as a linear combination $\mathbf{f} = \sum_{j=1}^n a_j V_j = V\mathbf{a}$, and our goal is to find such coefficients $a_1, \ldots, a_n$ that $\mathbf{f}$ becomes a minimum norm solution.

Set $\mathbf{m}' = U^T\mathbf{m} \in \mathbb{R}^k$ and compute

$$\begin{aligned} \|A\mathbf{f} - \mathbf{m}\|^2 &= \|UDV^T V\mathbf{a} - U\mathbf{m}'\|^2 \\ &= \|D\mathbf{a} - \mathbf{m}'\|^2 \\ &= \sum_{j=1}^r (d_j a_j - \mathbf{m}'_j)^2 + \sum_{j=r+1}^k (\mathbf{m}'_j)^2, \end{aligned} \quad (4.4)$$

where we used the orthogonality of U (namely, $\|U\mathbf{z}\| = \|\mathbf{z}\|$ for any vector $\mathbf{z} \in \mathbb{R}^k$). Now since d_j and $\mathbf{m}'_j$ are given and fixed, the expression (4.4) attains its minimum when $a_j = \mathbf{m}'_j/d_j$ for $j = 1, \ldots, r$. So any $\mathbf{f}$ of the form

$$\mathbf{f} = V \begin{bmatrix} d_1^{-1}\mathbf{m}'_1 \\ \vdots \\ d_r^{-1}\mathbf{m}'_r \\ a_{r+1} \\ \vdots \\ a_n \end{bmatrix}$$

is a least-squares solution. The smallest norm $\|\mathbf{f}\|$ is clearly given by the choice $a_j = 0$ for $r < j \leq n$, and so the minimum norm solution is uniquely determined by the formula $\mathbf{a} = D^+\mathbf{m}'$. □

Definition 4.1.2. *The matrix A^+ is called the* pseudoinverse, *or the* Moore–Penrose inverse *of A.*

How does the pseudoinverse take care of Hadamard's existence and uniqueness conditions H_1 and H_2? First of all, in the case of nonexistence, $\text{Coker}(A)$ is nontrivial, and any vector $\mathbf{m} \in \mathbb{R}^k$ can be written as the sum $\mathbf{m} = \mathbf{m}_A + (\mathbf{m}_A)^\perp$, where $\mathbf{m}_A \in \text{range}(A)$ and $(\mathbf{m}_A)^\perp \in \text{Coker}(A)$ and $\mathbf{m}_A \cdot (\mathbf{m}_A)^\perp = 0$. Then A^+ simply maps $(\mathbf{m}_A)^\perp$ to zero. Second, in the case of nonuniqueness, $\text{Ker}(A)$ is nontrivial, and we need to choose the reconstructed vector from a whole linear subspace of candidates. Using A^+ chooses the candidate with smallest norm.

4.2 Truncated SVD

After the analysis in Section 4.1 it remains to discuss Hadamard's continuity condition H_3. Recall from Section 3.5 that we may run into problems if d_r is much smaller than d_1. In that case even the use of the pseudoinverse results in numerical instability because the diagonal element d_r^{-1} appearing in D^+ is much larger than d_1^{-1}. We can overcome this by using the truncated SVD.

Definition 4.2.1. *For any $\alpha > 0$ define the truncated SVD (TSVD) by $A_\alpha^+ = V D_\alpha^+ U^T$, where*

$$D_\alpha^+ = \begin{bmatrix} 1/d_1 & 0 & \cdots & 0 & \cdots & 0 \\ 0 & 1/d_2 & & & & \vdots \\ \vdots & & \ddots & & & \\ & & & 1/d_{r_\alpha} & & \\ & & & & 0 & \\ \vdots & & & & \ddots & \vdots \\ 0 & \cdots & & & \cdots & 0 \end{bmatrix} \in \mathbb{R}^{n \times k}$$

and

$$r_\alpha = \min\left\{r, \max\{j \mid 1 \leq j \leq \min(k,n), d_j > \alpha\}\right\}. \tag{4.5}$$

We can then define a reconstruction function $\mathcal{L}_\alpha$ by the formula

$$\mathcal{L}_\alpha(m) = V D_\alpha^+ U^T m. \tag{4.6}$$

Then all of Hadamard's conditions hold: $\mathcal{L}_\alpha : \mathbb{R}^k \to \mathbb{R}^n$ is a well-defined, single-valued linear mapping with norm

$$\|\mathcal{L}_\alpha\| = \|V D_\alpha^+ U^T\| \leq \|V\| \|D_\alpha^+\| \|U^T\| = \|D_\alpha^+\| = d_{r_\alpha}^{-1},$$

implying continuity. To be specific, while of course $\|\mathcal{L}_\alpha\| = d_{r_\alpha}^{-1} < \infty$ implies the linear mapping is continuous in the mathematical sense, equation (3.16) now takes the form

$$\mathcal{L}_\alpha(m) = V D_\alpha^+ U^T (A\mathbf{f} + \varepsilon) = V D_\alpha^+ D V^T \mathbf{f} + V D_\alpha^+ U^T \varepsilon. \tag{4.7}$$

The vector $VD_\alpha^+ DV^T \mathbf{f}$ is an approximation to $\mathbf{f}$, and the error term can be estimated as follows:
$$\|VD_\alpha^+ U^T \varepsilon\| \leq \|VD_\alpha^+ U^T\| \|\varepsilon\| = \|D_\alpha^+\| \|\varepsilon\| = d_{r_\alpha}^{-1} \|\varepsilon\|. \quad (4.8)$$
By the ordering (3.15) of singular values we have
$$d_1^{-1} \leq d_2^{-1} \leq \cdots \leq d_r^{-1},$$
and by (4.8) the noise gets amplified in the inversion less and less if we keep fewer singular values (or, equivalently, increase α).

We see from definition (4.6) and by denoting $\mathbf{a} := D_\alpha^+ U^T m$ that the reconstruction is a linear combination of the columns $V_1, \ldots, V_n$ of the matrix $V = [V_1 \quad V_2 \quad \cdots \quad V_n]$:
$$\mathcal{L}_\alpha(m) = V\mathbf{a} = a_1 V_1 + \cdots + a_n V_n.$$

Thus the columns $V_1, \ldots, V_n$, called *singular vectors*, are the building blocks of any reconstruction using TSVD.

4.3 Measuring the quality of reconstructions

It is important to have a quantitative measure available for the closeness of a reconstruction to the actual measured object. Of course, such a measure makes sense only with simulated data, when the original object is known.

Denote the quantity of interest by $\mathbf{f} \in \mathbb{R}^n$ and assume that a regularized reconstruction $T(\mathbf{m}) \in \mathbb{R}^n$ is computed from noisy data $\mathbf{m} = A\mathbf{f} + \varepsilon \in \mathbb{R}^k$. Throughout the book we use the following definition of relative error:
$$\frac{\|\mathbf{f} - T(\mathbf{f})\|_2}{\|\mathbf{f}\|_2} \cdot 100\%, \quad (4.9)$$
where $\|\mathbf{f}\|_2^2 = \sum_{j=1}^n |\mathbf{f}_j|^2$.

Sometimes it is more informative to study relative errors in norms other than the Euclidean norm. For example, we may consider
$$\frac{\|\mathbf{f} - T(\mathbf{f})\|_p}{\|\mathbf{f}\|_p} \cdot 100\%, \quad (4.10)$$
where the p-norm is defined for $1 < p < \infty$ by
$$\|\mathbf{f}\|_p^p = \sum_{j=1}^n |\mathbf{f}_j|^p$$
and for $p = \infty$ by
$$\|\mathbf{f}\|_\infty = \max_{j=1,\ldots,n} |\mathbf{f}_j|.$$

It is important to realize that the relative error measures (4.9) and (4.10) do not necessarily coincide with the visual quality of the reconstruction as judged by a human observer. The measures can be used for detecting and quantifying gross deviations from the desired outcome, but if one reconstruction has relative error 20% and another one 22%, the latter is not necessarily worse. Defining what is a good reconstruction depends on the application. However, it is always valuable to calculate a quantitative error such as (4.9) or (4.10) so that gross errors will not go undetected.

4.4 TSVD for the guiding examples

4.4.1 TSVD for deconvolution

Let us study TSVD in the case of one-dimensional deconvolution. The right-hand column of Figure 4.1 shows reconstructions computed from data corrupted by random noise with standard deviation $\sigma = 0.05 \cdot \max |f(x)|$. The highest indexed singular vector used in each reconstruction is shown in the left-hand column.

We next study the noise-robustness of TSVD reconstructions. In Figure 4.2, we used 50 realizations of random noise for computing TSVD reconstructions using the first 10 singular vectors, and another 50 realizations for computing reconstructions using the first 40 singular vectors. We see in the figure that there is much more variation in the set of reconstructions using 40 singular vectors, implying that the reconstructions with only 10 singular vectors are more robust against noise. However, the reconstructions from 10 singular vectors are also more erroneous, so there is a trade-off between accuracy and noise-robustness. This is a typical feature of solutions to inverse problems.

4.4.2 TSVD for backward heat propagation

To study the method of TSVD for the backward heat propagation problem (2.21)–(2.22), we apply the TSVD to the matrix defined by (3.6), and to avoid an inverse crime, we create data by the finite difference solution of the forward problem presented in Section 2.2.2. The results of this method applied to the example with solution $u(x,0) = 10 \sin 2x$, $0 < x < \pi$, are shown in Figure 4.3 using noise-free data and noisy data with $\sigma = 0.10 \cdot \max |u(x,T)|$. Notice that this method is considerably more robust than the method in Section 2.2.3, and we are able to use a much higher noise level and still obtain good reconstructions of the initial temperature. The reconstructions in Figure 4.3 are from data measured at $T = 0.1$. It is an exercise to reconstruct the initial temperature from later times such as $T = 0.4$ (see Exercise 4.4.1).

4.4.3 TSVD for X-ray tomography

We consider the tomographic measurement matrix A constructed in Section 2.3.5 for the resolution 50×50 and with 50 projection directions. We compute numerically the SVD of A, which is already quite intensive computationally as the size of A is 3750×2500. The singular values are plotted in Figures 3.6 and 4.4.

We use formula (4.6) to compute reconstructions of the 50×50 Shepp–Logan phantom from noisy data with no inverse crime. We choose the truncation indices 200, 300, 1500, and 2000; the first two were chosen because the singular vectors 200 and 300 are especially beautiful. The reason for including number 1500 is that it is one of the choices yielding the minimal relative error of 48% using formula (4.9). Namely, all choices between 1440 and 1680 yield relative error of 48%, and any other choice leads to a larger error. Number 2000 was included to show what happens to the reconstruction when we start to use the really small singular values. Figure 4.4 shows the reconstructions and some of the singular vectors.

Exercise 4.4.1. *Compute reconstructions using TSVD for the example of Section* 4.4.2 *from data measured at time* $T = 0.4$. *Use the finite difference method to create the data*

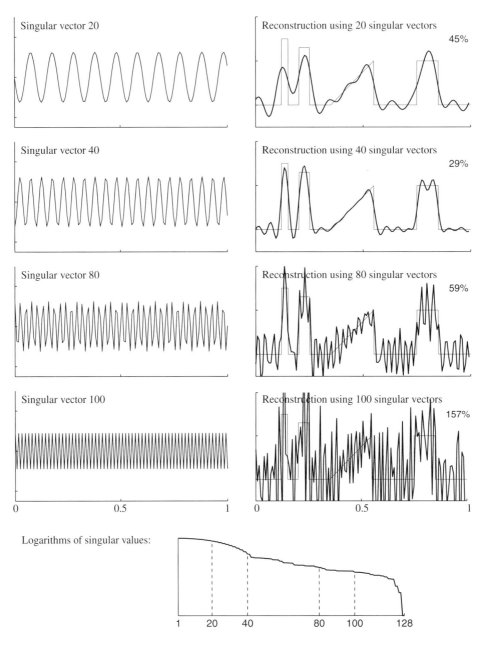

Figure 4.1. Left column: singular vectors related to the one-dimensional 128 × 128 convolution matrix. Right column: reconstructions (thick line) using all singular vectors up to the number shown on the left in the TSVD. Original signal is drawn with thin line for comparison. The percentages shown are relative errors of reconstructions. Bottom: logarithmic plot of the singular values.

4.4. TSVD for the guiding examples

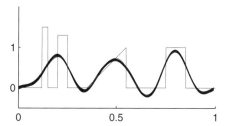 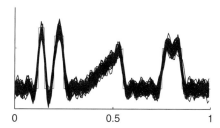

Figure 4.2. Study of noise-robustness of TSVD reconstructions. Here $\sigma = 0.05 \cdot \max |f(x)|$, and 50 independent realizations of random white noise was used for both of the two plots. Left: 50 reconstructions using the first 10 singular vectors. Right: 50 reconstructions using the first 40 singular vectors. The axis limits and scales are the same in both plots.

and compute reconstructions from noise-free data and data corrupted with noise with $\sigma = 0.10 \cdot \max |f(x)|$. *Produce plots as in Figure* 4.3.

Exercise 4.4.2. *Using the kernel*

$$f(x) = \begin{cases} 5x - 0.5 & \text{for } 0.1 \leq x \leq 0.2, \\ -5x + 1.5 & \text{for } 0.2 < x \leq 0.3, \\ 1 & \text{for } 0.5 \leq x \leq 0.6, \\ 0.2 & \text{for } 0.6 < x < 0.7, \\ 1.3 & \text{for } 0.7 \leq x \leq 0.8, \\ 0 & \text{otherwise}, \end{cases}$$

add noise with amplitude $\sigma = 0.05$ *and solve the one-dimensional deconvolution problem, reproducing plots as in Figure* 4.1.

Exercise 4.4.3. *Using the noisy kernel from Exercise* 4.4.2 *study the noise robustness of the TSVD reconstructions.*

Exercise 4.4.4. *Using data at resolution* 50×50 *and only* 12 *projection directions reproduce* Figure 4.4 *with the corresponding results.*

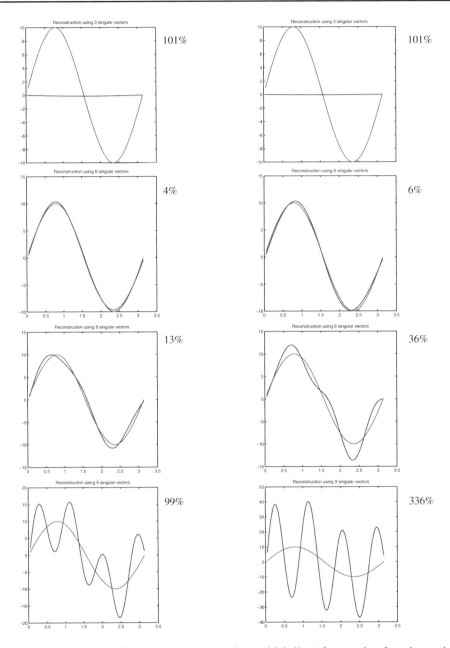

Figure 4.3. Left column: reconstructions (thick line) from noise-free data using all singular vectors up to the number shown. The actual solution is drawn with a thin line for comparison. The percentages shown are relative sup-norm errors of the reconstructions. Right column: reconstructions (thick line) from noisy data with $\sigma = 0.10 \cdot \max |u(x,T)|$ using all singular vectors up to the number shown. The actual solution is drawn with a thin line for comparison. The percentages shown are relative sup-norm errors of the reconstructions.

4.4. TSVD for the guiding examples

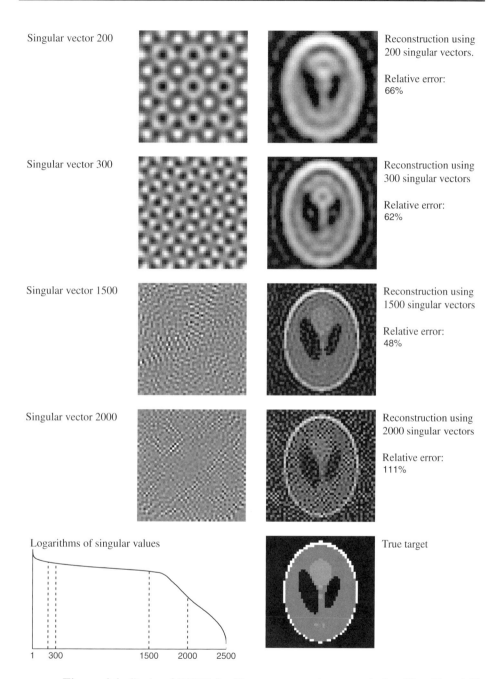

Figure 4.4. Study of TSVD for X-ray tomography at resolution 50×50 and 50 projection directions. Left column: singular vectors related to tomography matrix. Right column: reconstructions using all singular vectors up to the number shown on the left in the TSVD. The percentages shown are relative square norm errors of reconstructions.

Chapter 5
Tikhonov regularization

Tikhonov regularization is typically the method of first choice for linear problems. It provides some smoothing, and generalized Tikhonov regularization provides an opportunity to incorporate known properties of the solution into the solution method. As we shall see in this chapter, it is simple to implement, but introduces the classic question, "How can I choose the regularization parameter?" Two popular but not always reliable methods for selecting the regularization parameter, the Morozov discrepancy principle and the L-curve method, are explained in Section 5.4. In fact, there is no known method for choosing the regularization parameter that results in an optimal solution, but it is important to be aware of these widely used methods. The results of using Tikhonov regularization on our three guiding examples is demonstrated in this chapter and through the exercises.

5.1 Classical Tikhonov regularization

The Tikhonov regularized solution of equation $\mathbf{m} = A\mathbf{f} + \varepsilon$ is the vector $T_\alpha(\mathbf{m}) \in \mathbb{R}^n$ that minimizes the expression

$$\|AT_\alpha(\mathbf{m}) - \mathbf{m}\|^2 + \alpha \|T_\alpha(\mathbf{m})\|^2,$$

where $\alpha > 0$ is called a regularization parameter. We denote

$$T_\alpha(\mathbf{m}) = \arg\min_{\mathbf{z} \in \mathbb{R}^n} \left\{ \|A\mathbf{z} - \mathbf{m}\|^2 + \alpha \|\mathbf{z}\|^2 \right\}. \tag{5.1}$$

Tikhonov regularization can be understood as a balance between two requirements:

(i) $T_\alpha(\mathbf{m})$ should give a small residual $AT_\alpha(\mathbf{m}) - \mathbf{m}$.

(ii) $T_\alpha(\mathbf{m})$ should be small in L^2-norm.

The regularization parameter $\alpha > 0$ can be used to "tune" the balance.

Note that in inverse problems there are typically infinitely many choices of $T_\alpha(\mathbf{m})$ satisfying (i), and one of the roles of (ii) is to make the solution unique.

Theorem 5.1. *Let A be a $k \times n$ matrix. The Tikhonov regularized solution for equation $\mathbf{m} = A\mathbf{f} + \varepsilon$ is given by*

$$T_\alpha(\mathbf{m}) = V \mathcal{D}_\alpha^+ U^T \mathbf{m}, \tag{5.2}$$

where $A = UDV^T$ is the SVD, and

$$\mathcal{D}_\alpha^+ = \mathrm{diag}\left(\frac{d_1}{d_1^2 + \alpha}, \ldots, \frac{d_{\min(k,n)}}{d_{\min(k,n)}^2 + \alpha}\right) \in \mathbb{R}^{n \times k}. \tag{5.3}$$

Proof. Write $T_\alpha(\mathbf{m}) \in \mathbb{R}^n$ as linear combination of column vectors of the matrix V: $T_\alpha(\mathbf{m}) = \sum_{j=1}^n a_j V_j = V\mathbf{a}$. Set $\mathbf{m}' = U^T \mathbf{m}$ and compute

$$\|AT_\alpha(\mathbf{m}) - \mathbf{m}\|^2 + \alpha \|T_\alpha(\mathbf{m})\|^2$$
$$= \|UDV^T V\mathbf{a} - UU^T\mathbf{m}\|^2 + \alpha \|V\mathbf{a}\|^2$$
$$= \|D\mathbf{a} - \mathbf{m}'\|^2 + \alpha \|\mathbf{a}\|^2$$
$$= \sum_{j=1}^r (d_j a_j - \mathbf{m}'_j)^2 + \sum_{j=r+1}^k (\mathbf{m}'_j)^2 + \alpha \sum_{j=1}^n a_j^2$$
$$= \sum_{j=1}^r \left(d_j^2 + \alpha\right)\left(a_j^2 - 2\frac{d_j \mathbf{m}'_j}{d_j^2 + \alpha} a_j\right) + \alpha \sum_{j=r+1}^n a_j^2 + \sum_{j=1}^k (\mathbf{m}'_j)^2 \tag{5.4}$$
$$= \sum_{j=1}^r \left(d_j^2 + \alpha\right)\left(a_j - \frac{d_j \mathbf{m}'_j}{d_j^2 + \alpha}\right)^2 + \alpha \sum_{j=r+1}^n a_j^2 \tag{5.5}$$
$$- \sum_{j=1}^r \frac{(d_j \mathbf{m}'_j)^2}{d_j^2 + \alpha} + \sum_{j=1}^k (\mathbf{m}'_j)^2,$$

where completing the square in the leftmost term in (5.4) yields (5.5). Our task is to choose such values for the parameters $a_1, \ldots, a_n$ that (5.5) attains its minimum. Clearly the correct choice is

$$a_j = \begin{cases} \dfrac{d_j}{d_j^2 + \alpha} \mathbf{m}'_j, & 1 \le j \le r, \\ 0, & r+1 \le j \le n, \end{cases}$$

or, in short, $\mathbf{a} = \mathcal{D}_\alpha^+ \mathbf{m}'$. □

Recall from Section 4.2 that the TSVD solution is given by

$$\mathcal{L}_\alpha(\mathbf{m}) = V D_\alpha^+ U^T \mathbf{m} = \sum_{i=1}^{r_\alpha} \frac{\mathbf{u_i}^T \mathbf{m}}{d_i} \mathbf{v_i}, \tag{5.6}$$

where the $\mathbf{u_i}$ and $\mathbf{v_i}$ are the columns of the matrices U and V, respectively, while

$$T_\alpha(\mathbf{m}) = V \mathcal{D}_\alpha^+ U^T \mathbf{m} = \sum_{i=1}^r \left(\frac{d_i^2}{d_i^2 + \alpha}\right) \frac{\mathbf{u_i}^T \mathbf{m}}{d_i} \mathbf{v_i} = \sum_{i=1}^r \left(\frac{d_i}{d_i^2 + \alpha}\right) (\mathbf{u_i}^T \mathbf{m}) \mathbf{v_i}. \tag{5.7}$$

Recall that $r_\alpha \le r$. Thus, we see that the entries of $\mathcal{D}_\alpha^+$ effectively weight the contributions of the vectors in the SVD, and if the regularization parameter α is sufficiently small (smaller

5.1. Classical Tikhonov regularization

than the smallest singular value), the Tikhonov regularized solution is essentially the same as the solution obtained by the SVD. By increasing α, less weight is placed on the small singular values, which also correspond to the highly oscillatory right singular vectors.

5.1.1 Tikhonov regularization for the deconvolution problem

Let us apply Tikhonov regularization to our basic test problem of one-dimensional deconvolution. In Figure 5.1 we see the Tikhonov regularized solutions corresponding to four different choices of regularization parameter. To investigate the noise-robustness of the Tikhonov regularized solutions, reconstructions from 50 data sets with independent realizations of random white noise with $\sigma = 0.05$ were computed. The plots are superimposed in Figure 5.2.

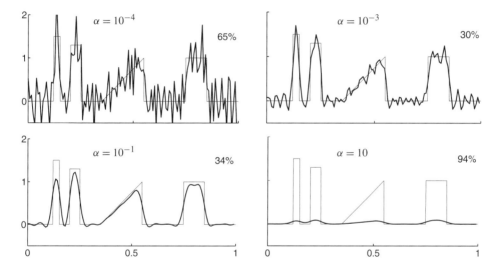

Figure 5.1. Tikhonov regularized reconstructions. The percentages shown are relative errors of reconstructions.

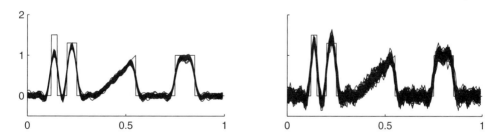

Figure 5.2. Study of noise-robustness of Tikhonov regularized reconstructions. Here the standard deviation of the noise level is $\sigma = 0.05 \cdot \max|f(x)|$, and 50 independent realizations of random white noise was used for both of the two plots. Left: regularization parameter $\alpha = 10^{-1}$. Right: regularization parameter $\alpha = 10^{-2}$.

5.1.2 Tikhonov regularization for backward heat propagation

We apply Tikhonov regularization to the backward heat propagation problem with final time data at $T = 0.1$, computing data **m** from the finite difference method described in Section 4.4.2 with noise added with standard deviation $\sigma = 0.10 \cdot \max |u(x,T)|$. The matrix A was constructed as in (3.6), and the minimization problem (5.1) was solved for regularization parameter α ranging from 10^{-5} to 10^{-2}. The results for a selection of regularization parameters are given in Figure 5.3. It is clear from these figures that a choice of α that is too large overdamps the solution, while a small choice that is too small results in wild oscillations. However, here the choice $\alpha = 10^{-2}$ is just right and results in a good reconstruction. Exercises applying Tikhonov regularization to the problem of backward heat propagation are found in Section 5.4 after we have discussed methods of choosing the regularization parameter.

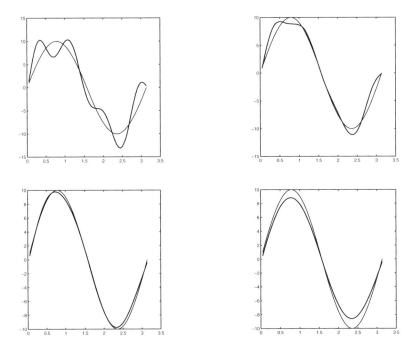

Figure 5.3. Tikhonov regularized reconstructions (dark line) with the actual solution superimposed (thin line) for the backward heat propagation problem of Section 2.2. The regularization parameter is $\alpha = 9 \times 10^{-5}$ (upper left), $\alpha = 4 \times 10^{-4}$ (upper right), $\alpha = 1 \times 10^{-2}$ (lower left), $\alpha = 7 \times 10^{-2}$ (lower right). The L^2-norm relative errors of reconstructions are (in the same order) 44%, 21%, 7%, and 14%.

Exercise 5.1.1. *Show that the matrix $A^T A + \alpha I$ is always invertible when $\alpha > 0$ and A is an arbitrary $k \times n$ matrix. Hint: use SVD.*

Exercise 5.1.2. *Using the kernel from Exercise 4.4.2 with added noise of amplitude $\sigma = 0.05$, compute deconvolutions using classical Tikhonov regularization with regularization* *parameter ranging from $\alpha = 10^0$ to $\alpha = 10^{-5}$ in increments of powers of* 10. *Create plots as in Figure* 5.1.

5.2 Normal equations and stacked form

Consider the quadratic functional $Q_\alpha : \mathbb{R}^n \to \mathbb{R}$ defined by

$$Q_\alpha(\mathbf{f}) = \|A\mathbf{f} - \mathbf{m}\|^2 + \alpha\|\mathbf{f}\|^2.$$

It can be proven that Q_α has a unique minimum for any $\alpha > 0$. The minimizer $T_\alpha(\mathbf{m})$ (i.e., the Tikhonov regularized solution of $\mathbf{m} = A\mathbf{f} + \varepsilon$) satisfies

$$0 = \frac{d}{dt}\left\{\|A(T_\alpha(\mathbf{m}) + t\mathbf{w}) - \mathbf{m}\|^2 + \alpha\|T_\alpha(\mathbf{m}) + t\mathbf{w}\|^2\right\}\Big|_{t=0}$$

for any $\mathbf{w} \in \mathbb{R}^n$.

Compute

$$\frac{d}{dt}\|A(T_\alpha(\mathbf{m}) + t\mathbf{w}) - \mathbf{m}\|^2\Big|_{t=0}$$
$$= \frac{d}{dt}\langle AT_\alpha(\mathbf{m}) + tA\mathbf{w} - \mathbf{m}, AT_\alpha(\mathbf{m}) + tA\mathbf{w} - \mathbf{m}\rangle\Big|_{t=0}$$
$$= \frac{d}{dt}\Big\{\|AT_\alpha(\mathbf{m})\|^2 + 2t\langle AT_\alpha(\mathbf{m}), A\mathbf{w}\rangle + t^2\|A\mathbf{w}\|^2$$
$$\quad - 2t\langle \mathbf{m}, A\mathbf{w}\rangle - 2\langle AT_\alpha(\mathbf{m}), \mathbf{m}\rangle + \|\mathbf{m}\|^2\Big\}\Big|_{t=0}$$
$$= 2\langle AT_\alpha(\mathbf{m}), A\mathbf{w}\rangle - 2\langle \mathbf{m}, A\mathbf{w}\rangle,$$

and

$$\frac{d}{dt}\alpha\langle T_\alpha(\mathbf{m}) + t\mathbf{w}, T_\alpha(\mathbf{m}) + t\mathbf{w}\rangle\Big|_{t=0}$$
$$= \alpha\frac{d}{dt}\Big\{\|T_\alpha(\mathbf{m})\|^2 + 2t\langle T_\alpha(\mathbf{m}), \mathbf{w}\rangle + t^2\|\mathbf{w}\|^2\Big\}\Big|_{t=0}$$
$$= 2\alpha\langle T_\alpha(\mathbf{m}), \mathbf{w}\rangle.$$

Thus, we have $\langle AT_\alpha(\mathbf{m}) - \mathbf{m}, A\mathbf{w}\rangle + \alpha\langle T_\alpha(\mathbf{m}), \mathbf{w}\rangle = 0$, and by taking the transpose,

$$\langle A^T AT_\alpha(\mathbf{m}) - A^T\mathbf{m}, \mathbf{w}\rangle + \alpha\langle T_\alpha(\mathbf{m}), \mathbf{w}\rangle = 0.$$

This results in the variational form

$$\langle (A^T A + \alpha I)T_\alpha(\mathbf{m}) - A^T\mathbf{m}, \mathbf{w}\rangle = 0. \tag{5.8}$$

Since (5.8) holds for any nonzero $\mathbf{w} \in \mathbb{R}^n$, we necessarily have $(A^T A + \alpha I)T_\alpha(\mathbf{m}) = A^T\mathbf{m}$. So the Tikhonov regularized solution $T_\alpha(\mathbf{m})$ satisfies

$$T_\alpha(\mathbf{m}) = (A^T A + \alpha I)^{-1} A^T\mathbf{m}, \tag{5.9}$$

and actually (5.9) can be used for computing $T_\alpha(\mathbf{m})$ defined in the basic situation (5.1).

In the generalized case of (5.14) we get by a similar computation,

$$T_\alpha(\mathbf{m}) = (A^T A + \alpha L^T L)^{-1} A^T \mathbf{m}. \tag{5.10}$$

Next we will derive a computationally attractive *stacked form* version of (5.2).

We rethink problem (5.2) so that we have two measurements on $\mathbf{f}$ that we minimize simultaneously in the least-squares sense. Namely, we consider both equations $A\mathbf{f} = \mathbf{m}$ and $\mathbf{f} = 0$ as independent measurements of the same object $\mathbf{f}$, where $A \in \mathbb{R}^{k \times n}$. Now we stack the matrices and right-hand sides so that the regularization parameter $\alpha > 0$ is involved correctly:

$$\begin{bmatrix} A \\ \sqrt{\alpha} \end{bmatrix} \mathbf{f} = \begin{bmatrix} \mathbf{m} \\ 0 \end{bmatrix}. \tag{5.11}$$

We write (5.11) as $\widetilde{A}\mathbf{f} = \widetilde{\mathbf{m}}$ and solve for $T_\alpha(\mathbf{m})$ defined in (5.10) in MATLAB by

$$\mathbf{f} = \widetilde{A} \backslash \widetilde{\mathbf{m}}, \tag{5.12}$$

where $\backslash$ stands for finding the least-squares solution. This is a good method for medium-dimensional inverse problems, where n and k are of the order $\sim 10^3$. Formula (5.12) is applicable to higher-dimensional problems than formula (5.2) since there is no need to compute the SVD for (5.12).

Why would (5.12) be equivalent to (5.9)? In general, a computation similar to the above shows that a vector $\mathbf{z}_0$, defined as the minimizer

$$\mathbf{z}_0 = \arg\min_\mathbf{z} \|B\mathbf{z} - \mathbf{b}\|^2,$$

satisfies the normal equations $B^T B \mathbf{z}_0 = B^T \mathbf{b}$. In this case the minimizing $\mathbf{z}_0$ is called the least-squares solution to equation $B\mathbf{z} = \mathbf{b}$. In the context of our stacked form formalism, the least-squares solution of (5.11) satisfies the normal equations

$$\widetilde{A}^T \widetilde{A} \mathbf{f} = \widetilde{A}^T \widetilde{\mathbf{m}}.$$

However,

$$\widetilde{A}^T \widetilde{A} = \begin{bmatrix} A^T & \sqrt{\alpha} I \end{bmatrix} \begin{bmatrix} A \\ \sqrt{\alpha} I \end{bmatrix} = A^T A + \alpha I$$

and

$$\widetilde{A}^T \widetilde{\mathbf{m}} = \begin{bmatrix} A^T & \sqrt{\alpha} I \end{bmatrix} \begin{bmatrix} \mathbf{m} \\ 0 \end{bmatrix} = A^T \mathbf{m},$$

so it follows that $(A^T A + \alpha I)\mathbf{f} = A^T \mathbf{m}$.

As an example, consider the tomographic measurement matrix A constructed in Section 2.3.5 for the resolution 50×50 and with 50 projection directions. We can use the stacked form approach (5.11) and (5.12) to avoid the expensive computation of the SVD of A. The results are found in Figure 5.4.

5.3. Generalized Tikhonov regularization

Original: 50×50 Shepp–Logan phantom

Too small parameter $\alpha = 0.1$, error 56%

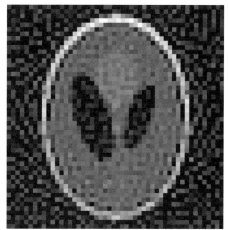

Medium parameter $\alpha = 10$, error 43%

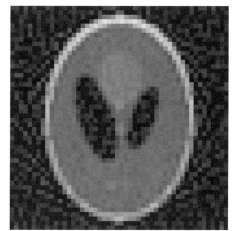

Too large parameter $\alpha = 100$, error 59%

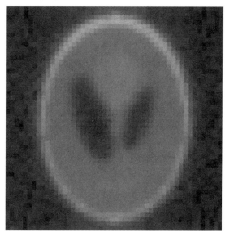

Figure 5.4. Tikhonov regularized tomographic reconstructions with $L = I$. The resolution is 50×50, and 50 projection directions were used evenly distributed along a full circle. The percentages shown are relative errors of reconstructions computed using formula (4.9).

5.3 Generalized Tikhonov regularization

Sometimes we have a priori information about the solution of the inverse problem. For example, we may know that $\mathbf{f}$ is close to a vector $\mathbf{f}_* \in \mathbb{R}^n$; then we minimize

$$T_\alpha(\mathbf{m}) = \arg\min_{\mathbf{z} \in \mathbb{R}^n} \left\{ \|A\mathbf{z} - \mathbf{m}\|^2 + \alpha \|\mathbf{z} - \mathbf{f}_*\|^2 \right\}. \tag{5.13}$$

Another typical situation is that **f** is known to be smooth. Then we minimize

$$T_\alpha(\mathbf{m}) = \arg\min_{\mathbf{z}\in\mathbb{R}^n}\left\{\|A\mathbf{z}-\mathbf{m}\|^2 + \alpha\|L\mathbf{z}\|^2\right\} \tag{5.14}$$

or

$$T_\alpha(\mathbf{m}) = \arg\min_{\mathbf{z}\in\mathbb{R}^n}\left\{\|A\mathbf{z}-\mathbf{m}\|^2 + \alpha\|L(\mathbf{z}-\mathbf{f}_*)\|^2\right\}, \tag{5.15}$$

where L is a discretized differential operator.

For example, in dimension one, representing the vector **f** as a continuous function f with $f(s_j) = \mathbf{f_j}$, we can discretize the derivative of the continuum by the difference quotient

$$\frac{df}{ds}(s_j) \approx \frac{f(s_{j+1}) - f(s_j)}{\Delta s} = \frac{\mathbf{f}_{j+1} - \mathbf{f}_j}{\Delta s}.$$

This leads to the discrete differentiation matrix

$$L = \frac{1}{\Delta s}\begin{bmatrix} -1 & 1 & 0 & 0 & 0 & \cdots & 0 \\ 0 & -1 & 1 & 0 & 0 & \cdots & 0 \\ 0 & 0 & -1 & 1 & 0 & \cdots & 0 \\ \vdots & & & \ddots & & & \\ \vdots & & & & \ddots & & \\ 0 & \cdots & & 0 & -1 & 1 & 0 \\ 0 & \cdots & & 0 & 0 & -1 & 1 \\ 1 & \cdots & & 0 & 0 & 0 & -1 \end{bmatrix}. \tag{5.16}$$

5.3.1 Generalized Tikhonov regularization for the deconvolution problem

Let us apply generalized Tikhonov regularization to our basic test problem of one-dimensional deconvolution. We take L as in formula (5.16) and use stacked form (5.12) for computing reconstructions with various choices of regularization parameter α. See Figure 5.5.

Compare Figure 5.5 to the classical Tikhonov regularization case $L = I$ shown in Figure 5.1. For large α the case $L = I$ leads to the reconstruction being close to zero, whereas in the case of the discrete derivative L given by (5.16) we see that large α takes the reconstruction close to a constant function, not necessarily zero. Also, using (5.16) seems to promote smoothness in the reconstructions, as expected.

Also, we study numerically the noise-robustness of the reconstructions; see Figure 5.6.

Exercise 5.3.1. *Show that the variational form corresponding to the minimization problem*

$$T_\alpha(m) = \arg\min_{\mathbf{z}\in\mathbb{R}^n}\{\|A\mathbf{z}-\mathbf{m}\|^2 + \alpha\|L\mathbf{z}\|^2\}$$

is given by

$$\left\langle (A^T A + \alpha L^T L)T_\alpha(\mathbf{m}) - A^T\mathbf{m}, \mathbf{w} \right\rangle = 0 \qquad \text{for all } \mathbf{w} \in \mathbb{R}^n.$$

5.3. Generalized Tikhonov regularization

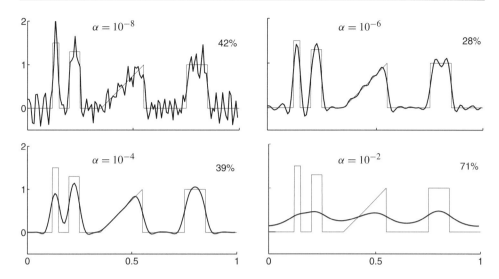

Figure 5.5. Generalized Tikhonov regularized reconstructions. The percentages shown are relative errors of reconstructions.

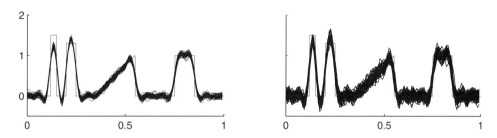

Figure 5.6. Study of noise-robustness of generalized Tikhonov regularized reconstructions. Here 50 independent realizations of random white noise was used for both of the two plots. The standard deviation of independent noise components was $\sigma = 0.05$. Left: regularization parameter $\alpha = 0.1$. Right: regularization parameter $\alpha = 0.01$.

Exercise 5.3.2. *Write the generalized Tikhonov problem*

$$T_\alpha(m) = \arg\min_{\mathbf{z} \in \mathbb{R}^n} \{\|A\mathbf{z} - \mathbf{m}\|^2 + \alpha \|L(\mathbf{z} - \mathbf{f}_\star)\|^2\}$$

in stacked form.

Exercise 5.3.3. *Experiment with generalized Tikhonov regularization on the problem of backward heat propagation.*

Exercise 5.3.4. *Choose L to be the Laplace operator and repeat the X-ray tomography example with generalized Tikhonov regularization and this choice of L. Note that in matrix* *form*

$$L = \begin{bmatrix} 0 & 1 & 0 \\ 1 & -4 & 1 \\ 0 & 1 & 0 \end{bmatrix}. \tag{5.17}$$

5.4 Choosing the regularization parameter

How does one choose the regularization parameter $\alpha > 0$ optimally? This is a difficult question and in general unsolved. There are some methods for choosing α that are based on the noise level in the data, for example, Morozov's discrepancy principle: If we have an estimate on the magnitude of error in the data, then any solution that produces a measurement with error of the same magnitude is acceptable. Other methods, such as the L-curve method, are based on striking a balance between the norm of a penalty term and the norm of the residual. We discuss both methods here.

Some methods, such as the generalized cross-validation method [165] and recent methods with inexact knowledge of the noise level [188], are not discussed here due to space restrictions. We remark that a very recent sparsity-based choice rule is explained below in Section 6.3.

5.4.1 Morozov's discrepancy principle

Assume that $\mathbf{m} = A\mathbf{f} + \varepsilon$ and that we know the size of noise: $\|\varepsilon\| = \delta > 0$. Then $T_\alpha(\mathbf{m})$ is an acceptable reconstruction if

$$\|AT_\alpha(\mathbf{m}) - \mathbf{m}\| \leq \delta.$$

For example, if the elements of the noise vector $\varepsilon \in \mathbb{R}^k$ satisfy $\varepsilon_j \sim N(0, \sigma^2)$, then we can take $\delta = \sqrt{k}\sigma$ since the expectation of the size is $\mathrm{E}(\|\varepsilon\|) = \sqrt{k}\sigma$.

The idea of Morozov's discrepancy principle is to choose $\alpha > 0$ such that

$$\|AT_\alpha(\mathbf{m}) - \mathbf{m}\| = \delta.$$

Theorem 5.2. *The Morozov discrepancy principle gives a unique choice for $\alpha > 0$ if and only if δ satisfies*

$$\|P\mathbf{m}\| \leq \delta \leq \|\mathbf{m}\|,$$

where P is orthogonal projection to the subspace Coker(A).

Proof. From the proof of Theorem 5.1 we find the equation

$$AT_\alpha(\mathbf{m}) = UDV^T V \mathcal{D}_\alpha^+ U^T \mathbf{m} = UD\mathcal{D}_\alpha^+ \mathbf{m}',$$

so we have

$$\|AT_\alpha(\mathbf{m}) - \mathbf{m}\|^2 = \|D\mathcal{D}_\alpha^+ \mathbf{m}' - \mathbf{m}'\|^2$$

$$= \sum_{j=1}^{\min(k,n)} \left(\frac{d_j^2}{d_j^2 + \alpha} - 1\right)^2 (\mathbf{m}'_j)^2 + \sum_{j=\min(k,n)+1}^{k} (\mathbf{m}'_j)^2$$

$$= \sum_{j=1}^{r} \left(\frac{\alpha}{d_j^2 + \alpha}\right)^2 (\mathbf{m}'_j)^2 + \sum_{j=r+1}^{k} (\mathbf{m}'_j)^2.$$

5.4. Choosing the regularization parameter

From this expression we see that the mapping
$$\alpha \mapsto \|AT_\alpha(\mathbf{m}) - \mathbf{m}\|^2$$
is monotonically increasing and thus, noting the formal identity
$$\sum_{j=r+1}^{k} (\mathbf{m}'_j)^2 = \|AT_0(\mathbf{m}) - \mathbf{m}\|^2,$$
we get
$$\sum_{j=r+1}^{k} (\mathbf{m}'_j)^2 \leq \|AT_\alpha(\mathbf{m}) - \mathbf{m}\|^2 \leq \lim_{\alpha \to \infty} \|AT_\alpha(\mathbf{m}) - \mathbf{m}\|^2 = \sum_{j=1}^{k} (\mathbf{m}'_j)^2,$$
and the claim follows from orthogonality of U. □

Numerical implementation of Morozov's method is now simple. Just find the optimal α as the unique zero of the function
$$f(\alpha) = \sum_{j=1}^{r} \left(\frac{\alpha}{d_j^2 + \alpha} \right)^2 (\mathbf{m}'_j)^2 + \sum_{j=r+1}^{k} (\mathbf{m}'_j)^2 - \delta^2. \tag{5.18}$$

Let us try Morozov's method for our one-dimensional deconvolution model problem. We take $n = 128 = k$ and simulate convolution data avoiding inverse crime as explained in Section 2.1.4. Next we add a noise vector $\varepsilon \in \mathbb{R}^k$ whose elements satisfy $\varepsilon_j \sim N(0, \sigma^2)$ with $\sigma = 0.01$. Then $\delta = \sqrt{k}\sigma \approx 0.113$. See Figure 5.7 for the resulting reconstruction.

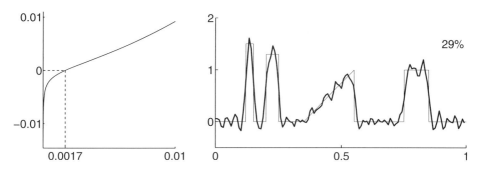

Figure 5.7. Morozov discrepancy principle and Tikhonov regularization for the one-dimensional deconvolution problem. Left: plot of the function $f(\alpha)$ defined in (5.18). Right: Tikhonov regularized deconvolution using parameter $\alpha = 0.0017$. Relative square norm error is 29%.

5.4.2 The L-curve method

As the method of Morozov does not apply to the generalized regularization formulas (5.13)–(5.15), we need to discuss alternative approaches. One possibility is to use the so-called L-curve method.

The idea of the L-curve method is to choose a collection of candidates for regularization parameter,
$$0 < \alpha_1 < \alpha_2 < \cdots < \alpha_M < \infty,$$
and compute $T_{\alpha_j}(\mathbf{m})$ for each $1 \leq j \leq M$. Then the points
$$(\log \|A T_\alpha(\mathbf{m}) - \mathbf{m}\|, \log \|L T_\alpha(\mathbf{m})\|) \in \mathbb{R}^2 \tag{5.19}$$
are plotted in the plane, forming approximately a smooth curve. This curve has typically the shape of the letter L with a smooth corner. The optimal value of α is thought to be found as near the corner as possible. See Figure 5.8 for an illustration of the L-curve.

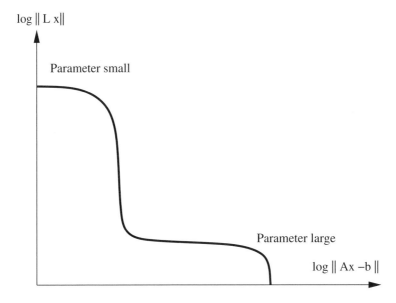

Figure 5.8. An idealized illustration of an L-curve formed by plotting a continuum of points defined by (5.19).

This method is applicable to generalized Tikhonov regularization as well as the case where $L = I$. To understand the L-curve method in a bit more detail, we introduce the generalized singular value decomposition of the matrix pair (A, L).

Definition 5.4.1. *Let A be an $m \times n$ matrix and L a $p \times n$ matrix with $m \geq n \geq p$, where $\ker(A) \cap \ker(L) = \{0\}$ and L has full row rank $(\mathrm{rank}(L) = p)$. The generalized singular value decomposition (GSVD) of the matrix pair (A, L) is*
$$A = U \begin{bmatrix} \Sigma & 0 \\ 0 & I_{n-p} \end{bmatrix} W^{-1} \quad \text{and} \quad L = V \begin{bmatrix} M & 0 \end{bmatrix} W^{-1},$$
where U, $m \times n$, and V, $p \times p$, are orthonormal matrices, W is an $n \times n$ nonsingular matrix, $\Sigma = \mathrm{diag}(\sigma_1, \sigma_2, \ldots, \sigma_p)$ with $0 \leq \sigma_1 \leq \sigma_2 \leq \cdots \leq \sigma_p \leq 1$, $M = \mathrm{diag}(\mu_1, \mu_2, \ldots, \mu_p)$

5.4. Choosing the regularization parameter

with $1 \geq \mu_i \geq \mu_2 \geq \cdots \geq \mu_p \geq 0$ and $\sigma_i^2 + \mu_i^2 = 1$. The generalized singular values are defined as the ratios $\gamma_i = \frac{\sigma_i}{\mu_i}$. Since the σ_i are increasing and μ_i is decreasing we find

$$\gamma_1 \leq \gamma_2 \leq \gamma_3 \leq \cdots \leq \gamma_p.$$

Similarly to formula (5.7) the generalized Tikhonov regularized solution can be expressed in terms of the generalized singular values by

$$T_\alpha(\mathbf{m}) = \sum_{i=1}^{r} \frac{\gamma_i^2}{\gamma_i^2 + \alpha} \frac{\mathbf{u_i}^T \mathbf{m}}{\gamma_i} \mathbf{v_i}. \tag{5.20}$$

This allows us to write down an explicit expression for $\|AT_\alpha(\mathbf{m}) - \mathbf{m}\|_2^2$ and $\|LT_\alpha(\mathbf{m})\|^2$. That is,

$$\|AT_\alpha(\mathbf{m}) - \mathbf{m}\|_2^2 = \left\| \begin{bmatrix} \Sigma & 0 \\ 0 & I_{n-p} \end{bmatrix} W^{-1} T_\alpha(\mathbf{m}) - U^T \mathbf{m} \right\|^2 - \|U^T \mathbf{m}\|^2 + \|\mathbf{m}\|^2$$

$$= \sum_{i=1}^{p} \left(\frac{\gamma_i^2}{\gamma_i^2 + \alpha^2} - 1 \right)^2 (\mathbf{u_i}^T \mathbf{m})^2 - \|U^T \mathbf{m}\|^2 + \|\mathbf{m}\|^2$$

$$= \sum_{i=1}^{p} \left(\frac{\alpha^2}{\gamma_i^2 + \delta^2} \right)^2 (\mathbf{u_i}^T \mathbf{m})^2 - \|U^T \mathbf{m}\|^2 + \|\mathbf{m}\|^2 \tag{5.21}$$

and

$$\|LT_\alpha(\mathbf{m})\|^2 = \left\| \begin{bmatrix} M & 0 \end{bmatrix} W^{-1} T_\alpha(\mathbf{m}) \right\|^2$$

$$= \left\| [M0] \begin{bmatrix} D & 0 \\ 0 & I_{n-p} \end{bmatrix} U^T \mathbf{m} \right|^2$$

$$= \sum_{i=1}^{p} \left(\frac{\gamma_i}{\gamma_i^2 + \alpha^2} \right)^2 (\mathbf{u_i}^T \mathbf{m})^2. \tag{5.22}$$

The above formula shows that $\|LT_\alpha(\mathbf{m})\|$ is a monotonically decreasing function and

$$\|LT_0(\mathbf{m})\|^2 = \sum_{i=1}^{p} \frac{1}{\gamma_i^2} (\mathbf{u_i}^T \mathbf{m})^2 = \|L\mathcal{L}_\alpha(\mathbf{m})\|^2,$$

$$\|LT_\alpha(\mathbf{m})\|^2 \to 0 \text{ as } \alpha \to \infty.$$

This explains the ordering of points along the L-curve: as the regularization parameter increases, the points move downward and to the right. Also, the L-curve is bounded from above and bounded to the left and right.

From [194] any point (a, b) on the curve $(\|AT_\alpha(\mathbf{m}) - \mathbf{m}\|, \|LT_\alpha(\mathbf{m})\|)$ is a solution of the two constrained least-squares problems:

$$a = \min\{\|A\mathbf{f} - \mathbf{m}\|\} \quad \text{subject to} \quad \|L\mathbf{f}\| \leq b, \quad 0 \leq b \leq \|LT_0(\mathbf{m})\|,$$
$$b = \min\{\|L\mathbf{f}\|\} \quad \text{subject to} \quad \|A\mathbf{f} - \mathbf{m}\| \leq a, \|P\mathbf{m}\| \leq a \leq \|\mathbf{m}\|.$$

As a consequence, all approximate solutions of the regularization problem by other computations will generate a point in Figure 5.8 above the L-curve.

To understand the shape of the L-curve, assumptions are needed on the Fourier coefficients $c_i = u_i^T \mathbf{m}$. One can show that the faster the c_i decay to 0, the sharper the L-shaped corner. It can also be shown that a log-log scale emphasizes the L-shape. The short horizontal section of the L-curve in Figure 5.8 arises because when α is extremely small,

$$\|A\,T_\alpha(\mathbf{m}) - \mathbf{m}\|_2^2 \approx \sum_{i=1}^{p} \left(\frac{\alpha^2}{\gamma_i^2}\right)^2 (\mathbf{u_i}^T\mathbf{m})^2 - \|U^T\mathbf{m}\|^2 + \|\mathbf{m}\|^2,$$

$$\|L\,T_\alpha(\mathbf{m})\|^2 \approx \|L\,T_0(\mathbf{m})\|^2 = \|L\,\mathcal{L}_\alpha(\mathbf{m})\|^2,$$

and so $\|L\,T_\alpha(\mathbf{m})\|^2$ is constant while $\|A\,T_\alpha(\mathbf{m}) - \mathbf{m}\|_2^2$ is increasing. For α extremely large we find a vertical line downward at $\|A\,T_\alpha(\mathbf{m}) - \mathbf{m}\| \approx \|\mathbf{m}\|$.

For more information about the L-curve method, see the book by Hansen [194] and the references therein.

The L-curve method applied to our guiding examples can be found in Figures 5.9, 5.10, and 5.11 for the problems, respectively, of one-dimensional deconvolution, the problem of backward heat propagation from data at $T = 0.1$, and X-ray tomography for the resolution 50×50 and with 50 projection directions. It is clear from the results that the regularization parameter α corresponding to the corner of the L does not always result in the optimal reconstruction. This is explored further in the exercises, and the reader is referred to the articles [462, 190] for analysis of the nonconvergence of the L-curve method.

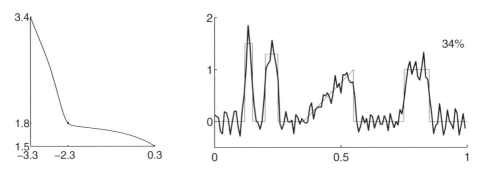

Figure 5.9. L-curve for the one-dimensional deconvolution problem.

Exercise 5.4.1. *Consider the regularized solution $T_\alpha(\mathbf{m})$ of equation $\mathbf{m} = A\mathbf{x} + \varepsilon$ using TSVD with truncation index $p(\alpha)$ for $\alpha > 0$. Assume that the noise level is $\delta = \|\varepsilon\|$. Show that the discrepancy condition $\|AT_\alpha(\mathbf{m}) - \mathbf{m}\| \leq \delta$ can be written in the form*

$$\sum_{j=p(\alpha)+1}^{m} (y'_j)^2 \leq \delta^2.$$

(This is the equivalent of Morozov's discrepancy condition for TSVD.)

Exercise 5.4.2. *Use the Morozov discrepancy principle for the backward heat propagation problem from $T = 0.1$ with added noise of 1%. Does the Morozov discrepancy principle predict an optimal parameter?*

5.4. Choosing the regularization parameter

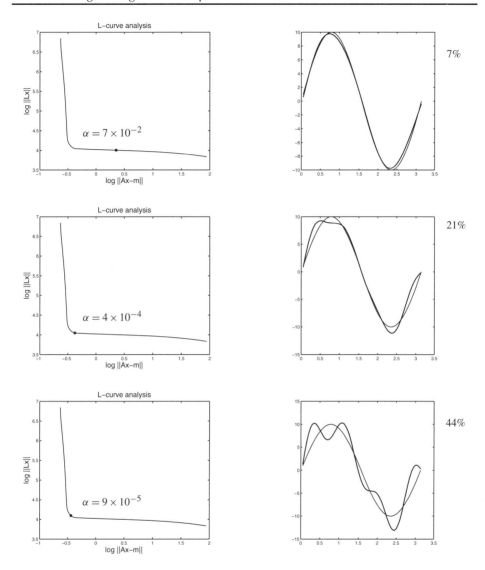

Figure 5.10. L-curves and the corresponding reconstructions for the backward heat problem. Notice that this is an example of nonconvergence of the L-curve method. The best reconstruction does not correspond to a regularization parameter α coinciding with the corner of the L. The percentages represent the percent relative error.

Exercise 5.4.3. *Use Tikhonov regularization for the backward heat propagation problem from $T = 0.4$. Plot the L-curves and determine whether the corner corresponds to the best reconstruction in this case. What regularization parameter gives you the best reconstruction?*

Exercise 5.4.4. *Find the parameter for the one-dimensional deconvolution problem in Figure 5.9 that corresponds to the corner of the L. Find the parameter that results in the*

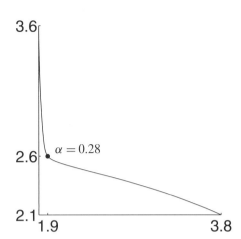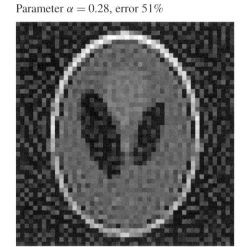

Figure 5.11. L-curve method for Tikhonov regularized tomographic reconstructions with $L = I$. The resolution is 50×50, and 50 projection directions were used evenly distributed along a full circle. Compare to Figure 5.4.

best reconstruction and plot its location on the L-curve. What does its position tell you about the balance between a small norm residual and a small penalty term?

Exercise 5.4.5. *Repeat Exercise 5.4.4 using the kernel defined in Exercise 4.4.2.*

Exercise 5.4.6. *Use instead the Morozov discrepancy principle to choose a regularization parameter in Exercise 5.4.5. Does it predict an optimal regularization parameter?*

Exercise 5.4.7. *Determine whether the parameter $\alpha = 0.28$ results in the best reconstruction for X-ray tomographic reconstructions of the Shepp–Logan phantom with resolution 50×50 and data from 50 projection directions, as shown in Figure 5.4. Clarify whether you are using a norm criterion to decide which reconstruction is best or whether you are choosing it visually. If $\alpha = 0.28$ is not best, determine what value of α is and where it lies on the L-curve.*

5.5 Large-scale implementation

The formulation (5.10) of Tikhonov regularization is remarkable because it allows matrix-free implementation. Namely, assume that we have available computational routines called `Amult` and `Lmult` that take an arbitrary vector $\mathbf{z} \in \mathbb{R}^n$ as argument and return

$$\texttt{Amult}(\mathbf{z}) = A\mathbf{z} \in \mathbb{R}^k, \qquad \texttt{Lmult}(\mathbf{z}) = L\mathbf{z} \in \mathbb{R}^{k'},$$

respectively. Further, since the transposes $A^T : \mathbb{R}^k \to \mathbb{R}^n$ and $L^T : \mathbb{R}^{k'} \to \mathbb{R}^n$ appear in (5.10) as well, we need computational routines called `ATmult` and `LTmult` that take

5.5. Large-scale implementation

vectors $\mathbf{v} \in \mathbb{R}^k$ and $\mathbf{w} \in \mathbb{R}^{k'}$ as arguments and return

$$\mathtt{ATmult}(\mathbf{v}) = A^T \mathbf{v} \in \mathbb{R}^n, \qquad \mathtt{LTmult}(\mathbf{w}) = L^T \mathbf{w} \in \mathbb{R}^n.$$

Now we can solve the linear equation $(A^T A + \alpha L^T L)\mathbf{f} = A^T \mathbf{m}$ without actually constructing any of the matrices $A, A^T, L,$ or L^T! The trick is to use an iterative solution strategy, such as the conjugate gradient method.

We will demonstrate large-scale inversion in Chapter 9 in the context of X-ray tomography with realistic resolution.

5.5.1 Conjugate direction methods

The conjugate gradient method belongs to a class of methods known as conjugate direction methods. These were invented for the quadratic optimization problem

$$\text{minimize } \frac{1}{2}\mathbf{x}^T Q \mathbf{x} - \mathbf{b}^T \mathbf{x}, \tag{5.23}$$

where Q is an $n \times n$ symmetric positive definite matrix.

Definition 5.5.1. *Given a symmetric matrix Q, then two vectors $\mathbf{d_1}$ and $\mathbf{d_2}$ are said to be Q-orthogonal, or conjugate with respect to Q, if $\mathbf{d_1}^T Q \mathbf{d_2} = 0$. A finite set of vectors is said to be a Q-orthogonal set if $\mathbf{d_i}^T Q \mathbf{d_j} = 0$ for all $i \neq j$.*

Theorem 5.3. *If Q is positive definite and the set of nonzero vectors $\mathbf{d_0}, \mathbf{d_1}, \mathbf{d_2}, \ldots, \mathbf{d_k}$ are Q-orthogonal, then these vectors are linearly independent.*

Proof. Suppose there are constants $\alpha_i, i = 0, 1, \ldots, k$ such that $\alpha_0 \mathbf{d_0} + \cdots + \alpha_k \mathbf{d_k} = 0$. Then, multiplying by Q and $\mathbf{d_i}^T$,

$$\alpha_0 \mathbf{d_i}^T Q \mathbf{d_0} + \cdots + \alpha_i \mathbf{d_i}^T Q \mathbf{d_i} + \cdots + \alpha_k \mathbf{d_i}^T Q \mathbf{d_k} = 0.$$

By Q-orthogonality, all terms are 0 except the ith, so we must also have $\alpha_i \mathbf{d_i}^T Q \mathbf{d_i} = 0$. But this means $\alpha_i = 0$ since Q is positive definite. Since i was arbitrary, we can then conclude that $\alpha_i = 0$ for $i = 0, 1, \ldots, k$. □

Let us consider why Q-orthogonality is so useful. Corresponding to the $n \times n$ matrix Q, let $\mathbf{d_0}, \mathbf{d_1}, \ldots, \mathbf{d_{n-1}}$ be n nonzero Q-orthogonal vectors. Then they are linearly independent, so the solution $\mathbf{x}^*$ of (5.23) can be written

$$\mathbf{x}^* = \sum_{i=0}^{n-1} \alpha_i \mathbf{d_i}.$$

Then

$$Q\mathbf{x}^* = \sum_{i=0}^{n-1} \alpha_i Q \mathbf{d_i}$$

and

$$\mathbf{d_j}^T Q \mathbf{x}^* = \sum_{i=0}^{n-1} \alpha_i \mathbf{d_j} Q \mathbf{d_i} = \alpha_j \mathbf{d_j}^T Q \mathbf{d_j}.$$

Recall that the solution to (5.23) satisfies $Q\mathbf{x}^* = \mathbf{b}$. Solving for α_j,

$$\alpha_j = \frac{\mathbf{d_j}^T Q \mathbf{x}^*}{\mathbf{d_j}^T Q \mathbf{d_j}} = \frac{\mathbf{d_j}^T \mathbf{b}}{\mathbf{d_j}^T Q \mathbf{d_j}}.$$

So

$$\mathbf{x}^* = \sum_{i=0}^{n-1} \frac{\mathbf{d_j}^T \mathbf{b}}{\mathbf{d_j}^T Q \mathbf{d_j}} \mathbf{d_j}.$$

The Q-orthogonal set allowed us to find an expression for $\mathbf{x}^*$ only in terms of $Q, \mathbf{b}$ and the $\mathbf{d_i}$'s. The expansion for $\mathbf{x}^*$ can be viewed as the result of an iterative process of n steps where at the ith step $\alpha_i \mathbf{d_i}$ is added. This gives us the basic conjugate direction method.

Theorem 5.4 (conjugate direction theorem). *Let $\{\mathbf{d_i}\}_{i=0}^{n-1}$ be a set of nonzero Q-orthogonal vectors. For any $\mathbf{x_0} \in \mathbb{R}^n$ the sequence $\{\mathbf{x_k}\}$ generated by*

$$\mathbf{x_{k+1}} = \mathbf{x_k} + \alpha_k \mathbf{d_k}, \quad k \geq 0, \quad \text{with} \tag{5.24}$$

$$\alpha_k = -\frac{\mathbf{g_k}^T \mathbf{d_k}}{\mathbf{d_k}^T Q \mathbf{d_k}} \quad \text{and} \quad \mathbf{g_k} = Q\mathbf{x_k} - \mathbf{b} \tag{5.25}$$

converges to the unique solution $\mathbf{x}^*$ of $Q\mathbf{x} = \mathbf{b}$ after n steps. That is, $\mathbf{x_n} = \mathbf{x}^*$.

Proof. Since the $\mathbf{d_i}$'s are linearly independent, we can write $\mathbf{x}^* - \mathbf{x_0} = \alpha_0 \mathbf{d_0} + \alpha_1 \mathbf{d_1} + \cdots + \alpha_{n-1} \mathbf{d_{n-1}}$, multiply by Q and then $\mathbf{d_k}^T$ to get

$$\alpha_k = \frac{\mathbf{d_k}^T Q (\mathbf{x}^* - \mathbf{x_0})}{\mathbf{d_k}^T Q \mathbf{d_k}}. \tag{5.26}$$

Following the iterative process (5.24),

$$\mathbf{x_1} = \mathbf{x_0} + \alpha_0 \mathbf{d_0},$$
$$\mathbf{x_2} = \mathbf{x_1} + \alpha_1 \mathbf{d_1} = \mathbf{x_0} + \alpha_0 \mathbf{d_0} + \alpha_1 \mathbf{d_1},$$
$$\mathbf{x_k} = \mathbf{x_0} + \alpha_0 \mathbf{d_0} + \cdots + \alpha_{k-1} \mathbf{d_{k-1}},$$

we see that

$$\mathbf{d_k}^T Q (\mathbf{x_k} - \mathbf{x_0}) = \sum_{i=0}^{k-1} \alpha_i \mathbf{d_k}^T Q \mathbf{d_i},$$

which implies by the Q-orthogonality of the $\mathbf{d_i}$'s that $\mathbf{d_k}^T Q \mathbf{x_k} = \mathbf{d_k}^T Q \mathbf{x_0}$. Substituting this into (5.26) gives

$$\alpha_k = \frac{\mathbf{d_k}^T Q \mathbf{x}^* - \mathbf{d_k}^T Q \mathbf{x_k}}{\mathbf{d_k}^T Q \mathbf{d_k}} = \frac{\mathbf{d_k}^T \mathbf{b} - \mathbf{d_k}^T Q \mathbf{x_k}}{\mathbf{d_k}^T Q \mathbf{d_k}}$$
$$= \frac{-\mathbf{d_k}^T (Q \mathbf{x_k} - \mathbf{b})}{\mathbf{d_k}^T Q \mathbf{d_k}} = \frac{\mathbf{g_k}^T \mathbf{d_k}}{\mathbf{d_k}^T Q \mathbf{d_k}}. \quad \square$$

5.5.2 Conjugate gradient method

In the conjugate gradient method the directions are not specified beforehand, but rather are determined at each step in the iteration. At step k, one evaluates the current negative gradient vector and adds to it a linear combination of the previous direction vectors to obtain a new conjugate direction vector along which to move. Advantages of the conjugate gradient method include the following:

1. Unless the solution is attained in less than n steps, the gradient is always nonzero and linearly independent of all previous direction vectors. In fact, the gradient $\mathbf{g_k}$ is orthogonal to the subspace generated by $\mathbf{d_0}, \ldots, \mathbf{d_{k-1}}$.

2. The conjugate gradient method has a simple formula for determining the new direction vectors.

3. Since the directions are based on gradients, the method makes good progress toward the solution at every step. Note that in arbitrary sequences of conjugate directions, progress may be poor until the final few steps. While this is not significant for quadratic problems, it is for the generalization to nonquadratic problems.

Note that with roundoff error, the conjugate gradient method does not terminate in a finite number of steps.

The Conjugate Gradient Algorithm
For any $\mathbf{x_0} \in \mathbb{R}^n$, define $\mathbf{d_0} = -\mathbf{g_0} = \mathbf{b} - Q\mathbf{x_0}$ and

$$\mathbf{x_{k+1}} = \mathbf{x_k} + \alpha_k \mathbf{d_k},$$

$$\alpha_k = \frac{-\mathbf{g_k}^T \mathbf{d_k}}{\mathbf{d_k}^T Q \mathbf{d_k}},$$

$$\mathbf{d_{k+1}} = -\mathbf{g_{k+1}} + \beta_k \mathbf{d_k},$$

$$\beta_k = \frac{-\mathbf{g_{k+1}}^T Q \mathbf{d_k}}{\mathbf{d_k}^T Q \mathbf{d_k}},$$

where $\mathbf{g_k} = Q\mathbf{x_k} - \mathbf{b}$. Note that the first step is a Steepest Descent step.

5.5.3 Preconditioning

Consider the unregularized minimization problem

$$\text{minimize} \|A\mathbf{f} - \mathbf{m}\|_2^2 \tag{5.27}$$

or the regularized problem

$$\text{minimize} \|A\mathbf{f} - \mathbf{m}\|_2^2 + \alpha \|\mathbf{f}\|_2^2. \tag{5.28}$$

Each is a quadratic minimization problem with a unique minimum. We showed that the Tikhonov regularized solution $T_\alpha(\mathbf{m})$ to (5.28) satisfies $T_\alpha(\mathbf{m}) = (A^T A + \alpha I)^{-1} A^T \mathbf{m}$ or the *normal equations*

$$(A^T A + \alpha I) T_\alpha(\mathbf{m}) = A^T \mathbf{m}.$$

Preconditioning is a technique to speed up potentially slow convergence of the minimization problem solved by an iterative method. The preconditioner is a matrix S chosen so that AS^{-1} has a more favorable spectrum than A, and in place of the unregularized minimization problem (5.27), we instead minimize

$$\text{minimize} \| (AS^{-1})\mathbf{y} - \mathbf{m} \|_2^2, \quad S\mathbf{f} = \mathbf{y}. \tag{5.29}$$

This is an equivalent problem since

$$AS^{-1}\mathbf{y} - \mathbf{m} = AS^{-1}S\mathbf{f} - \mathbf{m} = A\mathbf{f} - \mathbf{m}.$$

Thus, we have the normal equations

$$(S^{-1})^T A^T A S^{-1} \mathbf{y} = (S^{-1})^T A \mathbf{m}.$$

The desired (and partially contradictory) properties of S are as follows:

- AS^{-1} should be better conditioned than A and/or have only a few distinct singular values.

- S should have about the same number of nonzero entries as A.

- It should be cheap to solve equations with matrices S and S^T.

Note that there is an extra cost associated with preconditioning, and that is solving $S\mathbf{f} = \mathbf{y}$. One example of a cheap and simple preconditioner is a diagonal scaling of the columns of A:

$$S = \text{diag}\,(s_1^{1/2},\ldots,s_n^{1/2}), \quad s_j = \|a_j\|_2^2.$$

For further study on conjugate direction methods, the conjugate gradient method, and preconditioning, we recommend going to a text on optimization methods. While such references abound, we referred to [49, 119, 263, 325] in writing these sections.

Exercise 5.5.1. *Explain how the conjugate gradient method can be used to solve the linear equation $(A^T A + \alpha L^T L)\mathbf{f} = A^T \mathbf{m}$ without actually constructing any of the matrices A, A^T, L or L^T.*

Exercise 5.5.2. *Implement the conjugate gradient method to solve the linear equation $(A^T A + \alpha L^T L)\mathbf{f} = A^T \mathbf{m}$ for the deconvolution problem.*

Exercise 5.5.3. *Implement the conjugate gradient method to solve the linear equation $(A^T A + \alpha L^T L)\mathbf{f} = A^T \mathbf{m}$ for the backward heat equation.*

Chapter 6
Total variation regularization

Many applications of computational inversion methods call for reconstructions with sharp features. For example, natural scenes contain edges between areas of different colors. Therefore, a method recovering ideal sharp photographs from misfocused noisy snapshots should be *edge-preserving*.

Using TSVD or Tikhonov regularization, especially with derivative penalty, generally results in smooth reconstructions. An alternative method was introduced in [372, 399] with edge-preservation in mind. The basic idea is to replace the 2-norm by 1-norm in the penalty term of generalized Tikhonov regularization: find a vector $\mathbf{f} \in \mathbb{R}^n$ that minimizes the expression

$$\|A\mathbf{f} - \mathbf{m}\|_2^2 + \alpha \sum_{j=1}^{n} |(L\mathbf{f})_j|, \qquad (6.1)$$

where the finite difference matrix L is given by (5.16). This resulted in a rich literature of fruitful applications of this so-called *total variation regularization*. In statistical literature, this type of method goes by the name *lasso*; see [437].

In this book we concentrate on two computational approaches for minimizing the expression (6.1): medium-scale constrained quadratic programming [304, 210, 163, 283] and large-scale gradient-based minimization methods for an approximate version of (6.1) with smoothed absolute value function similarly to [289, 104, 237, 202]. These methods have the advantage that nonnegativity constraints can be easily implemented.

Many other computational methods for minimizing the expression (6.4) have been introduced as well, including a lagged diffusivity method [127], Lagrange multiplier methods [73], frame-based thresholding methods [112, 99, 77], domain decomposition methods [147, 146], Bregman distance methods [369, 484, 164, 61, 371, 493], primal-dual methods [75, 74, 135, 356], finite element methods [142, 202], and other methods [312, 126, 463, 457, 315, 183, 469, 184, 72]. Further treatments of total variation regularization and related methods can be found in the books [461, 370, 76, 405, 195].

6.1 What is total variation?

The total variation of a function in one dimension on an interval $[a,b]$ is defined in [398] as follows.

Definition 6.1.1. *Let f be a real-valued function defined on the interval $[a,b]$. The* total variation *of f, denoted by $TV(f)$, is defined to be*

$$TV(f) = \sup \sum_{i=1}^{k} |f(x_i) - f(x_{i-1})|,$$

where the supremum is over all partitions $a = x_0 < x_1 < \cdots < x_k = b$ of $[a,b]$.

Note that if f is piecewise constant with a finite number of jump discontinuities, then the total variation is the sum of the magnitude of the jumps. If f is differentiable, letting $\Delta x_i = x_i - x_{i-1}$,

$$TV(f) = \sup \sum_{i=1}^{k} \frac{|f(x_i) - f(x_{i-1})|}{|x_i - x_{i-1}|} |x_i - x_{i-1}|,$$

and taking the limit as $\Delta x_i \to 0$ results in

$$TV(f) = \int_a^b |f'(x)| dx. \tag{6.2}$$

In higher dimensions this generalizes to ($\Omega = [a,b]^n$)

$$TV(f) = \int_\Omega |\nabla f(x)| dx. \tag{6.3}$$

By constraining our least-squares minimization problem to minimize the total variation of the solution (in the discrete sense), we expect the solution to have fewer oscillations. We also expect it to be "blockier" since we are not minimizing the function amplitude or derivative directly.

One way to generalize Tikhonov regularization for the indirect measurement $\mathbf{m} = A\mathbf{f} + \varepsilon$ is to consider the minimization problem

$$T_\alpha(\mathbf{m}) = \underset{\mathbf{z} \in \mathbb{R}^n}{\arg\min} \left\{ \|A\mathbf{z} - \mathbf{m}\|_2^2 + \alpha \|L\mathbf{z}\|_p^p \right\}, \tag{6.4}$$

where the finite difference matrix L is given by (5.16). The parameter $1 \leq p < \infty$ in (6.4) is related to the so-called ℓ^p-*norm* defined for vectors $\mathbf{h} \in \mathbb{R}^n$ by

$$\|\mathbf{h}\|_p := \left(\sum_{j=1}^{n} |\mathbf{h}_j|^p \right)^{1/p}. \tag{6.5}$$

Note that taking $p = 2$ in formula (6.4) leads to generalized Tikhonov regularization as discussed in Section 5.3.

An important special case is $p = 1$; then (6.5) simplifies to the form

$$\|\mathbf{h}\|_1 := |\mathbf{h}_1| + \cdots + |\mathbf{h}_n|. \tag{6.6}$$

Inversion based on formula (6.4) with $p = 1$ is referred to as *total variation regularization*.

6.1. What is total variation?

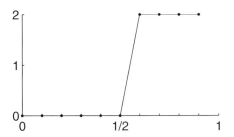 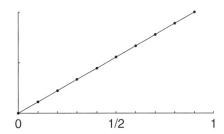

Figure 6.1. Two functions with interesting differences in their 1- and 2-norms. Left: **h**. Right: **f**.

Two functions that differ in their 1- and 2-norms are shown in Figure 6.1. The reader is encouraged to compute the norms.

Total variation regularization can be understood as a balance between two requirements:

(i) $T_\alpha(\mathbf{m})$ should give a small residual $AT_\alpha(\mathbf{m}) - \mathbf{m}$.

(ii) $LT_\alpha(\mathbf{m})$ should be small in ℓ^1-norm.

Taking $p = 2$ in formula (6.4) strongly favors smooth reconstructions over discontinuous ones, whereas $p = 1$ allows piecewise constant reconstructions.

Let us give an intuitive reason why taking $p = 1$ in formula (6.4) is so different from taking $p = 2$. We define two functions in the interval $0 \leq x \leq 1$. The first one is the linear function $f(x) = ax$ with $a = 20/9$, and the second one is the discontinuous function

$$h(x) = \begin{cases} 0 & \text{for } 0 \leq x \leq 1/2, \\ 2 & \text{for } 1/2 < x \leq 1. \end{cases}$$

We choose $n = 10$ and discretize the functions using the points

$$x_1 = 0, \ x_2 = \frac{1}{10}, \ldots, x_{10} = \frac{9}{10},$$

and setting $\mathbf{f} = [f(x_1), f(x_2), \ldots, f(x_{10})]^T$ and $\mathbf{h} = [h(x_1), h(x_2), \ldots, h(x_{10})]^T$. Furthermore, denote $\Delta x := x_2 - x_1 = 1/10$.

Note that both functions climb up from zero to two in the same interval ($f(0) = 0 = h(0)$ and $f(9/10) = 2 = h(9/10)$), but f does so smoothly and h with a single jump discontinuity.

Let L_0 be the 9×10 matrix achieved by removing the last row in matrix L given by formula (5.16). Then

$$L_0 \mathbf{h} = 10 \begin{bmatrix} \mathbf{h}_2 - \mathbf{h}_1 \\ \vdots \\ \mathbf{h}_{10} - \mathbf{h}_9 \end{bmatrix},$$

and similarly for $L_0 \mathbf{f}$.

A simple calculation shows that

$$\|L_0 \mathbf{f}\|_2^2 = 44.44\ldots, \qquad \|L_0 \mathbf{f}\|_1 = 20,$$

and
$$\|L_0\mathbf{h}\|_2^2 = 400, \qquad \|L_0\mathbf{h}\|_1 = 20.$$

The point is that the ℓ^2-norm penalizes the jump discontinuity much more than a smooth change, whereas the ℓ^1-norm gives exactly the same penalty for both.

Exercise 6.1.1. *Consider the function $f(x) = \sin 2\pi x$ on the interval $0 \leq x \leq 1$ and define $h(x)$ to be the "square wave"*

$$h(x) = \begin{cases} 0 & \text{for } x = 0, \\ 1 & \text{for } 0 < x < 0.5, \\ 0 & \text{for } x = 0.5, \\ -1 & \text{for } 0.5 < x < 1 \\ 0 & \text{for } x = 1. \end{cases}$$

Choose $n = 11$ and discretize the functions using the points

$$x_1 = 0, \; x_2 = \frac{1}{10}, \ldots, x_{10} = \frac{9}{10}, x_{11} = 1.$$

Using this discretization, compute $\|L_0\mathbf{f}\|_2^2$, $\|L_0\mathbf{f}\|_1$, $\|L_0\mathbf{h}\|_2^2$, and $\|L_0\mathbf{h}\|_1$. How does this compare to the results of the example of this section in terms of the net change of the function? Does the result change if you use 21 discretization points?

6.2 Quadratic programming

We want to find a vector $\mathbf{f} \in \mathbb{R}^n$ that solves (6.4) with $p = 1$. We write the vector $L\mathbf{f} \in \mathbb{R}^n$ in the form
$$\mathbf{v}_+ - \mathbf{v}_- = L\mathbf{f},$$
where $\mathbf{v}_\pm$ are nonnegative vectors: $\mathbf{v}_\pm \in \mathbb{R}_+^n$, or $(\mathbf{v}_\pm)_j \geq 0$ for all $j = 1, \ldots, n$. Now minimizing (6.4) with $p = 1$ is equivalent to minimizing

$$\|A\mathbf{f}\|_2^2 - 2\mathbf{m}^T A\mathbf{f} + \alpha \mathbf{1}^T \mathbf{v}_+ + \alpha \mathbf{1}^T \mathbf{v}_-,$$

where $\mathbf{1}$ is the vector with all elements equal to one: $\mathbf{1} = \begin{bmatrix} 1 & 1 & \cdots & 1 \end{bmatrix}^T \in \mathbb{R}^n$, and the minimization is taken over $y \in \mathbb{R}^{3n}$ defined by

$$\mathbf{y} = \begin{bmatrix} \mathbf{f} \\ \mathbf{v}_+ \\ \mathbf{v}_- \end{bmatrix}, \quad \text{where} \quad \begin{array}{l} \mathbf{f} \in \mathbb{R}^n, \\ \mathbf{v}_+ \in \mathbb{R}_+^n, \\ \mathbf{v}_- \in \mathbb{R}_+^n. \end{array}$$

Note the identity $\|A\mathbf{f}\|_2^2 = \mathbf{f}^T A^T A\mathbf{f}$ and write

$$H = \begin{bmatrix} 2A^T A & 0 & 0 \\ 0 & 0 & 0 \\ 0 & 0 & 0 \end{bmatrix}, \qquad \mathbf{h} = \begin{bmatrix} -2A^T \mathbf{m} \\ \alpha \mathbf{1} \\ \alpha \mathbf{1} \end{bmatrix}.$$

6.2. Quadratic programming

We then have the quadratic optimization problem in standard form,

$$\arg\min_{\mathbf{y}} \left\{ \frac{1}{2}\mathbf{y}^T H \mathbf{y} + \mathbf{h}^T \mathbf{y} \right\}, \tag{6.7}$$

with the constraints

$$L \begin{bmatrix} y_1 \\ \vdots \\ y_n \end{bmatrix} = \begin{bmatrix} y_{n+1} \\ \vdots \\ y_{2n} \end{bmatrix} - \begin{bmatrix} y_{2n+1} \\ \vdots \\ y_{3n} \end{bmatrix} \tag{6.8}$$

and

$$y_j \geq 0 \quad \text{for } j = n+1, \ldots, 3n. \tag{6.9}$$

Several software packages (such as the `quadprog.m` routine in MATLAB's Optimization Toolbox) exist that can deal with a problem of the form (6.7) with constraints of type (6.8).

One downside of the above approach is that the optimization problem (6.7) has $3n$ degrees of freedom, whereas the original problem (6.4) has only n. Numerical optimization becomes harder in higher dimensions. However, the advantage is that (6.7) is in a well-understood standard form.

Let us apply the above method to our one-dimensional deconvolution test problem. In Figures 6.2 and 6.3 the blockiness, or edge-preserving nature, of the reconstructions is evident. One also sees that too large a choice of α damps out the amplitude of the reconstructions, while too small of a choice results in a highly oscillatory reconstruction. The challenge remains to find just the right choice. The noise-robustness of the reconstruction is investigated in Figure 6.3. The robustness diminishes with the size of α.

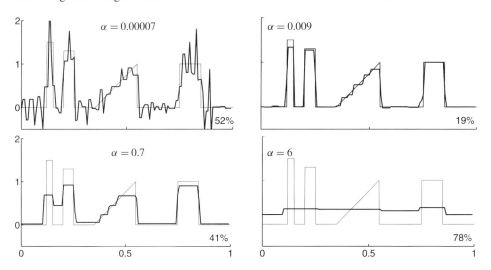

Figure 6.2. Total variation regularized reconstructions. The percentages shown are relative errors of reconstructions. Note the staircasing effect in the linear ramp part of the signal; this is a typical artefact of total variation inversion. Here $n = 128$.

The two-dimensional case is slightly more complicated since we need to discretize the gradient of the unknown with respect to two directions. One possibility is to write

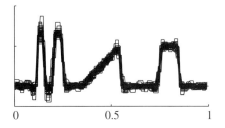 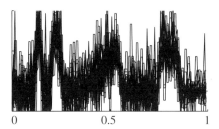

Figure 6.3. Study of noise-robustness of total variation regularized reconstructions. Here the noise level is $\sigma = 0.05 \cdot \max|f(x)|$, and 50 independent realizations of random white noise was used for both of the two plots. Left: regularization parameter $\alpha = 0.02$. Right: regularization parameter $\alpha = 0.005$.

horizontal and vertical difference quotients in the form of two matrices L_H and L_V and look for a vector $\mathbf{f} \in \mathbb{R}^n$ that minimizes the expression

$$\|A\mathbf{f} - \mathbf{m}\|_2^2 + \alpha \sum_{j=1}^n |(L_H \mathbf{f})_j| + \alpha \sum_{j=1}^n |(L_V \mathbf{f})_j|. \qquad (6.10)$$

We remark that modifying the formulation (6.10) as above leads to an optimization problem with $5n$ degrees of freedom.

Exercise 6.2.1. *Experiment with total variation regularization on the deconvolution problem with kernel defined in Exercise 4.4.2. Consider noise levels with standard deviation $\sigma = 0.01 \cdot \max|f(x)|$, $\sigma = 0.05 \cdot \max|f(x)|$, and $\sigma = 0.1 \cdot \max|f(x)|$. Try to find an optimal regularization parameter.*

Exercise 6.2.2. *Experiment with total variation regularization using the kernel*

$$f(x) = \begin{cases} 1 & \text{for } 0.1 \leq x \leq 0.2, \\ -2.5x + 1.55 & \text{for } 0.3 \leq x \leq 0.5, \\ 3x - 1.2 & \text{for } 0.5 < x < 0.6, \\ 1.3 & \text{for } 0.8 \leq x \leq 0.9, \\ 0 & \text{otherwise} \end{cases}$$

for the deconvolution problem, and reproduce plots as in Figure 6.2.

6.3 Sparsity-based parameter choice

Total variation regularization can be seen as an example of the more general concept of *compressed sensing*, whose study started with the articles [69, 128]. Compressed sensing

6.3. Sparsity-based parameter choice

is an efficient framework for recovering sparse signals from noisy and corrupted data. The term "sparse" means here that a continuous signal can be represented by a finite number of coefficients in a suitable basis, for example, using Fourier or wavelet transform coefficients.

A new sparsity-based method for choosing the regularization parameter was introduced in [283]. Suppose that we know a priori the finite number of nonzero coefficients in the true signal. We then compute reconstructions from the noisy data using a large collection of regularization parameters (similarly to the L-curve method of Section 5.4.2), count the number of nonzero transform coefficients of each reconstruction, and choose the parameter so that the corresponding reconstruction has approximately the correct number of nonzero transform coefficients. This parameter choice rule is called the *S-curve method*.

Why would the above method work in general? The thinking goes as follows. When the regularization parameter tends to infinity, the reconstruction converges typically to a constant, often zero. Thus there are no nonzero coefficients (or at most a small and known maximum number of them) in the limit case. On the other hand, when the parameter tends to zero, there is no regularization and the reconstruction is typically very erratic and oscillatory. Most of the transform coefficients are then needed to represent the reconstruction. Thus there is a more or less continuous transition from most nonzero coefficients (small regularization parameter) to no nonzero coefficients (large regularization parameter). Roughly speaking, somewhere in between those extremes there is a parameter value yielding approximately the desired level of sparsity.

Before turning to a numerical example, let us comment on the possibility of having the above kind of a priori information. How would we know the number of nonzero coefficients in the unknown target? There are at least two typical sources of such information. First, we may know some technical properties of the true signal, as in the recovery of corrupted dual-tone multifrequency signals used for telecommunications over analog telephone lines. There one knows that the clean signal consists of exactly two pure frequencies. Second, we may have available a collection of typical signals: for example, in medical imaging one can use a set of full-data CT slice images for measuring typical sparsity values and then use this knowledge as prior knowledge in limited-data tomography.

Let us now see how the sparsity-based choice rule works in practice. Let us compute the same example as in Figure 6.2 with $n = 128$. First of all, we compute the number of nonzero coefficients in our actual signal. Let the matrix L be given by formula (5.16), and compute $L\mathbf{f}$ with $\mathbf{f}$ being the true signal evaluated at the points x_j defined in (2.6). It turns out that the difference vector $L\mathbf{f}$ has exactly 32 nonzero entries. In other words, the signal $\mathbf{f}$ has 32 jumps.

Next we choose 30 parameters $\alpha_1, \ldots, \alpha_{30}$ in the interval $[10^{-6}, 100]$ with logarithmic spacing. We solve the minimization problem (6.4) for each α_j using quadratic programming as explained in Section 6.2. Figure 6.4(a) shows the numbers of jumps in each reconstruction. We want to choose a value of α that yields approximately 32 jumps in the reconstruction. However, the curve shown in Figure 6.4(a) is not monotonically decreasing, so there is in principle a possibility of nonuniqueness in the choice. We use the following simple idea: let α_{j_0} be the largest parameter used in the computation that yields *more than* 32 jumps in the reconstruction. Choose $\alpha := \alpha_{j_0+1}$. See Figure 6.4(b) for the resulting reconstruction.

Section 9.1.3 presents a further example of the sparsity-based choice rule in the context of X-ray tomography and wavelet-based regularization.

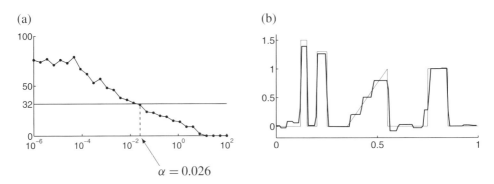

Figure 6.4. Sparsity-based choice of regularization parameter for total variation regularization. Here $n = 128$. (a) Number of jumps in the reconstruction as function of regularization parameter α. Note the logarithmic scale in the horizontal α-axis. (b) Reconstruction corresponding to the choice $\alpha = 0.026$ (thick line) and original signal (thin line).

6.4 Large-scale implementation

We apply the gradient descent minimization method of Barzilai and Borwein [31] to (approximate) large-scale total variation regularization. Our aim is to minimize

$$G(\mathbf{f}) = \|A\mathbf{f} - \mathbf{m}\|_2^2 + \alpha \|L\mathbf{f}\|_1$$
$$= \|A\mathbf{f} - \mathbf{m}\|_2^2 + \alpha \sum_{j=1}^n |\mathbf{f}_j - \mathbf{f}_{j-1}|, \qquad (6.11)$$

where we use the convention $f_0 = f_n$ according to the periodic boundary condition. Now the *objective functional* $G : \mathbb{R}^n \to \mathbb{R}$ is not continuously differentiable because of the absolute values appearing in (6.11), so we cannot apply any derivative-based optimization method.

Let us replace the absolute value function $|t|$ by an approximation:

$$|t|_\beta := \sqrt{t^2 + \beta},$$

where $\beta > 0$ is small. (Another suitable choice is $|t|_\beta = \frac{1}{\beta}\log(\cosh(\beta t))$.) Then the modified objective functional

$$G_\beta(\mathbf{f}) = \|A\mathbf{f} - \mathbf{m}\|_2^2 + \alpha \sum_{j=1}^n |\mathbf{f}_j - \mathbf{f}_{j-1}|_\beta \qquad (6.12)$$

is continuously differentiable and we can apply gradient-based optimization methods.

We need to compute the gradient of G_β. Write

$$\nabla G_\beta(\mathbf{f}) = \nabla \|A\mathbf{f} - \mathbf{m}\|_2^2 + \alpha \nabla \left(\sum_{j=1}^n |\mathbf{f}_j - \mathbf{f}_{j-1}|_\beta \right)$$

6.4. Large-scale implementation

and calculate for the first term

$$\nabla \|Af - m\|_2^2 = 2A^T A f - 2A^T m. \tag{6.13}$$

Computing the gradient of the second term involves keeping track of the following kind of calculations:

$$\frac{\partial}{\partial \mathbf{f}_\nu} \left(\sum_{j=1}^n |\mathbf{f}_j - \mathbf{f}_{j-1}|_\beta \right)$$

$$= \sum_{j=1}^n \frac{\partial}{\partial \mathbf{f}_\nu} \left((\mathbf{f}_j - \mathbf{f}_{j-1})^2 + \beta \right)^{1/2}$$

$$= \frac{\mathbf{f}_\nu - \mathbf{f}_{\nu-1}}{\left((\mathbf{f}_\nu - \mathbf{f}_{\nu-1})^2 + \beta \right)^{1/2}} - \frac{\mathbf{f}_{\nu+1} - \mathbf{f}_\nu}{\left((\mathbf{f}_{\nu+1} - \mathbf{f}_\nu)^2 + \beta \right)^{1/2}}. \tag{6.14}$$

The details are left as an exercise.

The *steepest descent method* was introduced by Cauchy in 1847. It is an iterative method where the initial guess $\mathbf{f}^{(1)}$ is chosen some way (e.g., $\mathbf{f}^{(1)} = 0$) and the next iterates are found inductively by

$$\mathbf{f}^{(\ell+1)} = \mathbf{f}^{(\ell)} - \kappa_\ell \nabla G_\beta(\mathbf{f}^{(\ell)}),$$

where the steplength parameter κ_ℓ is determined from

$$\kappa_\ell = \arg\min_{\kappa \geq 0} G_\beta(\mathbf{f}^{(\ell)} - \kappa \nabla G_\beta(\mathbf{f}^{(\ell)})).$$

Note that κ_ℓ is called a *steplength parameter* instead of just *steplength* because the length of the gradient is not necessarily one in general; thus the actual length of the step is $|\kappa_\ell \nabla G_\beta(\mathbf{f}^{(\ell)})|$. The steepest descent method is known to converge very slowly.

In 1988, Barzilai and Borwein [31] introduced the following optimization strategy which differs from the steepest descent method only by the choice of the steplength parameter:

$$\mathbf{f}^{(\ell+1)} = \mathbf{f}^{(\ell)} - \delta_\ell \nabla G_\beta(\mathbf{f}^{(\ell)}),$$

where δ_ℓ is given by setting $y_\ell := \mathbf{f}^{(\ell)} - \mathbf{f}^{(\ell-1)}$ and $g_\ell := \nabla G_\beta(\mathbf{f}^{(\ell)}) - \nabla G_\beta(\mathbf{f}^{(\ell-1)})$ and

$$\delta_\ell = \frac{y_\ell^T y_\ell}{y_\ell^T g_\ell}.$$

This method converges faster and is less affected by ill-conditioning than the steepest descent method (especially for quadratic objective functionals.) There are some practical problems with the method of Barzilai and Borwein:

(i) How to choose $\beta > 0$? Too large a value leads to smooth reconstructions, and too small a value may cause convergence problems.

(ii) How to choose the first steplength δ_1?

(iii) The objective function is not guaranteed to get smaller with each step. What to do in case it becomes bigger?

Usually (i) can be solved by some experimentation. The quick-and-dirty solution to (ii) is just choosing δ_1 to be small, for example, $\delta_1 = \frac{1}{10\,000}$. Another practical way is to choose δ_1 by line minimization.

One simple way to deal with (iii) is to check if f increases and, if so, halve the steplength. However, this is not the best possible way to ensure the convergence of the method, since just the increasing steps have turned out to be essential for the local convergence properties of the Barzilai–Borwein method. It is often advisable to just let the method run in spite of occasionally increasing objective function values.

Strategies to guarantee the global convergence of the Barzilai–Borwein method can be found, for instance, in [393, 105]. Constrained optimization, such as enforcing nonnegativity, using Barzilai–Borwein method is discussed in [105, 467].

Note that the storage need of the Barzilai–Borwein method is of the order n instead of n^2 typical for many other methods. If **f** is a large $M \times N$ size image, then $n^2 = M^2 N^2$ is too large for most computer memories!

Exercise 6.4.1. *Use formulas* (6.13) *and* (6.14) *to determine the gradient of the objective functional G_β defined in* (6.12).

Exercise 6.4.2. *Plot $|t|_\beta = \frac{1}{\beta} \log(\cosh(\beta t))$ for various small values of $\beta > 0$ and explain why this is a suitable choice to approximate the absolute value function $|t|$.*

6.4.1 TV regularization for tomography

Consider the tomographic measurement matrix A constructed in Section 2.3.5 for the resolution 50×50 and with 50 projection directions. We simulate data avoiding inverse crime as explained in Section 2.3.6.

Next we apply the approximate large-scale total variation algorithm described in Section 6.4. We take $\beta = 0.0001$ and $\delta_1 = 0.0001$ and let the Barzilai–Borwein method run for 200 iteration steps. The initial guess is the all-zero image. See Figure 6.5 for the results using various regularization parameters.

Chapter 9 contains further tomographic examples based on total variation regularization, including reconstructions from measured X-ray data.

6.4. Large-scale implementation

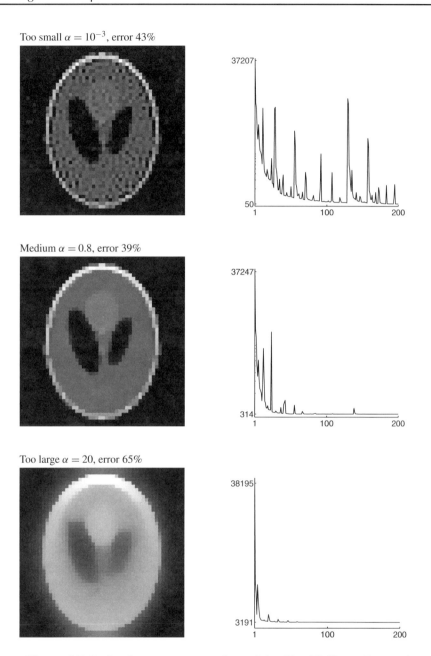

Figure 6.5. Left column: reconstructions of the 50×50 Shepp–Logan phantom using approximate total variation regularization and Barzilai–Borwein minimization with nonnegativity constraint. Note the edge-preserving nature of the total variation penalty. Right column: evolution of the objective functional during the minimization process. Note that the Barzilai–Borwein minimization does not decrease the value of the objective function at every iteration.

Chapter 7
Besov space regularization using wavelets

The Besov space norm provides another means of promoting solutions with sharp edges and introducing sparsity in the penalty function. It has great potential for capturing fine details in the reconstruction due to the *multiresolution* properties of wavelets. The reader is referred to texts such as [2, 86, 111, 248, 262, 335, 441] for a more thorough introduction to wavelets than what we provide here. Our introduction is mainly structured as in [111]. Examples using Besov space regularization for tomographic data are found in Section 9.1.3. For experimentation with wavelets, the reader may wish to try MATLAB's Wavelet Toolbox.

7.1 An introduction to wavelets

The first example of wavelets we will consider is the Haar wavelet basis. This basis for $L^2(\mathbb{R})$ has been known since 1910 and provides an example that is both intuitive and easy to compute. Define the Haar function

$$\psi(x) = \begin{cases} 1 & \text{for } 0 \leq x < 1/2, \\ -1 & \text{for } 1/2 \leq x < 1, \\ 0 & \text{otherwise} \end{cases} \tag{7.1}$$

and $\psi_{m,n}(x) = 2^{-m/2}\psi(2^{-m}x - n)$. The set $\{\psi_{m,n}\}$ is an orthonormal basis for $L^2(\mathbb{R})$. One proof of this result is found in [111], where it is shown that the $\psi_{m,n}$ are orthonormal and any L^2 function f can be approximated to arbitrarily small precision by a finite linear combination of the $\psi_{m,n}$.

A key concept to wavelet bases is the idea of multiresolution analysis.

Definition 7.1.1. *A multiresolution analysis is a sequence of closed subspaces V_j that satisfy the following five properties:*

1. $\cdots V_2 \subset V_1 \subset V_0 \subset V_{-1} \subset V_{-2} \subset \cdots$.

2. $\overline{\bigcup_{j \in \mathbb{Z}} V_j} = L^2(\mathbb{R})$ *and* $\bigcap_{j \in \mathbb{Z}} V_j = \{0\}$.

3. $f \in V_j \iff f(2^j \cdot) \in V_0$.

4. $f \in V_0 \implies f(\cdot - n) \in V_0$ *for all* $n \in \mathbb{Z}$.

5. *There exists a function* $\phi \in V_0$ *such that* $\{\phi_{0,n} : n \in \mathbb{Z}\}$, *where* $\phi_{j,n} = 2^{-j/2}\phi(2^{-j}x - n)$ *is an orthonormal basis in* V_0.

The function ϕ *is called the* scaling function *of the multiresolution analysis.*

The third property is the one that epitomizes the idea of multiresolution: All of the spaces V_j are scaled versions of V_0.

Here one choice for the scaling function ϕ is

$$\phi(x) = \begin{cases} 1 & \text{for } 0 \leq x \leq 1, \\ 0 & \text{otherwise}, \end{cases}$$

related to the Haar multiresolution analysis with spaces

$$V_j = \left\{ f \in L^2(\mathbb{R}) : f|_{[2^j k, 2^j(k+1))} = \text{constant for all } k \in \mathbb{Z} \right\}. \tag{7.2}$$

The usefulness of a multiresolution analysis is that whenever a collection of closed subspaces satisfies the five properties of Definition 7.1.1, then there exists an orthonormal wavelet basis $\{\psi_{j,k}\}$ of $L^2(\mathbb{R})$ such that for all $f \in L^2(\mathbb{R})$,

$$P_{j-1}f = P_j f + \sum_{k \in \mathbb{Z}} \langle f, \psi_{j,k} \rangle \psi_{j,k},$$

where $P_j f$ is the orthogonal projection of f onto V_j. Moreover, the wavelet can be constructed explicitly. One possibility for the construction of ψ is to define its Fourier transform by

$$\hat{\psi}(\xi) = e^{i\xi/2} \overline{m_0(\xi/2 + \pi)} \hat{\phi}(\xi/2), \tag{7.3}$$

where

$$m_0(\xi) = \frac{1}{\sqrt{2}} \sum_n h_n e^{-in\xi}, \qquad h_n = \langle \phi, \phi_{-1,n} \rangle, \quad \text{and} \quad \sum_{n \in \mathbb{Z}} |h_n|^2 = 1. \tag{7.4}$$

Equivalently,

$$\psi(x) = \sum_n (-1)^{n-1} \overline{h_{-n-1}} \phi_{-1,n}(x) = \sqrt{2} \sum_n (-1)^{n-1} \overline{h_{-n-1}} \phi(2x - n). \tag{7.5}$$

This ψ is known as the *mother wavelet*. Note that this construction is not unique.

Exercises 7.1.3 and 7.1.4 illustrate that while the Haar wavelet basis is easy to use and has the attribute that the wavelets are compactly supported, it does not have very good time-frequency localization. Other examples of wavelet bases can be found in the references at the beginning of this chapter. Those constructed by the method described in this section lead to wavelets with infinite support. An important breakthrough in wavelet analysis was the construction of a more general family of compactly supported wavelets [110]. Note that in general they cannot be written in closed form, and so we refer the reader to [111] for their construction, and we will use them in the remainder of this section.

With the deconvolution problem in mind, we construct a wavelet representation for 1-periodic functions on $\mathbb{R}$ following Daubechies [111, Section 9.3]. In other words, we

7.1. An introduction to wavelets

will give a construction of a wavelet basis for functions on the one-dimensional torus $\mathbb{T}$ constructed by identifying the endpoints of the interval $[0,1]$.

Let ψ^C and ϕ^C be a compactly supported mother wavelet and scaling function, respectively, suitable for multiresolution analysis in $\mathbb{R}$. Set

$$\phi_{j,k}(x) = \sum_{\ell \in \mathbb{Z}} \phi^C(2^j(x+\ell) - k), \tag{7.6}$$

$$\psi_{j,k}(x) = \sum_{\ell \in \mathbb{Z}} \psi^C(2^j(x+\ell) - k), \tag{7.7}$$

and define spaces $V_j := \overline{\text{span}\{\phi_{j,k} \,|\, k \in \mathbb{Z}\}}$ and $W_j := \overline{\text{span}\{\psi_{j,k} \,|\, k \in \mathbb{Z}\}}$. From [111] the V_j are spaces of constant functions for $j \leq 0$. Thus we have a ladder $V_0 \subset V_1 \subset V_2 \subset \cdots$ of multiresolution spaces satisfying $\overline{\cup_{j \geq 0} V_j} = L^2(\mathbb{T})$. Notice that we have flipped notation from [111] in which the ladder goes in the opposite direction and the V_j are constant for negative j. We have replaced j by $-j$ in the notation of [111].

The successive orthogonal complements of V_j in V_{j+1} turn out to be W_j for $j \geq 0$. Each space W_j has the orthonormal basis $\{\psi_{j,k} \,|\, k = 0, \ldots, 2^j - 1\}$, and we can represent functions as follows:

$$f(x) = c_0 + \sum_{j=0}^{\infty} \sum_{k=0}^{2^j-1} w_{j,k} \psi_{j,k}(x), \tag{7.8}$$

where the coefficients are defined by

$$c_0 := \langle f, 1 \rangle = \int_{\mathbb{T}} f(x)\,dx, \tag{7.9}$$

$$w_{j,k} := \langle f, \psi_{j,k} \rangle = \int_{\mathbb{T}} f(x) \psi_{j,k}(x)\,dx. \tag{7.10}$$

Given a vector

$$\mathbf{f} = [\mathbf{f}_1 \quad \mathbf{f}_2 \quad \cdots \quad \mathbf{f}_{n-1}]^T \in \mathbb{R}^n,$$

it is quite easy and computationally effective to evaluate approximately the wavelet coefficients c_0 and $w_{j,k}$. Namely, two discrete functions $g = [g_0 \cdots g_L]$ and $h = [h_0 \cdots h_L]$, are related to any fixed pair of mother wavelet $\psi_{0,0}$ and scaling function $\phi_{0,0}$. Denote the discrete periodic convolutions between the two discrete functions and the vector $\mathbf{f}$ by

$$G\mathbf{f} := g * \mathbf{f} \in \mathbb{R}^n \qquad \text{and} \qquad H\mathbf{f} := h * \mathbf{f} \in \mathbb{R}^n.$$

Furthermore, define a *downsampling operator* $S : \mathbb{R}^n \to \mathbb{R}^{n/2}$ by

$$S\mathbf{v} = S[v_1 \; v_2 \; v_3 \cdots v_{n-1} \; v_n]^T := [v_1 \; v_3 \cdots v_{n-1}]^T.$$

The wavelet coefficients corresponding to the finest scale basis functions are then given by

$$SH\mathbf{f} = [w_{(N-1),0} \; w_{(N-1),1} \cdots w_{(N-1),(2^{N-1}-1)}]^T.$$

The next coarser level is given by

$$SHSG\mathbf{f} = [w_{(N-2),0} \; w_{(N-2),1} \cdots w_{(N-2),(2^{N-2}-1)}]^T,$$

and finally the process stops by the computation of the scalars $c_0 = (SG)^N \mathbf{f}$ and $w_{0,0} = SH(SG)^{N-1}\mathbf{f}$.

We organize the wavelet coefficients of $\mathbf{f} \in \mathbb{R}^n$ into a column vector $w \in \mathbb{R}^n$:

$$w = [c_0 \quad w_{0,0} \quad w_{1,0} \quad w_{1,1} \quad \cdots \quad w_{(N-1),(2^{N-1}-1)}]^T.$$

Furthermore, we denote the periodic wavelet reconstruction and decomposition by

$$\mathbf{f} = Bw \quad \text{and} \quad w = B^{-1}\mathbf{f}. \tag{7.11}$$

Exercise 7.1.1. *Verify that the sets V_j defined by (7.2) satisfy conditions 1–4 of Definition 7.1.1.*

Exercise 7.1.2. *Show that the Haar mother wavelet computed from (7.5) agrees with (7.1).*

Exercise 7.1.3. *Consider the truncated sine function defined by*

$$f(x) = \begin{cases} 0 & \text{for } x < 0, \\ \sin 2\pi x & \text{for } 0 \le x \le 1, \\ 0 & \text{for } x > 1. \end{cases}$$

Express f in the Haar wavelet basis. Plot your results on the same axis as f for several choices of N in the expansion. How many wavelets are needed to approximate f to an accuracy of 10^{-6} in ℓ^2-norm?

Exercise 7.1.4. *Consider the more oscillatory function defined by*

$$f(x) = \begin{cases} 0 & \text{for } x < 0, \\ \sin 8\pi x & \text{for } 0 \le x \le 1, \\ 0 & \text{for } x > 1. \end{cases}$$

Express f in the Haar wavelet basis. Plot your results on the same axis as f for several choices of N in the expansion. How many wavelets are needed to approximate f to an accuracy of 10^{-6} in ℓ^2-norm?

7.2 Besov spaces and wavelets

The Besov spaces are the natural function spaces to work with for wavelet representations of functions. Besov spaces are function spaces denoted by $B^s_{pq}(\mathbb{T})$, where $s \in \mathbb{R}$ is a smoothness index and $1 \le p < \infty$ and $1 \le q < \infty$ are integrability exponents. Roughly speaking, the greater s, the more derivatives a function $f \in B^s_{pq}(\mathbb{T})$ has with finite $L^p(\mathbb{T})$-norm. The following definition is from [335].

Definition 7.2.1. *Suppose that a wavelet family is r-regular or C^r-regular (see [335]). A function f in $L^p(\mathbb{R})$, $1 \le p \le \infty$, represented as in (7.8) is in the Besov space $B^s_{pq}(\mathbb{T})$, $s < r$, $1 \le p, q < \infty$, if the following norm is finite:*

$$\|f\|_{B^s_{pq}(\mathbb{T})} := \left(|c_0|^q + \sum_{j=0}^{\infty} 2^{jq(s+\frac{1}{2}-\frac{1}{p})} \left(\sum_{k=0}^{2^j-1} |w_{j,k}|^p \right)^{\frac{q}{p}} \right)^{\frac{1}{q}}.$$

The choice $p = 1$ and $q = 1$ and $s = 1$ is interesting because it is related to the total variation norm; then the Besov norm takes the form

$$\|f\|_{B^1_{11}(\mathbb{T})} = |c_0| + \sum_{j=0}^{\infty} \sum_{k=0}^{2^j-1} 2^{j/2} |w_{j,k}|, \qquad (7.12)$$

provided that $\psi_{0,0}$ and $\phi_{0,0}$ are once continuously differentiable [441, Theorem 1.20].

7.3 Using B^1_{11} regularization to promote sparsity

The regularized inverse problem by Besov space regularization is

$$\text{minimize} \left\{ \|Af - \mathbf{m}\|_2^2 + \alpha \|f\|_{B^1_{11}(\mathbb{T})} \right\}. \qquad (7.13)$$

Such regularization was first suggested in [112], where the solution is based on a thresholding procedure. We will derive an alternative implementation following [283] based on constrained quadratic programming; the advantage of this approach is easy implementation of nonnegativity constraints.

We truncate the wavelet expansion to a finite number of the coarseness scales in the following way. We only consider $n = 2^N$ with some $N \geq 0$, and define projection operator T_n acting on functions f expanded as (7.8) by

$$(T_n f)(x) = c_0 + \sum_{j=0}^{N-1} \sum_{k=0}^{2^j-1} w_{j,k} \psi_{j,k}(x). \qquad (7.14)$$

Note that the right-hand side of (7.14) spans an n-dimensional subspace of $B^1_{11}(\mathbb{T})$.

Now we can use (7.14) to construct a truncated version of the norm in (7.12). Defining $f_n \equiv T_n f$, the minimization problem (7.13) becomes

$$\text{minimize} \left\{ \|Af_n - \mathbf{m}\|_2^2 + \alpha \|f_n\|_{B^1_{11}(\mathbb{T})} \right\}, \qquad (7.15)$$

where the Besov norm takes the finite form

$$\|f_n\|_{B^1_{11}(\mathbb{T})} = |c_0| + \sum_{j=0}^{N-1} \sum_{k=0}^{2^j-1} 2^{j/2} |w_{j,k}|. \qquad (7.16)$$

The Besov norm defined by (7.16) takes the computationally effective form

$$\|f_n\|_{B^1_{11}(\mathbb{T})} = |(WB^{-1}\mathbf{f})_\nu|, \qquad (7.17)$$

where W is a diagonal matrix containing the power-of-two weights that appear in formula (7.16).

The regularization task is an optimization problem with mixed ℓ^2- and ℓ^1-norms:

$$\underset{\mathbf{f} \in \mathbb{R}^n}{\arg\min} \left\{ \frac{1}{2\sigma^2} \|A\mathbf{f} - \mathbf{m}\|_2^2 + \alpha_n \sum_{\nu=1}^{n} |(WB^{-1}\mathbf{f})_\nu| \right\}, \qquad (7.18)$$

The minimization of (7.18) can be reformulated into a quadratic programming (QP) form as follows. Denote $WB^{-1}\mathbf{f} = \mathbf{u}^+ - \mathbf{u}^-$, where $\mathbf{u}^+, \mathbf{u}^- \geq 0$. Now the problem (7.18) can be written as

$$\arg\min_{\mathbf{z}} \left\{ \frac{1}{2\sigma^2} \mathbf{f}^T A^T A \mathbf{f} - \frac{1}{\sigma^2} \mathbf{f}^T A^T \mathbf{m} + \alpha_n e^T \mathbf{u}^+ + \alpha_n e^T \mathbf{u}^- + \frac{1}{2\sigma^2} \mathbf{m}^T \mathbf{m} \right\},$$

where $\mathbf{e} \in \mathbb{R}^n$ is vector of all ones, and we denote

$$\mathbf{z} = \begin{bmatrix} \mathbf{f} \\ \mathbf{u}^+ \\ \mathbf{u}^- \end{bmatrix}, \quad Q = \begin{bmatrix} \frac{1}{\sigma^2} A^T A & 0 & 0 \\ 0 & 0 & 0 \\ 0 & 0 & 0 \end{bmatrix}, \quad \mathbf{c} = \begin{bmatrix} -\frac{1}{\sigma^2} A^T \mathbf{m} \\ \alpha_n \mathbf{e} \\ \alpha_n \mathbf{e} \end{bmatrix},$$

and $\alpha = \frac{1}{2\sigma^2} \mathbf{m}^T \mathbf{m}$. Now the minimization of (7.18) becomes

$$\min_{\mathbf{z}} \quad \frac{1}{2} \mathbf{z}^T Q \mathbf{z} + \mathbf{c}^T \mathbf{z} + \alpha$$

$$\text{such that } \mathcal{A}\mathbf{z} = b \quad \text{and} \quad \mathbf{z} \geq \Lambda = \begin{bmatrix} -M \\ 0 \\ 0 \end{bmatrix}, \quad (7.19)$$

where Λ is a lower bound for the primal variable z related to the following three inequalities: $u^+ \geq 0$ and $u^- \geq 0$ and $\mathbf{f} > -M$ (with some $M \gg 0$ so that the constraint on $\mathbf{f}$ is practically ineffective). The matrix $\mathcal{A}$ is an equality constraint matrix related to $WB^{-1}\mathbf{f} = u^+ - u^-$:

$$\mathcal{A} = \begin{bmatrix} WB^{-1} & -I & I \end{bmatrix}.$$

The regularized solution can then be computed using QP optimization methods, such as primal-dual path-following interior-point methods [478, 330, 451, 143, 365]. See Figure 7.1 for regularized deconvolutions computed as explained in [283].

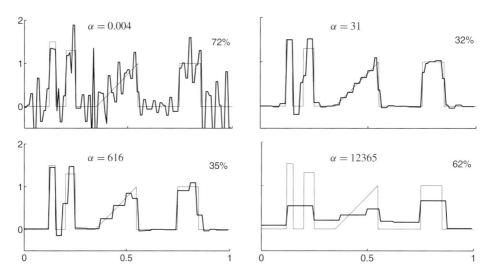

Figure 7.1. Besov space regularization using the Haar wavelet basis and $n = 128$. Note the edge-preserving nature of the reconstruction similar to total variation regularization.

7.3. Using B_{11}^1 regularization to promote sparsity

See Section 9.1.3 for real-data tomographic reconstructions using Besov space regularization and sparsity-based parameter choice.

Exercise 7.3.1. *Using formula* (7.17), *compute the Besov norm of the approximation to the function $f(x)$ in Exercise 7.1.3 for several values of N. What happens as N increases?*

Chapter 8
Discretization-invariance

In numerical analysis it is standard practice to analyze the error in the computational result as a function of the size of the grid. For example, the approximation in numerical integration quadrature is shown to be better when the number of quadrature points increases. Also, in finite element methods (FEMs) it is proven that if the diameter of the triangles in the mesh is proportional to h, then the solution of a PDE is approximated better by an FEM when the mesh is refined, or $h \to 0$.

Inverse problems often arise from continuous physical models. Such models must be discretized, or approximated by a finite-dimensional computational model, for calculating practical reconstructions using a computer. Usually there is no canonical computational grid, but instead it can be rather freely chosen. As we saw in Section 3.3, the continuous model for the inverse problem $\mathcal{A}f = \mathbf{m}$ is ill-posed when $\mathcal{A}$ is a compact operator, and so finite-dimensional discretizations of this problem will result in increasingly ill-conditioned matrices as the mesh is refined and the discrete problem more closely approximates the continuous problem. However, this is not necessarily the case when we study the regularized problem.

Assume that there are k indirect measurements available in an inverse problem, and we choose to use n degrees of freedom in the computational representation of the unknown. It is natural to ask what happens to the regularized reconstruction when $n \to \infty$. Surprisingly little research has so far been devoted to this question by the inverse problems research community. In this section we use numerical examples to illustrate the concept of *discretization-invariance* related to the above question.

Consider an ill-posed inverse problem where a periodic function $f : \mathbb{T} \to \mathbb{R}$ needs to be recovered from an indirect measurement $\mathbf{m} \in \mathbb{R}^k$ modeled by

$$\mathbf{m} = \mathcal{A}f + \varepsilon, \tag{8.1}$$

where $\mathcal{A}$ is a linear operator modeling the measurement and $\varepsilon \in \mathbb{R}^k$ is a measurement error, here assumed to be white noise.

Define a grid of n points in the interval $[0, 1)$:

$$x_n(j) = \frac{j-1}{n} \quad \text{for } j = 1, 2, \ldots, n, \tag{8.2}$$

and piecewise linear rooftop functions $\theta_j^n : \mathbb{T} \to \mathbb{R}$ by the requirement $\theta_\nu^n(x_n(j)) = \delta_{j\nu}$ for $j = 1,\ldots,n$ and $\nu = 1,\ldots,n$, where $\delta_{j\nu}$ is the Kronecker delta:

$$\delta_{j\nu} = \begin{cases} 1 \text{ when } j = \nu, \\ 0 \text{ when } j \neq \nu. \end{cases}$$

To discretize the inverse problem we use vectors

$$\mathbf{f} = [\mathbf{f}_1 \quad \mathbf{f}_2 \quad \cdots \quad \mathbf{f}_{n-1}]^T \in \mathbb{R}^n \tag{8.3}$$

and consider piecewise linear functions $f_n : \mathbb{T} \to \mathbb{R}$ of the form

$$f_n(x) = \sum_{\nu=1}^n \mathbf{f}_\nu \theta_\nu^n(x). \tag{8.4}$$

Now functions of the form (8.4) form a linear n-dimensional vector space. Also, it follows from (8.3) and (8.4) that

$$\mathbf{f} = [f_n(x_n(1)) \quad f_n(x_n(2)) \quad \cdots \quad f_n(x_n(n))]^T. \tag{8.5}$$

Let us return to the inverse problem of recovering the function f approximately from the measurement $\mathbf{m}$ defined by (8.1). Fix n and assume that we have computational algorithms available for computing $\mathcal{A} f_n \in \mathbb{R}^k$ approximately for any given $\mathbf{f} \in \mathbb{R}^n$ and f_n defined by (8.4). Then we can compute regularized reconstructions by solving the following optimization problem:

$$\underset{\mathbf{f} \in \mathbb{R}^n}{\arg\min} \{\|\mathcal{A} f_n - \mathbf{m}\|_2^2 + \alpha_n \Psi(f_n)\}, \tag{8.6}$$

where $0 < \alpha_n < \infty$ is a regularization parameter and $\Psi(f_n)$ is a suitable penalty term. Note that Ψ takes as an argument a function defined on $\mathbb{T}$, not a vector. The idea behind this is to have a well-defined penalty term for functions that is naturally discretized by restricting it to finite-dimensional function spaces.

The concept of discretization-invariance arises from the observation that the dimension n in the minimization problem (8.6) can be chosen arbitrarily, in particular independently of the number k of measurements. It is natural to require that computational reconstructions (8.6) converge as the grid is refined arbitrarily, in other words as $n \to \infty$. Choices of Ψ and α_n in (8.6) satisfying this condition are called *discretization-invariant*.

Exercise 8.0.2. *Prove formula* (8.5).

8.1 Tikhonov regularization and discretizations

In the case of Tikhonov regularization, recall that the penalty term in (8.6) is defined as follows:

$$\Psi(f_n) = \|f_n'\|_{L^2(\mathbb{T})}^2 = \int_\mathbb{T} \left|\frac{df_n(x)}{dx}\right|^2 dx. \tag{8.7}$$

We set $\Delta_n := x_n(2) - x_n(1)$ and proceed to discretize the integral in (8.7) as a Riemann sum, and we use a finite difference approximation for the derivative. These approximations

8.1. Tikhonov regularization and discretizations

are actually exact for the piecewise linear functions of the form (8.4), resulting in

$$\int_{\mathbb{T}} \left| \frac{df_n}{dx}(x) \right|^2 dx = \Delta_n \sum_{\nu=1}^{n} \left| \frac{f_n(x_n(\nu+1)) - f_n(x_n(\nu))}{\Delta_n} \right|^2$$

$$= \frac{1}{\Delta_n} \sum_{\nu=1}^{n} |\mathbf{f}_{\nu+1} - \mathbf{f}_\nu|^2$$

$$= \Delta_n \sum_{\nu=1}^{n} |(L\mathbf{f})_\nu|^2, \tag{8.8}$$

where we use the notations $x_n(n+1) = x_n(1)$ and $\mathbf{f}_{n+1} = \mathbf{f}_1$ in accordance with the periodic boundary condition, and the $n \times n$ difference matrix L is given by

$$L = \frac{1}{\Delta_n} \begin{bmatrix} -1 & 1 & 0 & 0 & 0 & \cdots & 0 \\ 0 & -1 & 1 & 0 & 0 & \cdots & 0 \\ 0 & 0 & -1 & 1 & 0 & \cdots & 0 \\ \vdots & & & \ddots & & & \\ \vdots & & & & \ddots & & \\ 0 & \cdots & & 0 & -1 & 1 & 0 \\ 0 & \cdots & & 0 & 0 & -1 & 1 \\ 1 & \cdots & & 0 & 0 & 0 & -1 \end{bmatrix}. \tag{8.9}$$

Let us consider Tikhonov regularized deconvolutions at different resolutions n. Problem (8.6) takes the discrete form

$$T_\alpha^{(n)}(\mathbf{m}) = \arg\min_{\mathbf{f} \in \mathbb{R}^n} \left\{ \|A\mathbf{f} - \mathbf{m}\|_2^2 + \alpha_n \Delta_n \sum_{\nu=1}^{n} |(L\mathbf{f})_\nu|^2 \right\}, \tag{8.10}$$

which we solve using the stacked form formulation

$$T_\alpha^{(n)}(\mathbf{m}) = \begin{bmatrix} A \\ (\alpha_n \Delta_n)^{1/2} L \end{bmatrix} \backslash \begin{bmatrix} \mathbf{m} \\ 0 \end{bmatrix}, \tag{8.11}$$

where the backslash denotes minimum square norm solution.

As an example, we consider the one-dimensional deconvolution problem, and we simulate the measurement vector $\mathbf{m} \in \mathbb{R}^k = \mathbb{R}^{64}$ by computing the convolution with a highly accurate numerical integration at the 64 sample points, thus avoiding an inverse crime. White noise with standard deviation 0.01 is added to each component of $\mathbf{m}$.

We build a 64×64 measurement matrix A and use the L-curve method to find an appropriate regularization parameter α_{64} at the discretization $n = 64$. A suitable value is found to be $\alpha_{64} = 3.1 \cdot 10^{-5}$. Note that in this case $k = n = 64$.

Next we repeat the same reconstruction task with greater values of n while keeping $\mathbf{m}$ and k fixed. Note that the inverse problem is now underdetermined. We keep things simple by choosing $n = 2^N$ with $N = 6, 7, 8, \ldots, 13$; then all of the computational grids we use contain the 64-point grid as a subset. Consequently we can construct the $k \times n$ measurement matrix A as follows. Given $N > 6$, take $n = 2^N$ and let $\widetilde{A}$ be the $n \times n$ convolution matrix defined analogously to (2.14). Then we get the $k \times n$ matrix A by taking every $\frac{n}{64}$th row of the larger matrix $\widetilde{A}$.

Define $\alpha_n = \alpha_{64}$ for all $n > 64$. Disregarding a slightly different boundary condition, [304, Theorem 4.1(i)] implies that the regularized solutions $T_\alpha^{(n)}(\mathbf{m})$ converge to a well-defined limit function as $n \to \infty$. Indeed, evaluating (8.11) numerically for $n = 64, 128, 256, \ldots, 8192$ leads to almost identical functions, as shown in Figures 8.1 and 8.2. This is discretization-invariance.

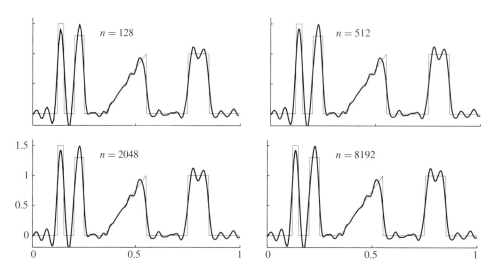

Figure 8.1. Discretization-invariant deconvolutions. Original signal (thin line) and Tikhonov regularized reconstructions (thick line). The number of measurements was $k = 64$ in each case. See also Figure 8.2.

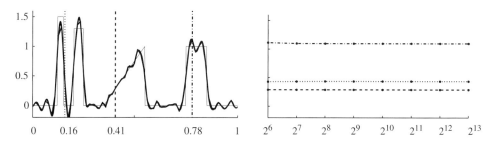

Figure 8.2. Discretization-invariant deconvolutions. Left: original signal (thin line) and Tikhonov regularized reconstructions (thick line) at resolutions $n = 64, 128, 256, \ldots, 8192$. The reconstructions shown in Figure 8.1 are plotted here as well. Right: convergence of estimates at three points when the resolution n ranges from $2^6 = 64$ to $2^{13} = 8192$.

Exercise 8.1.1. *Verify that the choice $\alpha_{64} = 3.1 \cdot 10^{-5}$ is a suitable choice for the example above based on the L-curve method.*

Exercise 8.1.2. *Investigate the property of discretization invariance for the backward heat equation with Tikhonov regularization.*

8.2 Total variation regularization and discretizations

For total variation regularization, the penalty term in (8.6) is defined as follows:

$$\Psi(f_n) = \|f_n'\|_{L^1(\mathbb{T})} = \int_{\mathbb{T}} \left|\frac{df_n(x)}{dx}\right| dx. \tag{8.12}$$

As above, we set $\Delta_n := x_n(2) - x_n(1)$, discretize the integral in (8.7) as a Riemann sum, and employ a finite difference approximation for the derivative. Similarly to (8.8) for functions of the form (8.4) the penalty term becomes

$$\int_{\mathbb{T}} \left|\frac{df_n}{dx}(x)\right| dx = \Delta_n \sum_{\nu=1}^{n} \left|\frac{f_n(x_n(\nu+1)) - f_n(x_n(\nu))}{\Delta_n}\right|$$

$$= \sum_{\nu=1}^{n} |\mathbf{f}_{\nu+1} - \mathbf{f}_\nu|. \tag{8.13}$$

Compare (8.13) with (8.8); note especially how the factor Δ_n cancels in (8.13).

We compute total variation regularized deconvolutions with various resolutions n. See Figures 8.3 and 8.4 for the results. The close agreement of the reconstructions computed at different resolutions reflects the fact that total variation regularization is discretization-invariant [304, Theorem 4.1(ii)].

However, the statistical (Bayesian) interpretation of total variation is shown in [304] not to be discretization-invariant. This observation started a quest for edge-preserving or

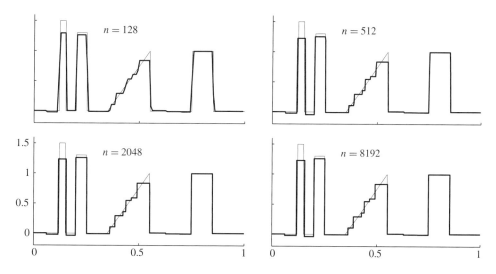

Figure 8.3. Discretization-invariant deconvolutions based on total variation regularization. Original signal is plotted using a thin line and regularized reconstructions are plotted using a thick line. Computations courtesy of Kati Niinimäki; see [283] for more details. See also Figure 8.4.

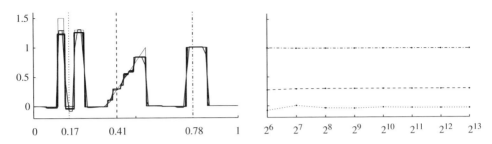

Figure 8.4. Discretization-invariant deconvolutions. Left: original signal (thin line) and total variation regularized reconstructions (thick line) at resolutions $n = 64, 128, 256, \ldots, 8192$. The reconstructions shown in Figure 8.3 are plotted here as well. Right: convergence of estimates at three points when the resolution n ranges from $2^6 = 64$ to $2^{13} = 8192$.

sparsity-promoting regularization methods that would allow a well-defined and discretization-invariant Bayesian interpretation. One possibility to achieve that is to use wavelets and Besov spaces, as shown theoretically in [303] and demonstrated numerically in [283].

8.3 Besov norm regularization and discretizations

The idea of using Besov space regularization for linear inverse problems was first introduced in [111]. Their Bayesian extensions are discretization-invariant; see [303, 283]. Discretization-invariance of deterministic Besov space regularization is proved in [170, 171, 283].

We define the penalty term in (8.6) as follows:

$$\Psi(f_n) = \|f_n\|_{B^1_{11}(\mathbf{t})}. \tag{8.14}$$

Here $n = 2^N$ with some $N \geq 0$ and

$$f_n(x) = c_0 + \sum_{j=0}^{N-1} \sum_{k=0}^{2^j-1} w_{j,k} \, \psi_{j,k}(x). \tag{8.15}$$

According to [441, Theorem 1.20], if the mother wavelet $\psi_{0,0}$ is continuously differentiable, then $B^1_{11}(\mathbf{t})$ functions can be written in the form (8.15). We choose the orthonormal Daubechies 7 wavelet basis since its mother wavelet satisfies that condition. See the right column in Figures 8.5 and 8.6 for the resulting reconstructions. The reconstructions seem to converge to a well-defined continuous limit, as predicted theoretically.

Since the wavelet expansion needs to be finite for computational reasons, the smoothness of the Daubechies 7 basis functions results in smooth reconstructions. Jumps in the signal are not recovered well, but the linear ramp in the middle of the interval is nicely recovered.

8.3. Besov norm regularization and discretizations

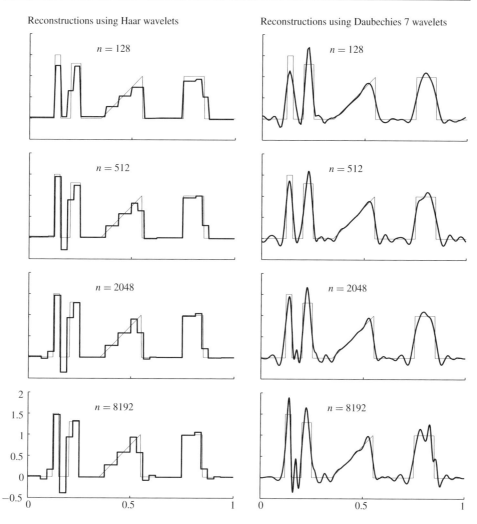

Figure 8.5. Discretization-invariant deconvolutions. Left column: original signal (thin line) and Besov norm regularized reconstructions (thick line) using Haar wavelet basis at various resolutions. See also Figure 8.7. Right column: original signal (thin line) and Besov norm regularized reconstructions (thick line) using Daubechies 7 wavelet basis at various resolutions. See also Figure 8.6. Computations illustrated in this figure were implemented by Kati Niinimäki; see [283] for more details.

To promote edge-preservation we try using the discontinuous Haar wavelet basis in (8.15). Then it is unclear whether we are working with a norm of the space $B_{11}^1(\mathbf{t})$, but computationally there is no problem in proceeding like this. See the left column in Figures 8.5 and 8.7 for reconstructions using the Haar wavelet basis. Numerical evidence suggests that these reconstructions are discretization-invariant as well.

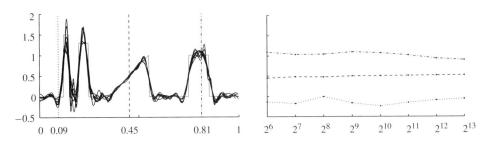

Figure 8.6. Discretization-invariant deconvolutions. Left: original signal (thin line) and Besov norm regularized reconstructions (thick line) using Daubechies 7 wavelet basis at resolutions $n = 64, 128, 256, \ldots, 8192$. The reconstructions shown in the right column of Figure 8.5 are plotted here as well. Right: convergence of estimates at three points when the resolution n ranges from $2^6 = 64$ to $2^{13} = 8192$.

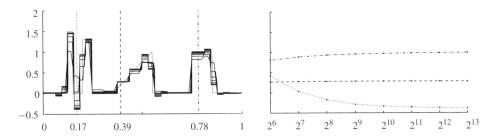

Figure 8.7. Discretization-invariant deconvolutions. Left: original signal (thin line) and Besov norm regularized reconstructions (thick line) using Haar wavelet basis at resolutions $n = 64, 128, 256, \ldots, 8192$. The reconstructions shown in the left column of Figure 8.5 are plotted here as well. Right: convergence of estimates at three points when the resolution n ranges from $2^6 = 64$ to $2^{13} = 8192$.

The Haar-based reconstructions are edge-preserving, which is similar to the total variation regularized reconstructions shown in Figures 8.3 and 8.4. Note also the staircasing effect along the ramp in the middle of the interval; this feature is shared by both total variation regularization and Haar-$B_{11}^1(\mathbf{t})$ regularization.

Chapter 9
Practical X-ray tomography with limited data

In many practical tomographic imaging situations there are restrictions on the number of available projection directions or on the angle of view. The reconstruction problem is more ill-posed when the data set is limited, as will be explained below in Sections 9.1 and 9.2. Recently there has been considerable interest in limited-data imaging, and even some commercial products are available. We will describe one such product, intended for dental implant planning, in Section 9.3.

Before discussing limited-data tomography, let us comment on full-data tomography. Currently almost all commercial tomographic imaging is based on very comprehensive data sets and on the use of filtered back-projection–type algorithms. Those algorithms are proven to recover the attenuation perfectly in the idealized continuum case, which is approximated closely by collecting data with very small angular increments. The effect of measurement noise can be typically kept very low using mild regularization, such as an appropriate filter in the frequency domain.

When dealing with comprehensive tomographic data, there is usually not much difference in the outcomes of various inversion algorithms. To demonstrate this, let us simulate tomographic data from a two-dimensional 512×512 Shepp–Logan phantom using 512 angles uniformly distributed in the interval $0° \leq \theta < 180°$, avoiding inverse crime by simulating on a twice finer grid first and then interpolating as explained in Section 2.3.6. We call the resulting sinogram $\mathbf{m} \in \mathbb{R}^k$; with the above choices we have $k = 373248$. We simulate measurement noise by adding an independent normally distributed noise component with standard deviation $0.01 \cdot \max_{1 \leq j \leq k} |\mathbf{m}_j|$ to each pixel in the sinogram. We recover the phantom using

1. filtered back-projection as explained in Section 2.3.3;

2. classical Tikhonov regularization with the matrix-free large-scale variant of Section 5.5; and

3. matrix-free large-scale approximate total variation regularization with nonnegativity constraint as explained in Section 6.4.1.

By *matrix-free* we mean the following. Note that $n = 512^2 = 262144$ and $k = 373248$, so the number of elements in A is $kn \approx 10^{11}$. Therefore, even storing A in computer memory is not advisable, not to mention the computation of the SVD. We do not construct the matrix A at all but instead use the MATLAB routines radon.m and iradon.m for computing $A\mathbf{f}$ and $A^T\mathbf{g}$ for given vectors $\mathbf{f} \in \mathbb{R}^n$ and $\mathbf{g} \in \mathbb{R}^k$. This is enough for implementing the large-scale methods described in Sections 5.5 and 6.4.1.

The reconstructions computed with the three different methods are shown in Figure 9.1. While the relative square norms errors are roughly 20% for all three methods, there is

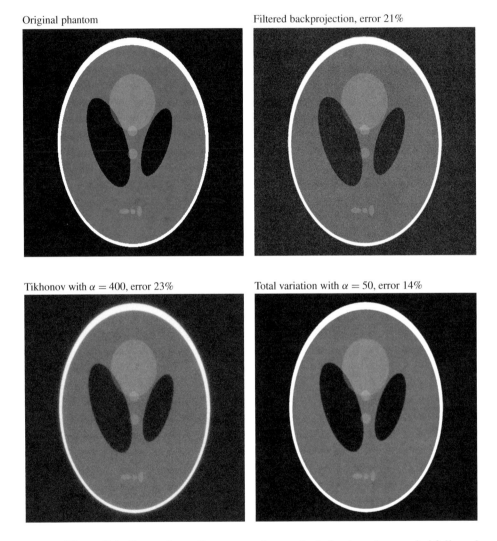

Figure 9.1. Comparison of reconstruction methods for densely sampled full-angle tomography. Here the phantom has size 512×512 and we use parallel beam geometry and 512 projection directions. The computation times of both regularization methods are more than 400 times longer than the time for the filtered back-projection algorithm, but there is no visible benefit for the extra investment in computational resources. In this kind of full-data situations filtered back-projection is the method of choice.

a big difference in computation times: filtered back-projection is done in less than half a second, but both matrix-free regularization methods take more than 200 seconds to run.

Our numerical example demonstrates the fact that whenever dealing with comprehensive, densely sampled tomographic data, it most efficient to use filtered back-projection. In dimension two, good references include [353, 249, 354, 133]. In dimension three, one typically uses helical trajectories for the X-ray source. Algorithms of the filtered back-projection flavor for various comprehensive three-dimensional data sets have been developed in [258, 261, 259]. For reviews of the mathematics of tomography, see [419, 140].

Let us now turn to limited-data imaging. There are many kinds of naturally appearing restricted data sets, for example, limited-angle data, sparse-angle data, and exterior or local tomographic data. Possible reasons for data limitations include the following:

- Geometrically limited angle of view. In medical imaging this happens, for example, in mammographic imaging with the breast in a stationary position between the detector surface and a compression plate and the X-ray source moving in an arc [360, 480, 392, 494]. Another example is electron microscope imaging [139].

- Impossibility of probing through a ball in the center of the target. These so-called limited-angle exterior data sets arise for instance in nondestructive testing [385, 384], reconstructing the structure of the solar corona [7], and in the recovery of ozone profiles around the Earth using spaceborne star occultation data [296].

- The desire to lower the radiation dose to a patient by taking fewer images. (There is considerable recent interest in CT dose reduction; see [444, 293, 328, 492].) We discuss such a case in three-dimensional dental X-ray imaging in Section 9.3.

- Cost of equipment in cases when each projection direction needs its own X-ray source. This is often the case with luggage screening at airports using one or two sources [395]. Another tomographic example (not based on X-rays) is magnetospheric imaging where each projection direction requires an own satellite orbiting the Earth [154].

The filtered back-projection algorithm is not well-suited for limited tomographic data [353]. This well-known fact is demonstrated numerically in Sections 9.1 and 9.2, where we show how Tikhonov regularization and nonnegativity constrained total variation regularization outperform filtered back-projection in the cases of sparse-angle and limited-angle tomography, respectively. See also the review article [375].

There are many successful computational approaches to sparse-data tomography discussed in the literature. Here is an incomplete list of results, based on Tikhonov regularization [218], total variation regularization [115, 377, 289, 410, 411, 290, 314, 412, 206, 433, 129, 46, 45, 239, 237, 436], strict piecewise-constant attenuation assumption [338, 425, 6, 397, 417, 316], level set methods [489, 141, 389, 288, 485, 272], modified algebraic reconstruction [379, 313], deformable models [197, 32, 198, 33, 337], variational methods [295, 122], sparsity-promoting methods [69, 68, 490, 491, 85, 238, 297], Bayesian methods [414, 247, 66], and multiresolution-sparsity methods [392, 359, 424, 452, 271]. See also the books [207, 208] for discrete tomography approaches.

We stress that any given solution method is probably not the best choice for *every* limited-data tomographic application. With severely limited data, each method produces errors and artifacts of its own kind, and it is up to the application area to choose the most suitable one. There are big differences in the computational burden of various methods, too.

9.1 Sparse full-angle tomography

In some tomographic applications it is possible to take images from all around the object, but the total number of images is restricted. We consider here taking images from full angle but with large uniform angular steps and call such data sets *sparse-angle data*.

One might think that it is impossible to compute reasonable reconstructions from sparse-angle data. Namely, it was noted already in [100], and later analyzed in [420, Theorem 4.2], that a finite number of line integrals does not determine the target uniquely since the measurement operator has a nontrivial nullspace. However, the "ghosts," or the objects invisible in the tomographic data, are known to be high-frequency functions [319, 321, 320, 326] and can be effectively suppressed by regularization.

Sparse-angle data violates the assumptions of filtered back-projection algorithms, and consequently filtered back-projection reconstructions typically contain a lot of artifacts as we will see below. In Sections 9.1.1–9.1.3 we demonstrate numerically Tikhonov, total variation, and Besov norm regularization for sparse-angle tomography and compare the results to those from filtered back-projection.

Exercise 9.1.1. *Compute and plot an invisible structure in tomographic imaging following the analysis in* [402].

9.1.1 Simulated data

We consider tomographic data taken from full angle of view but consisting only of 12 projection images with 30 degrees angular steps.

Note that here $n = 512^2 = 262144$, and the sinogram has $k = 128 \cdot 185 = 23680$ data points. The measurement matrix A has size 8748×262144, so the number of elements in A is approximately $2.3 \cdot 10^9$. We use the matrix-free algorithms described in Sections 5.5 and 6.4.1 for the Tikhonov and total variation regularization, respectively. See Figure 9.2 for the reconstructions.

Table 9.1 shows the relative square-norm errors in the reconstructions and the computation times of the three algorithms. The computation time of the filtered back-projection algorithm is very short, but the result has low quality. In this case it is beneficial to invest in computational resources and use total variation regularization, as the reconstruction error can be reduced from 126% to 19%. This is in contrast with the dense-data case shown in Figure 9.1.

9.1.2 Real data: Total variation regularization

We demonstrate the large-scale inversion methods introduced earlier in this book in the case of measured X-ray projection data. Our target is a walnut, and Keijo Hämäläinen and Aki Kallonen from University of Helsinki were kind enough to measure the projection data for us. The nut was measured using cone-beam geometry and collecting data from 90 directions. See Figure 9.3 for an illustration of the experimental setup and two examples of the resulting projection images.

The problem was reduced to a two-dimensional reconstruction task by choosing only the middle row of each projection image; this resulted in a fan-beam geometry. Each (one-dimensional) projection image consists of 512 measured attenuation values. We measured 1200 projection images from uniformly distributed angles around 360° and computed a filtered back-projection reconstruction using all this data. The resulting image is shown in

9.1. Sparse full-angle tomography

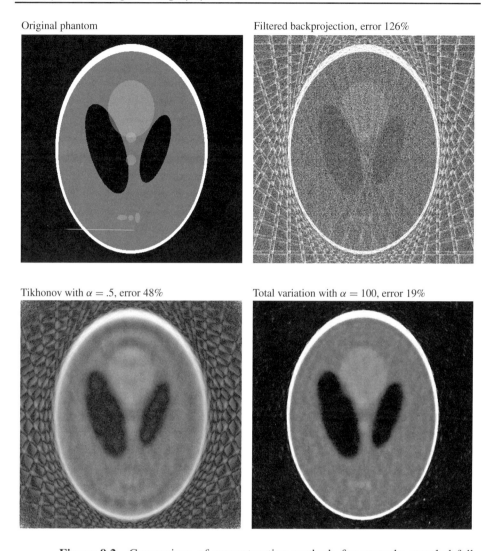

Figure 9.2. Comparison of reconstruction methods for sparsely sampled full-angle tomography. Here we simulated parallel-beam data from 20 uniformly spaced directions. The size of the reconstruction is 512×512. Approximate total variation regularization with nonnegativity constraint outperforms the competing methods clearly. See Table 9.2 for the computation times of the various methods. Also, compare these reconstructions with the dense-angle case shown in Figure 9.1; in the sparse-angle case there is much clearer difference between the reconstructions.

Figure 9.4 and is used as ground truth to which reconstructions from fewer projections can be compared.

We wish to study the performance of filtered back-projection and total variation regularization as the number of projection angles is reduced progressively.

Table 9.1. Computation times and relative square-norm errors of reconstructions from sparsely sampled full-angle tomographic data. See Figure 9.2 for the reconstructions.

Reconstruction method	Error	Computation time (s)
Filtered back-projection	126%	0.4
Tikhonov regularization	48%	69.5
Total variation regularization	19%	33.4

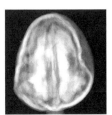
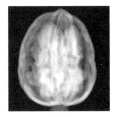

Figure 9.3. Left: Experimental setup for collecting tomographic X-ray data of a walnut. The detector plane is on the left and the X-ray source on the right in the picture. The walnut is attached to a computer-controlled rotator platform. Right: two examples of the resulting projection images.

The data was downsampled to 512 pixels per projection image, and rebinned to a parallel-beam geometry. This operation contains some interpolation error which is not involved in the computation of the ground truth image shown in Figure 9.4 as it used a built-in cone-beam reconstruction routine of the X-ray laboratory system. We pick three subsets from the original dense-angle dataset. The subsets consist of 10, 15, and 23 projections, respectively, each set spanning 180°. The number of data points in each case is $k = 10 \cdot 512 = 5120, k = 15 \cdot 512 = 7680$, and $k = 90 \cdot 512 = 11776$, respectively.

After rebinning, parallel-beam projections and back-projections were readily implemented for the three data sets using MATLAB's `radon.m` and `iradon.m` functions. The total variation regularized reconstruction is computed in all three cases on a 256×256 grid, so $n = 65536$. The computations were performed by Kati Niinimäki using a quadratic programming approach as explained in Section 6.2. See Figure 9.5 for the results.

It is clear from Figure 9.5 that filtered back-projection is not suitable for sparse-angle imaging, and that total variation regularization delivers useful piecewise-constant low-resolution reconstructions.

9.1. Sparse full-angle tomography 117

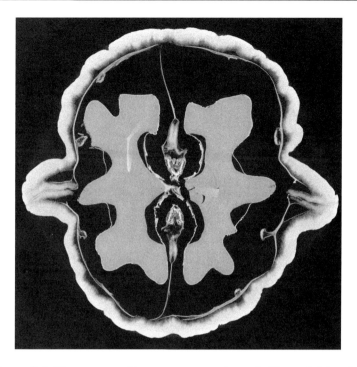

Figure 9.4. The tomographic measurement shown in Figure 9.3 involved recording projections from 1200 uniformly distributed angles. This image shows the filtered back-projection reconstruction from the full dense-angle data set. This is the ground truth to which the reconstructions computed using fewer projections can be compared.

The results and images presented in this section are due to a collaboration with Keijo Hämäläinen, Aki Kallonen, Ville Kolehmainen, Matti Lassas, and Kati Niinimäki. For more details, see [187].

9.1.3 Real data: Sparsity-promoting tomography

We demonstrate the sparsity-based parameter choice rule introduced in Section 6.3 here using X-ray tomography and Besov space regularization. The method requires a priori knowledge about the number of nonzero wavelet coefficients in the unknown attenuation coefficient. How would we know such a thing? In medical imaging applications, for example, we might have available an atlas of high-resolution CT slices of patients. We could then analyze the sparsity levels of those slices and use that knowledge to set the desired number of nonzero coefficients in the new (unknown) target.

We model the medical atlas scenario by sawing three walnuts in half and photographing the exposed slice. See Figure 9.6 for the photos. Further, we compute the wavelet transforms of the digital photographs and see how many of them are above a small threshold. It turns out that in size 128×128, in average 5936 wavelet coefficients out of the total of $128^2 = 16384$ are essentially nonzero.

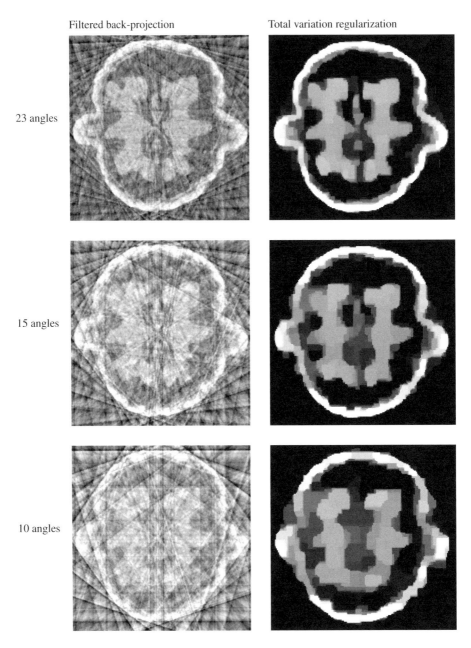

Figure 9.5. Sparse-angle tomographic reconstruction of a slice of a walnut using filtered back-projection (left column) and total variation regularization (right column). The number of equally distributed projections is indicated on the left. These images are computed by Kati Niinimäki. Compare the reconstructions to the ground truth shown in Figure 9.4.

9.2. Limited-angle tomography

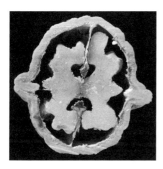
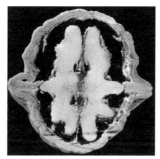
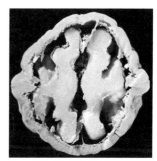

Figure 9.6. Photographs of walnuts sawn in half. These photos are used for estimating the expected number of nonzero wavelet coefficients in a two-dimensional tomographic reconstruction. Special thanks go to Esa Niemi for his careful job in sawing the walnuts.

We use the same 90-degree-angle data set as in Section 9.1.2. The reconstruction size is chosen to be $n = 128 \times 128$. We solve the minimization problem

$$\text{minimize}\left\{\|Af - \mathbf{m}\|_2^2 + \alpha \|f\|_{B^1_{11}(\mathbb{T}^2)}\right\}$$

using Haar wavelet basis and a quadratic programming code developed by Kati Niinimäki. The regularization parameter ranges between 10^{-6} and 10^4, and we compute reconstructions using 19 values of α in that interval. The result is readily expressed in wavelet transform domain, and we compute the number of (essentially) nonzero wavelet coefficients of each of the 19 reconstructions. Further, we fit a smooth interpolation curve to the numbers of nonzeros. See the left plot in Figure 9.7.

According to the sparsity-based choice rule for regularization parameter, we use the interpolation curve (the S-curve) in the left plot in Figure 9.7 to find the value of α that gives most closely the a priori known number 5936 of nonzero coefficients. The result is $\alpha = 0.024$. The resulting reconstruction is shown on the right in Figure 9.7.

The results and images presented in this section are due to a collaboration with Keijo Hämäläinen, Aki Kallonen, Ville Kolehmainen, Matti Lassas, Esa Niemi, and Kati Niinimäki. For more details, see [186].

9.2 Limited-angle tomography

In limited-angle tomography the projection images are available only from a restricted angle of view. From a purely theoretical point of view this may appear not to be a big problem. Namely, as discussed in [420, 353], having projection data available, for example, in dimension two from an angular range $\theta \in [-\varepsilon, \varepsilon]$ for any $\varepsilon > 0$ amounts to knowing the Fourier transform of the attenuation coefficient in a bowtie-shaped set in the frequency domain. In case the target has finite extent, the attenuation coefficient is compactly supported, and therefore its Fourier transform is real-analytic. Since we know the Fourier transform in a set containing an open disc, by analyticity we know it in the whole plane. Consequently we can recover the attenuation coefficient perfectly.

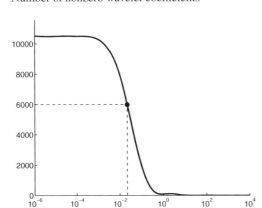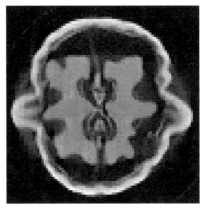

Figure 9.7. Reconstruction using Besov space regularization and sparsity-based choice of regularization parameter. Assume that we know a priori that the reconstruction should have 5936 nonzero wavelet coefficients (out of the maximal number of $128^2 = 16384$). Left: plot of a curve interpolating the numbers of nonzero wavelet coefficients in reconstructions computed using different regularization parameters α. Note that the horizontal axis is logarithmic. Right: reconstruction with the parameter $\alpha = 0.024$ leading to correct sparsity. Compare to the ground truth shown in Figure 9.4.

The above reasoning is based on the notoriously unstable theoretical device of analytic continuation. Practically, when the angle of view decreases from 180 degrees towards zero, the reconstruction problem becomes increasingly ill-posed. The sensitivity of the limited-angle reconstruction to measurement noise is analyzed quantitatively using singular value expansions in [432, 113, 322]. These studies show that limited-angle tomography is a very challenging inverse problem, especially with narrow angles of view.

We demonstrate Tikhonov and total variation regularization for limited-angle tomography with simulated data. We compute limited-angle tomographic data for the 512×512 Shepp–Logan phantom using 12 angles from 60 degrees angle of view. See Figure 9.8 for the reconstructions. Note the typical "stretching" of the reconstructions in the general direction of the X-rays. Table 9.2 lists relative errors in the reconstructions.

Despite the distorted nature of the reconstructions in Figure 9.8, some features are actually recovered quite reliably. According to the microlocal analysis presented in [178, 383], there are some boundaries of objects inside the target that are stably represented in the limited-angle data. Roughly speaking, those boundaries that are tangented by X-rays can be reliably reconstructed. See Figure 9.9 for horizontal profiles of the reconstructions in Figure 9.8 and note that the locations of the jumps are rather well reconstructed, as predicted by microlocal analysis. See Table 9.2 for relative errors in the reconstructed profiles.

Three-dimensional reconstructions computed from measured limited-angle X-ray data are shown in Section 9.3.4 below.

9.2. Limited-angle tomography

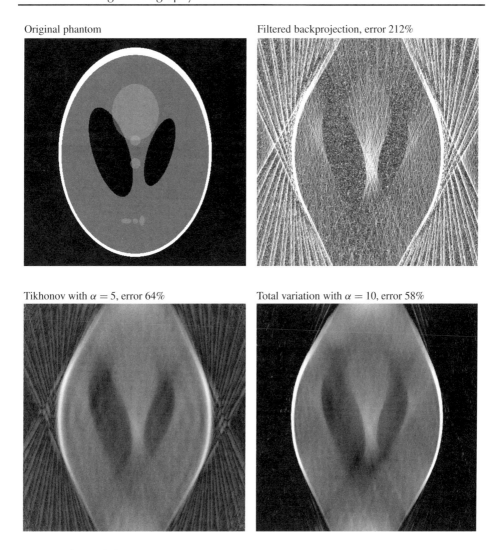

Figure 9.8. Comparison of reconstruction methods for limited-angle tomography. Note the typical "stretching" of the reconstructions along the directions of view. According to the microlocal analysis presented in [178, 383], some parts of the boundaries can be reconstructed better than others—in this case close to vertical boundaries. See Figure 9.9 for plots of horizontal profiles through the above reconstructions.

Table 9.2. Computation times and relative errors of reconstructions from limited-angle tomographic data. See Figures 9.8 and 9.9 for the reconstructions.

Reconstruction method	Error	Error (profile)	Computation time (s)
Filtered back-projection	212%	230%	0.1
Tikhonov regularization	64%	38%	10.2
Total variation regularization	58%	21%	14.7

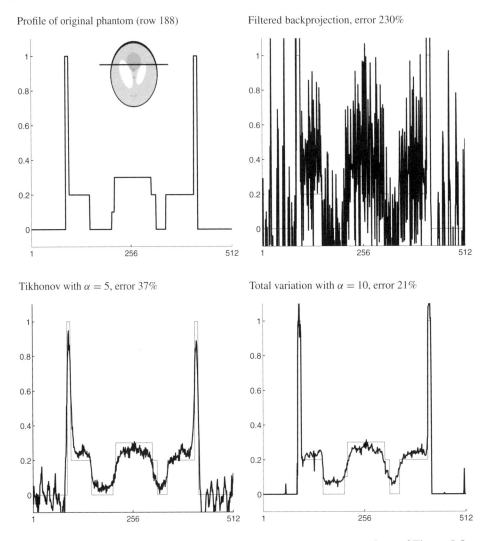

Figure 9.9. Horizontal profiles through the limited-angle reconstructions of Figure 9.8.

9.3 Low-dose three-dimensional dental X-ray imaging

In filtered back-projection the mathematics requires dense sampling of data, leading to high radiation doses. In low-dose three-dimensional X-ray imaging the thinking is reversed. Only the least possible amount of X-ray data is collected, and limitations of data are compensated for with tailored inversion algorithms. The reconstructions are not perfect, but they only need to be good enough for the clinical application at hand.

Now the mathematics does not dictate the dose, but instead the dose dictates the mathematics!

This section is dedicated to the description of the first commercial medical imaging apparatus based on limited-angle X-ray tomography. It is a novel application of a so-

9.3. Low-dose three-dimensional dental X-ray imaging

called panoramic imaging device that every dental clinic has nowadays. Programming new movements to the machine allows the collection of tomographic projection data. In essence, a two-dimensional imaging device that the clinic already has is reprogrammed, and suitable mathematics is developed for compensating for the incomplete data set. Here is an example where computational inversion is the core feature enabling a breakthrough in imaging technology.

Before discussing the new imaging device in Section 9.3.4 we need a couple of historical digressions.

9.3.1 Tomosynthesis

Tomosynthesis was invented (or at least first published) by a Dutch scientist called Ziedses des Plantes [120]. Later, tomosynthesis was named and developed further by David G. Grant [169]. The idea is to take X-ray images of a three-dimensional object from different directions and then use a simple shift-and-add algorithm to combine the images in such a way that details in a so-called *sharp layer* are more clearly visible than details away from that layer. We illustrate the shift-and-add process in Figures 9.10 and 9.11.

Mathematically, tomosynthesis is easily seen to be nothing else but *unfiltered* backprojection.

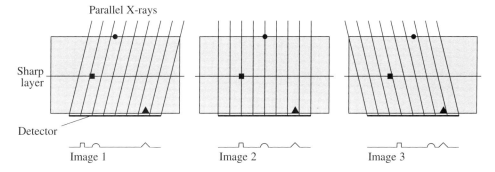

Figure 9.10. Image acquisition in tomosynthesis. X-ray images of the same object are taken from different directions. In this schematic two-dimensional illustration we use parallel-beam geometry for simplicity.

See the articles [180, 496, 166, 121, 124, 123, 440, 25] for more information on tomosynthesis and its applications.

9.3.2 Tuned-aperture computed tomography

Richard L. Webber patented in 1994 an imaging method called *tuned-aperture computed tomography* (TACT), where the idea is to record X-ray images of a target from different directions, calibrate the directions using a fiducial reference ball visible in each image, and use tomosynthesis for three-dimensional reconstruction. The reference ball is typically placed on top of the object to be imaged (as in Figure 9.10) and keep the object and detector in fixed positions relative to each other.

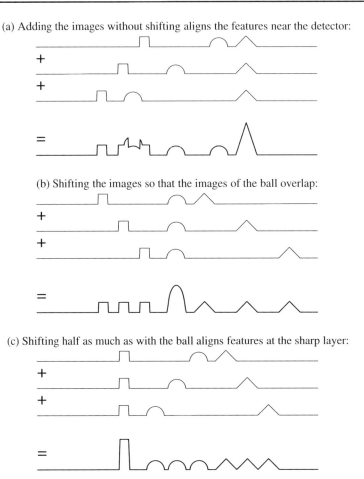

Figure 9.11. Computation of sharp layers in tomosynthesis. This is the so-called shift-and-add algorithm. See Figure 9.10 for the structure of the object under imaging.

The goal of Webber's work is to perform three-dimensional medical imaging with simple and cost-effective equipment. Any combination of an X-ray source and digital detector will do. This approach is in stark contrast with the trend of using big and expensive MRI and CT machines for three-dimensional imaging.

"Tuning" the aperture is related to the possibility of modifying the thickness of the sharp layer by choosing the imaging angles suitably. Roughly speaking, a wider angle of view results in narrower depth-of-field in the tomosynthetic slice.

The TACT imaging approach is useful for several applications in dental radiography; see [471, 472, 473, 474].

9.3.3 Panoramic dental imaging

Starting in the beginning of twentieth century there was a race to develop a device that could take an X-ray image of all the teeth simultaneously. The first approaches were based

9.3. Low-dose three-dimensional dental X-ray imaging 125

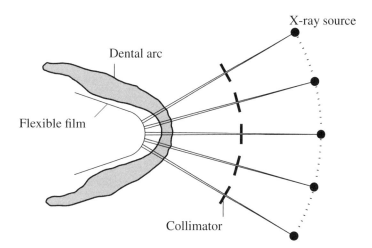

Figure 9.12. Principle of film-based intraoral (film inside the mouth) panoramic dental X-ray imaging. The X-ray beam is narrowed using a collimator, which is essentially a slit in a radiopaque lead plate.

on having a film inside the mouth and using a narrow X-ray beam as shown in Figure 9.12. The pioneering work of Hisatugu Numata [368] deserves special mention.

Yrjö Veli Paatero (1901–1963) defended his thesis in dentistry in 1939 and was nominated in 1945 to a teaching position in the Department of Dentistry at the University of Helsinki, Finland. He was responsible for X-ray diagnostics, and spent a lot of time in the tedious routine work of imaging the teeth of individual patients using several small intraoral films. He started to develop better methods for imaging all the teeth, starting with intraoral approaches [373]. Soon thereafter he came up with the idea of imaging a sharp layer following the dental arc using a geometric movement; in this approach both the film and the X-ray source are outside the mouth [374]. Paatero's invention received international attention, and he received an invitation to the University of Washington, Seattle, for the period 1950–1951. During that time he developed panoramic imaging further, and continued his work after returning to Finland in 1951.

The principle of the panoramic image formation is based on an ingenious combination of two simultaneous movements. A fixed arrangement of an X-ray source and film holder rotates with the patient's head located at the center of rotation. X-rays reach the film through a collimator (a narrow slit in a radiopaque lead plate). The film moves linearly with respect to the holder, effectively picking out features located in the sharp layer and blurring features away from the sharp layer. The faster the linear movement of the film, the larger the radius of the circular sharp layer. See Figure 9.13 for a simplified illustration.

The formation of the sharp layer is basically a continuum version of the tomosynthesis principle explained in Section 9.3.1 and Figures 9.10 and 9.11.

In modern digital panoramic devices the physical movement of the film is replaced by electronic movement of pixels in a narrow charge coupled device detecting the radiation. See Figure 9.14 for a picture of a modern panoramic imaging device and Figure 9.15 for a digital panoramic image.

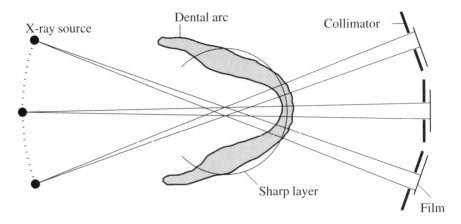

Figure 9.13. Principle of film-based extraoral (film outside the mouth) panoramic dental X-ray imaging, shown in a simplified cylindrical geometry and exaggerated angles for clarity. Note the crucial linear movement of the film with respect to the collimator slit. Due to the simplified geometry, the sharp layer does not follow the dental arc properly. In practice this problem is solved by varying the linear speed of the film and moving the center of rotation during the exposure.

In recognition of his groundbreaking work, Paatero received a professorship of Oral Radiology at the University of Turku, Finland, in 1961. However, he died tragically soon thereafter in 1963. His longtime collaborator Timo Nieminen founded a company called Palomex in 1964 and started manufacturing panoramic X-ray devices. The idea of panoramic imaging caught on in the 1970s, and since then a panoramic device has been considered a necessary tool for every dental clinic.

For more details on the history of panoramic imaging the reader is referred to [185].

9.3.4 The VT device for dental implant planning

The Palomex company mentioned in Section 9.3.3 was acquired in 1977 by Instrumentarium Corporation, a Finnish company founded in 1900. The Instrumentarium Imaging division concentrated on various medical X-ray technologies. The story of the VT device started in 1998, when the company licensed the patent for TACT imaging (discussed in Section 9.3.2).

Instrumentarium Imaging planned to use TACT technology for several applications, including mammography, surgical C-arm imaging and dental imaging. Research projects were initiated in 2001 together with academic partners and funding from the Finnish Technology Agency. Bayesian inversion was studied as a flexible framework of reconstruction instead of the outdated and rigid tomosynthesis approach. The first results were published in 2003 as the two-part article [414, 289].

Big changes followed on the corporate level. General Electric acquired Instrumentarium in 2003 and later sold the dental imaging division in 2005, resulting in the founding of PaloDEx Group. As a result, the originally quite general three-dimensional imaging project was refocused on dental imaging applications only. However, some three-dimensional re-

9.3. Low-dose three-dimensional dental X-ray imaging

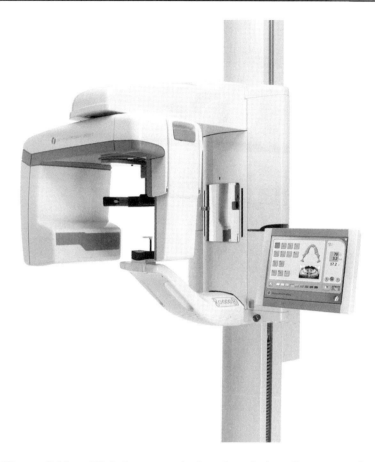

Figure 9.14. Digital panoramic imaging device (Instrumentarium Dental OP200D). Photograph courtesy of PaloDEx Group. See Figure 9.15 for a digital panoramic image.

sults concerning mammography were published; see [392]. Interestingly, pretty much the same team of researchers and engineers worked together regardless of the changing names of the parent corporation.

At the time of writing, these research and development projects during 2001–2007 have resulted in one commercial product: the so-called volumetric tomography (VT) device designed for dental implant planning.

Dental implants replace missing teeth. Implants are attached by drilling a hole into the bone, screwing a titanium screw to the bone, and attaching the implant to the screw. The hole must be deep enough for the implant to be sturdy, but not so deep that it damages nerves located inside the bone. See Figure 9.16 for an illustration. Planning the direction, width, and depth of an appropriate hole requires three-dimensional information on the tissue. Mere projection X-ray images or panoramic X-ray images are not enough, as they involve overlapping of structures and geometric distortions.

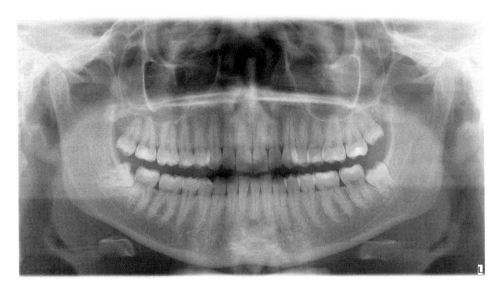

Figure 9.15. A typical digital panoramic image, taken with the Instrumentarium Dental OP200D device shown in Figure 9.14. Image courtesy of PaloDEx Group.

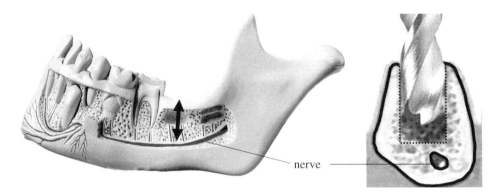

Figure 9.16. Left: typical situation for implant planning. A tooth is missing, and a hole should be drilled as indicated by the black arrow. However, it is important not to drill so deep that the nerve would be damaged. The photograph is taken from a plastic anatomical model. Right: cross-section of the mandible showing the location of the nerve. The appropriate depth of drilling can be measured from a reconstructed cross-sectional slice oriented like this.

The idea of the VT device is to use a digital panoramic imaging device, shown in Figure 9.14, in a new way for collecting limited-angle tomographic data. See Figure 9.17 for an illustration of the imaging geometry. These tailored movements of the panoramic device need to be specially designed because the detector is very narrow (as explained in Section 9.3.3, the detector is designed to mimic the moving film behind a narrow slit collimator). Thus two-dimensional projection images need to be formed by a linear movement,

9.3. Low-dose three-dimensional dental X-ray imaging

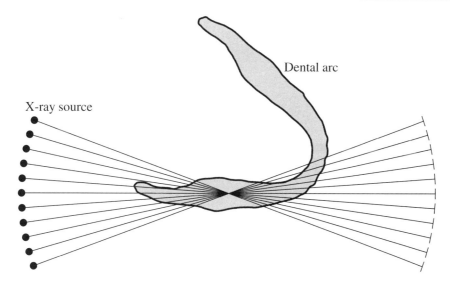

Figure 9.17. Imaging geometry of the VT device. A total of 11 projection images is taken from a ±20-degree angle of view. See Figure 9.18 for examples of actual projection images.

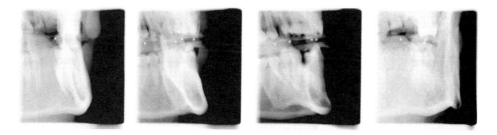

Figure 9.18. Some projection images of a dry scull taken with the VT device. The detector is very narrow in the horizontal direction, so the images are recorded using a scanning movement. Notice the fiducial reference balls used for calibrating the imaging geometry. Images courtesy of PaloDEx Group.

much as in a photocopier or a flatbed scanner. See Figure 9.18 for projection images of a dry skull.

The limited-angle data set is so incomplete that filtered back-projection–type algorithms do not give satisfactory results. Tikhonov regularization, iterative frequency-domain reconstruction, and total variation regularization do provide good enough reconstructions, as shown in [290, 291, 70, 218]. See Figure 9.19 for an example of actual *in vivo* imaging using the VT device. The computation is done according to [252].

It is also possible to augment the limited-angle projection data set by including a panoramic image. Now the panoramic image is not a projection image but rather a curved tomosynthetic slice. However, proper modeling of the sharp layer formation allows one to combine information from perpendicular directions, thus adding the most badly missing

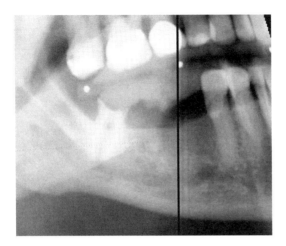 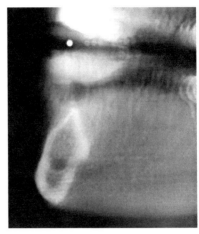

Figure 9.19. Slices from an *in vivo* three-dimensional reconstruction produced by the VT device from a real patient with missing teeth. Left: navigation slice showing the location of the cross-section on the right. Right: cross-section of the mandible with nerve canal clearly visible. Compare to Figure 9.16. Images courtesy of PaloDEx Group.

Table 9.3. X-ray radiation patient doses delivered by various devices used for three-dimensional dental imaging. The dose of the VT device is known to be 1–2 doses of a panoramic two-dimensional image, which is roughly 6.7 μSv [109, 477]. The data for the other devices is taken from [324].

Modality	Dose (μSv)
Head CT	2100
CB Mercury	558
NewTom 3G	59
VT device	13

information to the limited-angle data set. See [218] for reconstructions from such hybrid data sets.

The sparse-data VT product has the lowest radiation dose among its dense-data competitors, as shown in Table 9.3. The three-dimensional information provided by VT is, nevertheless, enough for the clinical task of dental implant planning. In this way, a device that the clinic already has can be upgraded by mathematical software to yield three-dimensional images.

Chapter 10
Projects

In this chapter we outline several computational projects related to linear inverse problems. Each project involves building a computational model for a practically relevant direct problem and applying some of the inversion methods developed in Part I of this book. The projects may serve as individual or group assignments in a course, or as starting points of academic theses. Also, successful execution and formal reporting of a project below serves as training for writing scientific articles or engineering reports.

Let us point out some basic principles related to this kind of projects combining theoretical and computational aspects as well as dealing with ill-posedness.

Every piece of software should be tested thoroughly until you can completely trust the results. It's a good idea to start with the simplest possible case (perhaps allowing analytical solution by an explicit formula), write an algorithm, and compare the outcome with an analytical formula or another algorithm which is as different as possible. Then add the slightest possible complication to the code and test it as well. This way you can work your way to the full complexity of the computation by taking as small, completely reliable steps as possible.

If you are working with simulated data (as opposed to measuring real data), be sure to avoid an inverse crime. Simulate the data with a different algorithm (or at least with a different grid and slightly perturbed parameters) from what was used in the inversion method; otherwise you may end up with computations that produce wonderfully accurate reconstructions from noise-free data but break down completely when noise is added. Since real data always contains noise such algorithms are utterly useless in practice. Let us stress once more: noise-robustness is the most important feature of practical inversion methods.

If your project involves data measured in the real world, start by building both the computational forward model (with accurate simulation of measurement noise) and the inversion algorithm. Then test the inversion thoroughly using simulated data and avoiding any inverse crime. Only then apply the method to real data. The reason for this procedure is efficient debugging. If you go right ahead and try to invert the measured data, chances are that something will go wrong (it always does), but you have no chance of knowing where the problem is: in the forward model, in the inversion algorithm, in the choice of parameters, in the representation of data, or something else. However, if you work your way from the simplest simulated case step by step to the most complicated simulated case, then compare the properties of the measured and simulated data, and only then attempt

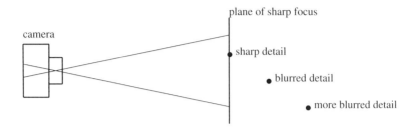

Figure 10.1. Schematic illustration of the focal plane of a camera. Details in the focal plane appear sharp, and details off the focal plane appear blurred. The blurring is stronger if the detail is farther away from the focal plane.

inversion of the real data, then you always know that the possible problem arises from the latest small addition to the so far reliable computational framework.

10.1 Image deblurring

The aim of this project is to build an algorithm that sharpens misfocused photographs. The data is measured using a digital camera, and consequently the reconstruction methods must be applicable to large-scale data. An illustration of the focal plane of a camera is found in Figure 10.1.

The optical construction of regular consumer cameras produces a plane of sharp focus. This means that details located on a two-dimensional plane in the three-dimensional scene under imaging show up sharp and crisp in the photograph, and details away from that plane are blurred. Moreover, the blur becomes worse as the distance between a detail and the plane of sharp focus grows. By "worse blur" we mean wider point spread function.

To keep the project simple, we choose a planar scene so that the point spread function is the same in all parts of the image. Take a printed paper with some text and an image; a newspaper page will do. Make sure that the page contains a small isolated black dot.

Now you need to take a misfocused photograph of the printed paper. The use of a tripod to keep the camera steady is recommended; it's also a good idea to use two light sources illuminating the paper with roughly 45 degree angles. If you have a camera with manual focus, you can simply focus perfectly on the plane of the paper and then turn the focus a little (or move the camera slightly away from the paper). A simple way to achieve the same result with an autofocus camera is to place the paper on the floor and put a book on top of the paper. (Of course, the book should not cover the important part of the paper.) Let the camera focus on the book cover; then the paper is out of focus. Varying the distance between the camera and the book will adjust the amount of blur.

It's a good idea to take a sharp picture for comparison and a few misfocused versions with different amount of blur. Also, take photographs with different sensitivities (ISO settings in the camera), resulting in various amounts of measurement noise.

Read the photograph into MATLAB using the `imread` command and pick out only one of the three color components (red, green, or blue). Then you have a grayscale pixel image that can be viewed as a discrete approximation of a real-valued light intensity distribution on the paper. You can read off the shape of the point spread function from the image of the isolated black dot.

10.2 Inversion of the Laplace transform

Let $f : [0,\infty) \to \mathbb{R}$. The Laplace transform F of f is defined by

$$F(s) = \int_0^\infty e^{-st} f(t)dt, \quad s \in \mathbb{C}, \tag{10.1}$$

provided that the integral converges. The direct problem is to find the Laplace transform for a given function f according to (10.1). The inverse problem is, given a Laplace transform F, find the corresponding function f.

Assume that we know the values of F at real points $0 < s_1 < s_2 < \cdots < s_n < \infty$. Then we may approximate the integral in (10.1) with, for example, the trapezoidal rule as

$$\int_0^\infty e^{-st} f(t)dt \approx \frac{t_k}{k} \left(\frac{1}{2} e^{-st_1} f(t_1) + e^{-st_2} f(t_2) + e^{-st_3} f(t_3) + \cdots \right.$$
$$\left. + e^{-st_{k-1}} f(t_{k-1}) + \frac{1}{2} e^{-st_k} f(t_k) \right), \tag{10.2}$$

where the vector $t = [t_1\ t_2\ \ldots\ t_k]^T \in \mathbb{R}^k$, $0 \leq t_1 < t_2 < \cdots < t_k$, contains the points at which the unknown function f will be evaluated. By denoting $x_l = f(t_l)$, $l = 1, \ldots, k$, and $m_j = F(s_j)$, $j = 1, \ldots, n$, and using (10.2), we get a linear model of the form $m = Ax + \varepsilon$ with

$$A = \frac{t_k}{k} \begin{bmatrix} \frac{1}{2}e^{-s_1 t_1} & e^{-s_1 t_2} & e^{-s_1 t_3} & \cdots & e^{-s_1 t_{k-1}} & \frac{1}{2}e^{-s_1 t_k} \\ \frac{1}{2}e^{-s_2 t_1} & e^{-s_2 t_2} & e^{-s_2 t_3} & \cdots & e^{-s_2 t_{k-1}} & \frac{1}{2}e^{-s_2 t_k} \\ \vdots & & & & & \vdots \\ \frac{1}{2}e^{-s_n t_1} & e^{-s_n t_2} & e^{-s_n t_3} & \cdots & e^{-s_n t_{k-1}} & \frac{1}{2}e^{-s_n t_k} \end{bmatrix}.$$

This problem is addressed in the article "The Bad Truth about the Laplace Transform" by Epstein and Schotland [134].

10.3 Backward parabolic problem

The backward heat equation introduced in Chapter 2 included the simplifying assumptions that the problem is posed in one spatial dimension and that the diffusion coefficient D is constant. Relaxing these assumptions leads to a more versatile model, although it will have more computational complexity. The solution can be approximated with a regularized least-squares approach or by methods of quasi-reversibility [307], in which the problem is transformed into a new problem that is well-posed in the backward direction, but ill-posed in the forward direction. The disadvantage of this approach is that the transformation is achieved by adding the biharmonic operator $\epsilon \Delta^2$ to the elliptic operator, resulting in a fourth-order PDE. A good project is to experiment with either of these approaches on a backward heat equation with one or both of these assumptions relaxed.

A more general model that includes a transport term, a source term, and a decay term is applicable not only to heat transfer [40], but also to contaminant transport in an underground aquifer [39]. In such a model, $u(x,y,t)$ represents the concentration of the contaminant at time t and point (x,y) in the aquifer. Consider the following backward parabolic problem for a two-dimensional convection-diffusion equation.

Find $\phi(x,y)$ such that $u(x,y,t_0) = \phi(x,y)$ is the solution to

$$u_t = (D_1(x,y)u_x)_x + (D_2(x,y)u_y)_y - (v(x,y)u)_x - \lambda(x,y)u + F(x,y),$$
$$u(x,y,t) \to 0 \quad \text{as} \quad x^2 + y^2 \to \infty, \tag{10.3}$$
$$u(x,y,T) = \psi(x,y) > 0, \quad T > t_0 > 0, \quad (x,y) \in \mathbb{R}^2.$$

The initial profile ϕ is assumed to have compact support. The coefficients $D_i(x,y), i = 1,2,$ are dispersion-diffusion coefficients, and the function $v(x,y)$ is the average linear velocity. The functions $D_1, D_2,$ and v are assumed to be positive piecewise continuous bounded functions in a domain Ω. The function $\lambda > 0$ is the decay rate, and $F(x,y)$ is the source of contaminant concentration per unit mass. The inverse problem is to determine the release history of the groundwater contaminant from knowledge of $\psi(x,y)$. The problem is ill-posed, as was the backward heat equation.

Note that the differential operator on the right-hand side of (10.3) is non-self-adjoint, and due to the presence of the transport term, the solution to the forward problem has a "wave-like" character. As a result, contrary to the self-adjoint case, the computational mesh depends on T, requiring the use of a forward solver with sufficient accuracy and stability for physically realistic time steps.

The continuous dependence of the solution to the forward problem on the initial data ϕ is proved in [200], where a stability estimate on the final time T is also proved that shows the decrease in stability as T increases. Using the stability result the existence of a least-squares solution to the backward parabolic problem is proved. Expressing the unknown initial profile in terms of a finite set of linearly independent basis functions $\xi_m(x,y), m = 1,\ldots,M,$

$$\phi(x,y) = \sum_{m=1}^{M} \beta_m \xi_m(x,y),$$

reduces the numerical solution of the backward parabolic problem to a finite number of well-posed forward problems with initial profile ξ_m. Expressing the solution at time T to these forward problems by $u[\xi_m]|_{t=T}$, the minimization functional J can be defined by

$$J(\beta) = \int_{\Omega} \left| \sum_{m=1}^{M} \beta_m u[\xi_m]|_{t=T} - \psi(x,y) \right|^2 dx\, dy.$$

The minimum occurs when

$$\frac{\partial}{\partial \beta_k} J(\beta_1,\ldots,\beta_M) = 0, \quad k = 1,\ldots,M,$$

resulting in an ill-posed system of linear algebraic equations $A\beta = b$, where $A = [a_{km}]$, $b = [b_1,\ldots,b_M]^T$, and

$$a_{km} = \int_{\Omega} u[\xi_k]|_{t=T} u[\xi_m]|_{t=T} dx\, dy,$$
$$b_k = \int_{\Omega} \psi(x,y) u[\xi_k]|_{t=T} dx\, dy.$$

A modified alternating-direction finite difference scheme for the solution of the intermediate forward problems is presented in [200], and another approach is the use of a finite element method. The minimization problem can be regularized by the methods of Part I of this book, resulting in many possibilities for a project solving the backward parabolic problem (10.3).

Part II

Nonlinear Inverse Problems

Chapter 11

Nonlinear inversion

We consider indirect measurements of the form

$$m = \mathcal{A}(f) + \varepsilon, \tag{11.1}$$

where f is the unknown function, ε models measurement error, and $\mathcal{A} : \mathcal{D}(\mathcal{A}) \subset X \to Y$ is the nonlinear forward map. Here the model space X and the data space Y are assumed to be Banach spaces. The only information we assume known about ε is an inequality $\|\varepsilon\|_Y \leq \delta$ with a known constant $\delta > 0$. Such models arise from various situations in technology and physics; the nonlinear operator $\mathcal{A}$ may be related to a PDE or to an integral equation. See Figure 11.1 for a schematic illustration of the model (11.1).

The inverse problem is

$$\begin{aligned}&\text{Given noisy measurement } m = \mathcal{A}(f) + \varepsilon \text{ and } \delta > 0, \\ &\text{extract information about } f.\end{aligned} \tag{11.2}$$

In Section 11.4 we introduce several motivating examples of nonlinear inverse problems. We wish to emphasize what the data is, what we recover, and why the forward map is nonlinear. Electrical impedance tomography (EIT), a fundamental nonlinear ill-posed inverse problem, will provide the guiding example that we treat with thoroughness and rigor in Chapters 12–16. We hope the work will serve as an inspiration for new methods in other applications to follow, and as a reference for researchers in EIT.

Ill-posed inverse problems of the form (11.2) are characterized by the failure of at least one of Hadamard's three requirements for a well-posed problem: (1) given the data m, there must be at least one $f \in X$ satisfying $\mathcal{A}(f) = m$ (existence); (2) if two objects $f, g \in X$ yield the same data $\mathcal{A}(f) = \mathcal{A}(g)$, then f must equal g (uniqueness); (3) the solution of the inverse problem must depend continuously on the data (stability).

In Part I of this book, the forward map $\mathcal{A}$ was assumed to be a linear operator between vector spaces. In that case, questions of uniqueness and continuous dependence on data could be analyzed in a straightforward manner by studying the kernel and singular system of the linear operator $\mathcal{A}$ or of its finite-dimensional matrix approximations. Nonlinear forward maps $\mathcal{A}$ require more complicated analysis. We discuss uniqueness, stability, and existence questions in Sections 11.1.1–11.1.3.

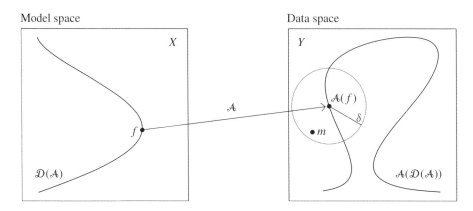

Figure 11.1. Nonlinear measurement model. The nonlinear forward map $\mathcal{A}$ is defined on the domain $\mathcal{D}(\mathcal{A}) \subset X$. The distance between the ideal measurement $\mathcal{A}(f)$ and noisy data m is known a priori to be less than δ.

Once Hadamard's conditions have been studied in the case of a given inverse problem, a noise-robust solution method should be constructed. Such constructions are called *regularization strategies* and are discussed below in Section 11.2.

The computational solution of inverse problems is much more complicated in the nonlinear case than in the linear case discussed in Part I of this book. We discuss the two main approaches, iterative and direct, in Section 11.3.

Actually it is sometimes possible to linearize a nonlinear inverse problem around an educated guess for the unknown, apply the methods from Part I, and achieve useful results. When starting research on a new nonlinear inverse problem, it is generally a good idea to start with the linearized version. However, if the problem is very nonlinear and a good guess is not available, linearization will not be a satisfactory solution approach.

Let us list some further reading about nonlinear inversion. The mathematical theory is discussed in general in the books [439, 340, 132, 236, 260, 445, 247, 254, 270, 269, 41], and more specifically inverse scattering in [358, 71, 91, 381, 62, 309, 63], and EIT and imaging in [354, 405]. A more practical perspective is offered by the application-oriented books [331, 211, 201, 355, 434, 215, 21, 403, 26].

11.1 Analysis of nonlinear ill-posedness

11.1.1 Uniqueness

Is it possible that two different targets of indirect measurement yield the same data? More precisely, do there exist $f, g \in X$ with $f \neq g$ and $\mathcal{A}(f) = \mathcal{A}(g)$? Such uniqueness questions can be extremely hard mathematical problems, as evidenced by the rich literature on the theory of inverse problems involving many areas of contemporary mathematics (geometry, real and complex analysis, PDEs, functional analysis). In general this is possible; see Figure 11.2 for an illustration.

If uniqueness does not hold, it is important to analyze the precise nature of the non-uniqueness. Then one can work with equivalence classes

$$[f] := \{g \in X \mid \mathcal{A}(g) = \mathcal{A}(f)\}$$

11.1. Analysis of nonlinear ill-posedness

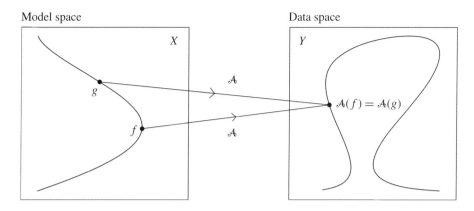

Figure 11.2. Nonlinear inverse problems may involve nonuniqueness, as illustrated here: two different objects $f \neq g$ give exactly the same measurement data.

of objects yielding the same data and consider a unique inverse problem in the quotient space $\{[f] \mid f \in X\}$. Often the smallest (in the sense of the $\|\cdot\|_X$-norm) representative of an equivalence class can be defined to be the reconstruction from ideal data.

If the nonuniqueness is really severe, or too many very different objects give the same data, there may be simply too little information to produce meaningful reconstructions. In such cases one should examine possibilities of (a) measuring more data or (b) finding a priori information about the unknown and complement the insufficient measurement data with that. The complementing process can be constructed into regularization strategies discussed in Section 11.2.

11.1.2 Continuous dependence on data

Assume that uniqueness holds (either due to injectivity of $\mathcal{A}$, or after constructing equivalence classes). There can still be a problem with Hadamard's third condition: the forward map $\mathcal{A}$ might not have a continuous inverse on $\mathcal{A}(\mathcal{D}(\mathcal{A}))$. This is actually the usual case when dealing with ill-posed inverse problems. The uniqueness analysis can be augmented with conditional stability results, which are inequalities of the form

$$\|f - g\|_X \leq \Phi(\|\mathcal{A}(f) - \mathcal{A}(g)\|_Y), \tag{11.3}$$

where $f, g \in \mathcal{D}(\mathcal{A})$ and $\Phi : \mathbb{R} \to \mathbb{R}$ is a nonnegative continuous function satisfying $\Phi(0) = 0$. Inequality (11.3) gives quantitative information about the ill-posedness of the inverse problem. See Figure 11.3 for a schematic illustration of conditional stability.

From a practical point of view inequalities of the form (11.3) are rarely useful, since often $m \notin \mathcal{A}(\mathcal{D}(\mathcal{A}))$.

11.1.3 Existence

Continuum inverse problems are modeled by a forward map $\mathcal{A}$ acting between some infinite-dimensional function spaces X and Y. As depicted in Figure 11.1, it is usually the case that $m \notin \mathcal{A}(\mathcal{D}(\mathcal{A}))$ so there is a problem with existence of a preimage $\mathcal{A}^{-1}(m)$. But why do

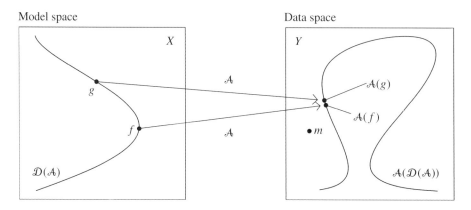

Figure 11.3. Conditional stability estimates bound the difference $\|f - g\|_X$ by the difference $\|\mathcal{A}(f) - \mathcal{A}(g)\|_Y$ in infinite-precision data. From a practical point of view conditional stability results are rarely applicable, as typically the measured data $m \notin \mathcal{A}(\mathcal{D}(\mathcal{A}))$.

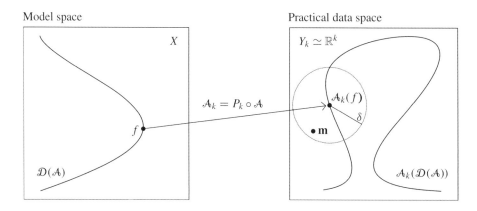

Figure 11.4. Existence failure in the practical nonlinear measurement model involving finite-dimensional data. The noisy data **m** is not the ideal measurement of any object in $\mathcal{D}(\mathcal{A})$.

we not define the data set to be $\mathcal{A}(\mathcal{D}(\mathcal{A}))$? This would take care of the lack of existence. However, the set $\mathcal{A}(\mathcal{D}(\mathcal{A}))$ is often most conveniently defined as a subset of a larger space Y, and the precise structure of $\mathcal{A}(\mathcal{D}(\mathcal{A}))$ is hard to analyze in case of most nonlinear and practically interesting inverse problems. Also, in finite-dimensional measurement models (that are necessary in practice), there is almost always lack of existence, as will be seen below. Thus it is reasonable to use a large data space in the infinite-dimensional limit case as well.

Using a practical device to measure values of the function $\mathcal{A}(f) \in Y$ can typically be modeled using a projection operator $P_k : Y \to Y_k \subset Y$ with a k-dimensional range $Y_k \simeq \mathbb{R}^k$, where $\simeq$ denotes isometry. The range $P_k(\mathcal{A}(\mathcal{D}(\mathcal{A})))$ is a proper subset of Y_k as indicated in Figure 11.4; it would be quite a coincidence to have the range equal Y_k in practice.

11.2. Nonlinear regularization

The measured data vector is then $\mathbf{m} = P_k(\mathcal{A}(f)) + \varepsilon$, where ε is a random vector taking values in $\mathbb{R}^k$. See Figure 11.4. The noisy data $\mathbf{m}$ is not the infinite-precision measurement of any object: $\mathbf{m} \notin \mathcal{A}(\mathcal{D}(\mathcal{A}))$. This can be seen as the failure of Hadamard's first condition of existence: the inverse of $\mathcal{A}$ is not defined for the element $m \in Y$.

Nevertheless, it may be that there is a characterization available for the set $\mathcal{A}(\mathcal{D}(\mathcal{A}))$. In that case one can project the noisy data $m = \mathcal{A}(f) + \varepsilon$ to some nearby element $P(m) \in \mathcal{A}(\mathcal{D}(\mathcal{A}))$ and use (11.3) for $g = P(m)$. However, the structure of the range $\mathcal{A}(\mathcal{D}(\mathcal{A}))$ is typically unknown, so there might not be useful characterizations available.

11.2 Nonlinear regularization

The goal related to an inverse problem is to recover information about the unknown quantity from the measurement data. Recall that the forward map does not allow a continuous inverse as the measurement data is perturbed by noise. Thus we need to define a *continuous* reconstruction function from the data space to the model space that comes as close to being the inverse of the forward map as possible. Such a process is called regularization.

A family of continuous maps $\mathcal{R}_\alpha : Y \to X$ parameterized by $0 < \alpha < \infty$ is called a *regularization strategy* if

$$\lim_{\alpha \to 0} \mathcal{R}_\alpha(\mathcal{A}(f)) = f$$

for every $f \in X$. Further, a choice of regularization parameter $\alpha = \alpha(\delta)$ as function of noise level $\delta > 0$ is called *admissible* if $\alpha(\delta) \to 0$ as $\delta \to 0$ and

$$\sup_m \left\{ \|\mathcal{R}_\alpha(m) - f\|_X : \|\mathcal{A}(f) - m\|_Y < \delta \right\} \longrightarrow 0, \quad \delta \to 0 \tag{11.4}$$

for every fixed $f \in X$. See Figure 11.5 for a schematic illustration of regularization.

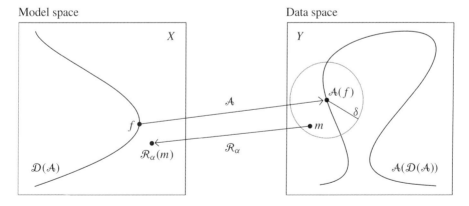

Figure 11.5. Regularized inversion. The forward map $\mathcal{A}$ does not have a continuous inverse. The idea of regularization is to construct a family $\mathcal{R}_\alpha$ of continuous maps from data space Y to model space X in such a way that f can be approximately recovered from noisy data m. For smaller noise level δ the approximation $\mathcal{R}_\alpha(m)$ is closer to f.

Let us comment on the above definition as it is quite technical. First of all, the requirement $\alpha(\delta) \to 0$ as $\delta \to 0$ is related to the idea that the more noise there is, the stronger regularization (bigger α) is needed. On the other hand, in the asymptotic limit $\delta \to 0$ of infinite-precision data, there is no need for regularization and "$\alpha = 0$." Of course, that limit is never actually achieved as there is always noise in the data and the lack of continuous inverse of $\mathcal{A}$ prevents taking α to zero.

Formula (11.4) can be interpreted as follows. The quantity $\|\mathcal{R}_\alpha(m) - f\|_X$ is the reconstruction error measured in the norm of the model space X. For fixed noise level $\delta > 0$, the supremum is taken over all data $m \in Y$ at most the distance δ away from the infinite-precision data: $\|\mathcal{A}(f) - m\|_Y < \delta$. Then (11.4) requires that in the asymptotic limit $\delta \to 0$ of noise-free data, the reconstruction error must vanish. Note that $f \in X$ is fixed in (11.4) and the speed of convergence is not necessarily uniform on X.

11.3 Computational inversion

Inverse problems arising from applications cannot be solved by theorems alone. On top of mathematical analysis, a computer program is needed for practical interpretation of noisy data. Furthermore, ill-posed inverse problems are characterized by extreme sensitivity to modeling and measurement errors, so the solution algorithms must be robust against such errors. The robustness is achieved through building the algorithm according to a regularization strategy in the sense of Section 11.2. As a by-product the regularization strategy provides a mathematically rigorous backing for the computational inversion method.

As shown in Part I of this book, linear inverse problems are all essentially alike and can be solved computationally using one of many types of general regularized solution strategies. In the case of nonlinear forward maps the situation is more complicated. There are two basic approaches to the computational regularized solution of nonlinear inverse problems: iterative and direct.

11.3.1 Iterative nonlinear regularization

The iterative approach is based on Tikhonov regularization or its generalizations. As in the linear case, we approximate the unknown function f by a finite-dimensional parameterized model $\mathbf{f} \in X_n \simeq \mathbb{R}^n$. Here $X_n \subset X$ is a finite-dimensional subset of the model space. Furthermore, we construct a numerical implementation $\mathcal{A}_{kn} : X_n \to Y_k$ that approximates the nonlinear map $\mathcal{A}_k = P_k \circ \mathcal{A} : X \to Y_k$. Then we define the regularized solution as the solution of the minimization problem of the form

$$T_\alpha(\mathbf{m}) = \arg\min_{\mathbf{g} \in X_n} \left\{ \|\mathcal{A}_{kn}(\mathbf{g}) - \mathbf{m}\|^2_{\mathbb{R}^k} + \alpha \|\mathbf{g}\|^2_{X_n} \right\}. \quad (11.5)$$

The solution $T_\alpha(\mathbf{m})$ of the nonlinear minimization problem (11.5) is then typically computed using an iterative optimization method.

The strength of the iterative approach lies in the generality of the formulation. The optimization routine need not be rewritten when going from one application to another; just a new code for $\mathcal{A}_{kn}$ is needed (unless the objective functional in (11.5) changes too much qualitatively, such as from smooth to nondifferentiable, in which case a different optimization routine may be required; however, even in this case the conceptual structure of the method stays the same).

However, the iterative approach has some drawbacks as well. Nonlinearity of the objective functional in (11.5) leads to the possibility of local minima. Without a good initial guess it may be difficult to get the algorithm to converge to the desired global minimum, and it may be hard to detect if the result is a local minimum or not. On the theoretical side, iterative methods can (so far) be proven to be regularization strategies only for inverse problems where the forward map is not too far from a linear map. Unfortunately, many important inverse problems are heavily nonlinear and not covered by existing theory on iterative regularization.

Further discussion about iterative regularization is outside the scope of this book. See the excellent book [254] and the references therein for detailed information about iterative regularization of nonlinear inverse problems.

11.3.2 Direct nonlinear regularization

By direct methods for nonlinear regularization we mean explicit construction of a nonlinear map $\mathcal{R}_\alpha$ without resorting to a minimization problem. The advantage of such an approach is that there is no iteration and therefore no possibility of the algorithm getting stuck in local minima. Also, sometimes it is possible to compute the reconstruction only at desired points, leading to natural region-of-interest imaging and trivial parallelization. Typically there is no need for a computational algorithm for $\mathcal{A}_{kn}$ in the reconstruction method; consequently there is no danger of inverse crimes even when working with computer-simulated data.

Direct methods have some drawbacks, too. Nonlinear inverse problems are quite different from each other, and the direct regularization method needs to be tailored to each inverse problem individually. Also, the algorithms may involve the solution of quite involved PDEs, leading to heavy computations. This is not always the case, though; for example, implementations of the enclosure method of Ikehata involve only simple integrations of the measured data and lead to fast computation. See Section 17.1.

There are currently only few regularized direct methods for nonlinear inverse problems, most notably the factorization method [270], the enclosure method and its variants [227, 223, 220, 447, 219], and D-bar methods [415, 234, 233, 277, 16]. So far only one of these method applies to a global PDE coefficient recovery problem *and* allows a full regularization analysis: the regularized D-bar method for EIT [277]; see Chapter 15.

Given the large body of theoretical work on uniqueness and conditional stability on various nonlinear inverse problems, there is great scope for research in designing and implementing new direct regularization strategies for practical applications.

11.4 Examples of nonlinear inverse problems

This section is devoted to the description of selected nonlinear inverse problems with great practical interest. We explain for each problem what is measured, what needs to be recovered, and why the forward map is nonlinear. We do not specify the model spaces or data spaces, as the precise definition of these is a nontrivial, application-dependent modeling task.

11.4.1 Glottal inverse filtering

The physiological process of human speech production can be modeled by two stages: excitation and filtering. The excitation, or *glottal flow* $f(t)$, is the air flow that streams from the lungs and oscillates due to vibrations of the vocal folds. Filtering corresponds to the effects of the *vocal tract* (the cavity between the vocal folds and lips), which is influenced, for example, by the positioning of the tongue. The air flow at the lips is then given by the convolution

$$\tilde{f}(t) = (v * f)(t) = \int_{\mathbb{R}} v(\tau) f(t-\tau) d\tau,$$

where v is the filter modeling the vocal tract.

When speech is recorded using a microphone, the resulting signal represents pressure $\tilde{p}(t) = \tilde{f}'(t)$, the derivative of the airflow. Properties of convolution allow us to write $\tilde{p} = v * f' = v * p$, where $p = f'$ denotes pressure at the glottis. The noisy microphone recording can then be modeled as

$$m = v * p + \varepsilon. \qquad (11.6)$$

The nonlinear inverse problem of glottal inverse filtering (GIF) is to reconstruct both the excitation pressure p and the vocal tract filter v from recorded voice. It is a variant of *blind deconvolution*, where the convolution kernel is also unknown. See [250, 251] for mathematical treatments of direct and inverse problem. The forward map $\mathcal{A} : (v, p) \mapsto m$ is nonlinear due to the convolution in (11.6).

Better models of speech production, based on GIF, help in addressing the grand challenge of automatic speech recognition in noisy real-world environments such as cafes, pubs, and traffic [253]. Also, good GIF leads to new speech synthesizers capable of modeling natural speech with a variety of affective factors [386]. These goals are important for people using hearing aids and reading devices and for minority language communities in need of speech recognizers and translators.

11.4.2 Inverse problems in hydrology

Several types of inverse problems arise in hydrology. The most common are backward parabolic problems to identify an initial state in a reaction-diffusion equation, as discussed in Section 10.3, and the problem of determining one or more unknown coefficients in a PDE. Sometimes the hydrologist needs to solve both problems. The unknown coefficient may be a function of the independent and/or dependent variables, and the underlying problem may be one of solute transport, or it may be to determine the hydraulic properties of the medium (or both). A thorough discussion of this type of problem from a hydrogeologist's point of view can be found in the classic text *Quantitative Hydrogeology* [118]. We discuss a few nonlinear examples of inverse problems in hydrology here.

The PDE (11.7) on the bounded domain $\Omega \in \mathbb{R}^3$ models the hydraulic head $u(x,t)$ in a confined inhomogeneous isotropic aquifer at time t and location $x \in \Omega$, where $p(x)$ is the hydraulic conductivity, $R(x,t)$ is known as the recharge rate, and $S(x)$ is the storativity. A *confined aquifer* is a saturated medium bounded from above by a region with zero or low permeability such that the hydraulic head of the water underneath is higher than the level of the aquifer, and hence the pressure of the water is higher than atmospheric pressure. A schematic of a confined aquifer is found in Figure 11.6. If the hydraulic head is at the same level of the free surface of the aquifer (and hence it is not under pressure), the aquifer is

11.4. Examples of nonlinear inverse problems

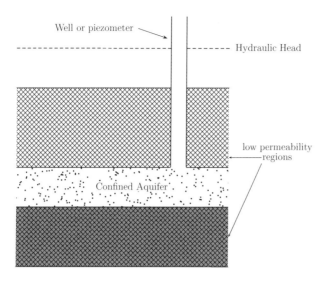

Figure 11.6. Schematic of a confined aquifer.

called *unconfined*.

$$\nabla \cdot (p(x)\nabla u(x,t)) = S(x)\frac{\partial u}{\partial t} + R(x,t), \quad x \in \Omega, \quad t > 0. \tag{11.7}$$

The steady-state problem is

$$\nabla \cdot (p(x)\nabla u(x,t)) = 0, \quad x \in \Omega, \tag{11.8}$$

also known as the conductivity equation, which we study in the context of EIT in the subsequent chapters. One inverse problem is to identify p from knowledge of the head throughout Ω. Since this data is difficult to measure, a more practical inverse problem is to identify p from $\nabla u(x)$ in Ω at discrete points.

The model for solute transport given in Section 10.3 assumes knowledge of the dispersion coefficients D_1 and D_2, but in practice these must often be calibrated before the model can be used to locate the source of a contaminant. This can be accomplished by injecting a tracer, taking measurements, and solving the problem with a known initial condition. Sometimes soil column samples are collected, and a one-dimensional diffusion tube is used, and the contaminant is injected on one end.

The dispersivity actually consists of the sum of a molecular diffusion and a kinematic dispersion, where the molecular diffusion is small in comparison with the kinematic dispersion. It is well known that the kinematic dispersion is directly related to the advective velocity, so a flow with primarily a longitudinal velocity component is also dominated by longitudinal dispersion. A good explanation of hydrodynamic dispersion can be found, for example, in [428]. Further evidence is found in [273], where it is shown through field experiments that the transversal dispersivity was 6 to 20 times smaller than the longitudinal dispersivity. This motivates the use of a model in one spatial dimension,

$$u_t = (D(x)u_x)_x - (v(x)u)_x - \lambda(x)u + F(x), \quad t > 0, \quad x > 0,$$
$$u(x,0) = g(x), \quad x > 0,$$

and a Fourier boundary condition to preserve mass balance [450],

$$-D(0)u_x(0,t) + v(0)u(0,t) = v(0)G(t), \quad t > 0,$$

where $G(t)$ is the concentration in the entrance reservoir. The dispersion coefficient may also be concentration dependent, so that $D = D(u)$. The introduction in [168] is a tutorial in which a nonlinear problem of this type is shown to result in a monotone coefficient to data map, and hence a theorem on unique identification of $D(u)$. Further instruction on exploiting and obtaining monotonicity is found in the tutorial in [168].

11.4.3 Inverse medium scattering in $\mathbb{R}^3$

Ultrasound waves are vibrations in a physical medium. They transmit energy in terms of pressure, density, temperature, and particle displacement. Unlike electromagnetic waves, sound waves cannot propagate in a vacuum. The distance traveled by the particles is called the particle displacement and is usually a few tenths of a nanometer in water. The particle velocity is usually a few centimeters per second in water. Note that this velocity is different from the rate of energy propagation through the medium, which is defined as the sound speed $c \approx 1.5 \times 10^5$ cm/s in water.

In ultrasound imaging, an electronic signal generator transmits a signal with frequency ω in the range of $2\pi * (1 \text{ to } 25)$ MHz. Transducers containing piezoelectric crystals convert the electrical impulse into acoustic vibrations which penetrate the tissue and reflect from object boundaries. The transducer may act as both a transmitter and a receiver with a Mylar separator between the transmitting crystals and receiving crystals, which may be made of natural quartz, barium titanate, or Rochelle salts, for example. The sound waves enter the subject, are reflected by the tissues in the interior, and are received by the transducer, where they are converted back to electrical energy through an analog-to-digital converter. In medical ultrasound, a pulse-echo system is typically used to solve the inverse obstacle problem to locate the boundaries of tissue. By displaying the amplitude of the reflected wave as a function of depth, a time of flight reconstruction can be used to identify tissue boundaries. However, the full inverse problem of determining the location and spatially dependent acoustic properties of the medium is nonlinear and demands more sophisticated and tailored techniques.

We describe here the PDE governing the propagation of acoustic waves of small amplitude in an isotropic medium in $\mathbb{R}^3$ and the data associated with the inverse medium problem. The scattered obstacle will be denoted by D and is surrounded by a homogeneous medium. See Figure 11.7. A thorough discussion of inverse acoustic scattering theory can be found in the classic text [91].

The velocity potential U of an acoustic wave satisfies the wave equation

$$\frac{1}{c^2}U_{tt} = \Delta U. \tag{11.9}$$

For time-harmonic waves $U(x,t) = \text{Re}(u(x)e^{-i\omega t})$ in a nonabsorbing, but possibly inhomogeneous, medium Ω with frequency ω, the acoustic pressure field $u(x)$ satisfies

$$\Delta u + \frac{\omega^2}{c^2}u = 0 \quad \text{in } \mathbb{R}^3,$$

11.4. Examples of nonlinear inverse problems

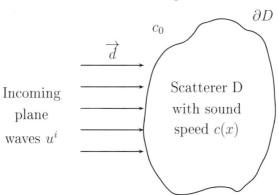

Figure 11.7. Illustration of the incident waves u^i for probing the inhomogeneous scatterer D.

where $c(x) = c_0$ outside of the bounded scatterer $D \subset \mathbb{R}^3$. If D is an impenetrable, or *sound-hard*, object, there is a Neumann boundary condition $\frac{\partial u}{\partial \nu} = 0$ on ∂D, where ν is the unit outward normal to ∂D. If D is a penetrable, or *sound-soft*, obstacle, then $u = 0$ on ∂D since the pressure of the total wave vanishes on the boundary. We will consider here the case of a sound-soft scatterer. Denote

$$\frac{\omega^2}{c^2} = k^2(1+q(x)),$$

where $k = \omega/c_0$ is the wave number, and define a compactly supported perturbation by

$$q(x) = \frac{c_0^2}{c(x)^2} - 1.$$

The ratio $n(x) := c_0^2/c(x)^2$ is called the *refractive index* of the medium, and q is the perturbation of the refractive index from unity.

The unknown scatterer will be probed by a transmitted wave from a certain direction. We assume that the *incoming field* is a plane wave of the form

$$u_0(x;d) = e^{ikx \cdot d}, \qquad d \in \mathbb{R}^3, |d| = 1,$$

where d is the direction of propagation. The resulting *total field* $u(x;d)$ can then be written as a sum of the incoming and *scattered field* u_{SC}:

$$u(x;d) = u_0(x;d) + u_{SC}(x;d) = e^{ikx \cdot d} + u_{SC}(x;d).$$

Let $r = |x|$. The scattered field u_{SC} satisfies the *Sommerfeld radiation condition*

$$\lim_{r \to \infty} r \left(\frac{\partial u_{SC}}{\partial r} - iku_{SC} \right) = 0$$

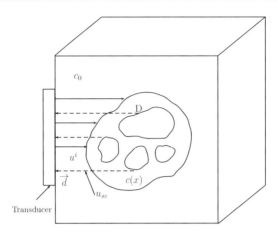

Figure 11.8. The inhomogeneous scatterer D with the incident and measured scattered waves in the inverse medium problem.

and guarantees that the scattered wave is outgoing. Using the fundamental solution $\Phi(x) = e^{ik|x|}/(4\pi|x|)$ to the Helmholtz equation, one can show that u satisfies the *Lippmann–Schwinger* integral equation

$$u(x;d) = u_0(x;d) - \frac{k^2}{4\pi} \int_D \frac{e^{ik|x-y|}}{|x-y|} q(y) u(y;d) \, dy. \tag{11.10}$$

One can then expand the kernel in (11.10) as a function of $1/r$ to find the asymptotic expansion

$$u_{\text{SC}}(x;d) = \frac{e^{ikr}}{r} \left(u_\infty(\hat{x};d) + O\left(\frac{1}{r}\right) \right), \quad \text{where} \quad \hat{x} = \frac{x}{|x|}. \tag{11.11}$$

The function u_∞ is known as the *far field pattern* and is given by

$$u_\infty(\hat{x};d) = -\frac{k^2}{4\pi} \int_D e^{-ik\hat{x}\cdot y} q(y) u(y;d) \, dy. \tag{11.12}$$

The inverse medium problem: Determine $q(x)$ from the knowledge of the far field pattern $u_\infty(\hat{x};d)$ for all (or some incomplete collection of) $\hat{x}, d \in S^2$. The forward map $\mathcal{A}: q \mapsto u_\infty$ is nonlinear since in (11.12) the function u depends on q via (11.10). An illustration of this problem is found in Figure 11.8.

It is shown in [348, 366, 390, 391] that $q(x)$, $x \in D \subset \mathbb{R}^3$, is uniquely determined by knowledge of the far field pattern $u_\infty(\hat{x};d)$ for $\hat{x}, d \in D$ and a fixed wave number k. The proof makes use of complex geometrical optics (CGO) solutions, as we discuss in Part II of this book in the context of EIT.

The far field pattern is an idealized measurement in the following sense. Ideally, we would choose the direction d and send a plane wave towards D. However, plane waves have infinite energy and cannot be practically applied, but fortunately using a point source far away from D results in an approximate plane wave near D. Furthermore, use the asymptotic expansion (11.11) to write

$$u_\infty(\hat{x};d) \approx r e^{-ikr} u_{\text{SC}}(r\hat{x};d), \tag{11.13}$$

11.4. Examples of nonlinear inverse problems

where equality holds at the asymptotic limit $r \to \infty$. Practically we can consider the following near field data. Measure the values of $u(x;d)$ far away from the bounded set D containing the inhomogeneity, say, at points ranging on the circle $|x| = R$. Then we can compute $u_{SC}(x;d) = u(x;d) - e^{ikx \cdot d}$, substitute the result to (11.13), and achieve an approximation to the far field pattern.

Optimization methods that seek a solution to the Lippmann–Schwinger equation (11.10) with (11.12) as a constraint are discussed in [91]. These include the dual space method [92, 89, 95, 91] and its modifications [93, 94, 91]. The variation in the refractive index between biological tissues is generally small. (This is due to the fact that the variation in sound speed is small; for example, see Table 11.1.) This motivates the use of the Born approximation in (11.12), and as a result the nonlinear inverse medium problem can be reduced to a set of linear integral equations of the first kind. Cancer cells, however, can cause a significant change in the refractive index in the bone marrow. A modified dual space method has been developed for the detection of leukemia in the bone marrow of the upper part of the lower leg using microwaves [96, 97].

Table 11.1. Density and sound speeds [409] of air, water, and human tissue.

Material	Density (g/cm^3)	Sound speed (m/s)
Air	0.001	330
Bone	1.85	3360
Muscle	1.06	1570
Fat	0.93	1480
Blood	1.0	1560
Water	0.99	1485

In Section 17.4 a project concerning inverse obstacle scattering is discussed. In this case the scatterer is assumed to be sound-hard, and the problem is to determine ∂D from knowledge of the far field pattern.

11.4.4 Spectral inversion

Can one hear the shape of a drum? was the title of the classic paper by Mark Kac in the American Mathematical Monthly in 1966 [244] that now serves as the prototypical inverse spectral problem. In this pleasant exposition, Kac begins by pointing out that a two-dimensional homogeneous membrane Ω with a fixed boundary perturbed in a direction normal to its surface satisfies the wave equation

$$F_{tt}(x,t) = c^2 \Delta F(x,t), \quad x \in \Omega, \quad t > 0, \quad (11.14)$$
$$F(x,t) = 0, \quad x \in \partial\Omega,$$

where c is a fixed constant that depends upon the composition of the membrane. Musically, solutions of the form $F(x,t) = U(x)e^{i\omega t}$ represent the pure tones the membrane is able to produce. Substituting solutions of this form into (11.14), we see that U must satisfy

$$\Delta U + \lambda U = 0, \quad x \in \Omega, \quad (11.15)$$
$$U(x) = 0, \quad x \in \partial\Omega, \quad (11.16)$$

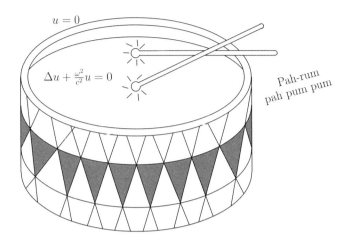

Figure 11.9. The classic inverse spectral problem: *Can you hear the shape of a drum?*

where $\lambda = \frac{\omega^2}{c^2}$. See Figure 11.9. For membranes Ω with a smooth boundary, there exists a sequence $\lambda_1 \leq \lambda_2 \leq \cdots$ of eigenvalues and corresponding eigenfunctions $\psi_n(x)$ such that

$$\Delta \psi_n + \lambda_n \psi_n = 0, \quad x \in \Omega,$$
$$\psi_n(x) \to 0, \quad \text{as } x \to \partial\Omega.$$

The inverse spectral problem is: given two eigenvalue problems of the form (11.15) on two distinct domains Ω_1 and Ω_2, if the sets of eigenvalues for the two problems are equal, are the domains Ω_1 and Ω_2 the same in the sense of Euclidean geometry? Since the eigenvalues determine the pure tones, the question is phrased more musically as, Do the pure tones uniquely determine the shape of the membrane? To study this question, the reader is referred to [244, 260]. More exotic "drums" such as those with fractal boundary have even been studied—see, for example, [298, 167]. Related studies are still ongoing.

The problem in one spatial dimension has been studied in [329, 88], where it has applications in speech production along a vocal tract [423], wave propagation in a duct [479], free, longitudinal vibrations in a thin straight rod [160], and transmission line synthesis [161]. Let us consider this problem with spatially varying coefficients in the PDE, as in [88]. The motion of one-dimensional waves in an inhomogeneous elastic medium is governed by

$$\rho(y)u_{tt} - (\mu(y)u_y)_y = 0, \quad (11.17)$$

where $u(y)$ denotes the shear displacement, $\rho(y)$ is the density of the medium, and $\mu(y)$ represents the shear modulus. This equation can be transformed to a generalized eigenvalue problem as follows. Let

$$x(y) = \int_0^y \left(\frac{\rho(s)}{\mu(s)}\right)^{1/2} ds,$$
$$A(x) = (\rho(y(x)) \cdot \mu(y(x)))^{1/2}.$$

11.4. Examples of nonlinear inverse problems

In the applications above, $A(x)$ represents a cross-sectional area. Then since

$$\frac{dx(y)}{dy} = \left(\frac{\rho(y)}{\mu(y)}\right)^{1/2} \quad \Longrightarrow \quad \frac{dy(x)}{dx} = \left(\frac{\mu(y)}{\rho(y)}\right)^{1/2},$$

one can differentiate A,

$$\frac{d}{dx} A(x) = \frac{1}{2} \frac{1}{(\mu(y(x)) \cdot \rho(y(x)))^{1/2}} \frac{dy(x)}{dx} = \frac{1}{2} \frac{1}{\rho(y(x))},$$

and using this in (11.17),

$$0 = \rho(y) u_{tt} - (\mu(y) u_y)_y$$
$$= \rho(y) u_{tt} - \frac{d}{dx}\left(\mu(y) u_x \frac{dx(y)}{dy}\right) \frac{dx(y)}{dy}$$
$$= \rho(y) u_{tt} - \frac{d}{dx}\left(\mu(y) u_x \left(\frac{\rho(y)}{\mu(y)}\right)^{1/2}\right) \left(\frac{\rho(y)}{\mu(y)}\right)^{1/2}$$
$$= \left(\frac{\rho(y(x))}{\mu(y(x))}\right)^{1/2} \left((\rho(y(x))\mu(y(x)))^{1/2} u_{tt} - \frac{d}{dx}\left((\rho(y(x))\mu(y(x)))^{1/2} u_x\right) \right),$$

equation (11.17) becomes

$$A(x) u_{tt} + (A(x) u_x)_x = 0.$$

With $u(x,t) = u(x) e^{-\sqrt{\lambda} t}$ and $p^2(x) = A(x)$, this is transformed into the generalized eigenvalue problem

$$\left(p^2(x) u'(x)\right)' + \lambda p^2(x) u(x) = 0$$

with Dirichlet boundary conditions. Now let $\alpha(x) = \frac{p'(x)}{p(x)}$ and the equation becomes

$$u''(x) + 2\alpha(x) u'(x) + \lambda u(x) = 0. \tag{11.18}$$

We can consider the forward problem and the inverse problem associated with (11.18):

The **Eigenvalue Problem** (EVP): For given $\alpha \in L^2([0,1])$ find $\lambda \geq 0$ and $u \in S = H^2([0,1]) \cap H^1_0([0,1])$ with $u \not\equiv 0$.

The **Inverse Eigenvalue Problem** (IEVP): Find $\alpha \in L^2([0,1])$ for given $\lambda_n \geq 0$ and $u_n \in S = H^2([0,1]) \cap H^1_0([0,1])$, $\bar{\alpha} = \int_0^1 \alpha(t)\, dt$, and $\kappa_n = \ln \frac{p(1) u'(1)}{p(0) u'(0)}$.

Another inverse spectral problem for vibrating systems is the Sturm–Liouville problem,

$$y''(x) + (\lambda - q(x)) y(x) = 0 \quad \text{on} \quad 0 < x < 1,$$

with $y(0) = y(1) = 0$ for $q \in L^2([0,1])$. Each eigenvalue λ is the square of a natural frequency of the system. One can perform the following experiment: Excite the vibrating system and measure the frequencies or the nodes of the eigenmodes. The corresponding inverse problem is: determine $q(x)$ from the position of the nodes. The function q represents, for example, the density of a vibrating string. The determination of the density of a vibrating string from spectral data was considered in [212]. The following result is from [329].

Theorem 11.1 (McLaughlin [329]). *Let $q_1, q_2 \in L^2(0,1)$. For each $n \geq 2$ find $k = 0, 1, 2, 3, \ldots$ and $m = 0, 1, 2, \ldots, 2^k - 1$ such that $n = 2^{k+1} - m$. Use this representation of n in terms of k and n and let $j(n) = m + 1$. Suppose the positions of the chosen zeros satisfy $x_n^{j(n)}(q_1) = x_n^{j(n)}(q_2)$ for $n = 2, 3, \ldots$ and $\int_0^1 q_1\, dx = \int_0^1 q_2\, dx$. Then $q_1 = q_2$, except possibly on a set of measure zero.*

Thus, the data for q is the position of one node of each eigenfunction and the average of the function q over the interval.

The literature includes several other uniqueness results for this problem; we include a few interesting ones here. Suppose $q_1, q_2 \in L^2([0,1])$ and $\lambda_n(q_1) = \lambda_n(q_2)$ for all n.

1. (Gel'fand [156].) If $\rho_n(q_1) = \rho_n(q_2)$ for all n, where

$$\rho_n(q) = \frac{\|y(\cdot, q, \lambda_n)\|_2^2}{|y'(0, q, \lambda_n)|^2},$$

then $q_1 = q_2$.

2. (See [235, 380].) If $\kappa_n(q_1) = \kappa_n(q_2)$ for all n, where

$$\kappa_n(q) = \ln \left| \frac{y'(1, q, \lambda_n)}{y'(1, q, \lambda_n)} \right|,$$

then $q_1 = q_2$.

Flux data at the origin was shown to be a sufficient set of data in [323] for the following problem.

Theorem 11.2 (Lowe and Rundell [323]). *If 0 is not a Dirichlet eigenvalue of $-\frac{d^2}{dx^2} + q(x)$ on $(0,1)$, then $q(x) \in L^2(0,1)$ is uniquely determined by the data $\{u_j'(0)\}_{j=1}^\infty$, where $u_j(x)$ solves*

$$-u_j''(x) + q(x) u_j(x) = \psi_j(x), \quad 0 < x < 1,$$
$$u_j(0) = 0,$$
$$u_j(1) = 0,$$

and $\{\psi_j(x)\}_{j=1}^\infty$ is a basis of $L^2(0,1)$.

11.4.5 Diffuse optical tomography

Optical tomography is an imaging technique in which a low-energy lightwave guided by fiber optics is sent into a highly scattering medium and the amplitude of the refracted waves are measured on sensors on the boundary. Figure 11.10 illustrates the scattered path of a photon beam through brain tissue. Applications include neonatal brain imaging for the

11.4. Examples of nonlinear inverse problems

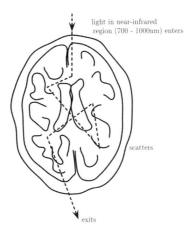

Figure 11.10. Simple illustration of the path of a photon beam through brain tissue.

detection of intraventricular hemorrhage and imaging changes in blood volume and oxygenation [203], breast imaging [483], and muscle imaging [209]. Mathematically, it is closely related to EIT. It is a nonlinear problem, and we introduce its mathematical formulation here. The reader is referred to other sources such as [11, 157] for a more thorough treatment, and to [205, 435, 12] for details of computational techniques related to optical tomography.

As observed by Arridge [11], the inverse problem of optical tomography has several interesting properties that distinguish it from other modalities, and provide a rich source of research problems:

- There are two types of data acquisition: time-varying intensities corresponding to the system response to an ultrashort impulse pulse or a steady-state complex-valued function.

- The forward problem can be formulated in terms of particles or waves. If a wave formulation is used, a variety of models are possible, ranging from a transport equation to a parabolic or elliptic PDE.

- Various coefficients (with different physical relevance) can arise in the formulation of the inverse problem. These affect the usefulness of the results.

We will introduce the inverse problems arising from a transport model and a diffusion model.

Transport model

Let $\Omega \in \mathbb{R}^3$ be a convex region containing the object of interest. Let $u(x,\theta,t)$ be the particle (photon) flux in a direction $\theta \in S^2$ at $x \in \Omega$. Here $\theta = [\sin\phi_1 \cos\phi_2, \sin\phi_1 \cos\phi_2]^T$, where ϕ_1 is the angle θ projected into the xy-plane, and ϕ_2 is the angle θ projected perpendicular to the xy-plane. Let c be a known and constant particle speed. The function $u(x,\theta,t)$

satisfies the transport equation

$$\frac{1}{c}u_t(x,\theta,t) + \theta \cdot \nabla u(x,\theta,t) + \mu(x)u(x,\theta,t)$$
$$= \mu_s(x)\int_{S^2} \eta(\theta \cdot \theta')u(x,\theta',t)d\theta', \quad (11.19)$$
$$u(x,\theta,0) = 0, \quad x \in \Omega, \quad \theta \in S^2, \quad (11.20)$$
$$u(x,\theta,t) = g^-(x,\theta,t), \quad x \in \partial\Omega,$$
$$\nu(x) \cdot \theta \leq 0, \, t \geq 0,$$

where ν is the exterior unit normal on $\partial\Omega$ and $\mu = \mu_a + \mu_s$ is the sum of the *absorption coefficient* μ_a and the *scattering coefficient* μ_s. The boundary condition models the incoming particle flux. The initial boundary-value problem has a unique solution under natural assumptions on μ, μ_s, and η. The function η is the *scattering kernel*, which is assumed to be normalized so that

$$\int_{S^2} \eta(\theta \cdot \theta')d\theta' = 1.$$

We will assume *Henyey–Greenstein* scattering,

$$\eta(s) = \frac{1}{4\pi}\frac{1-g^2}{(1+g^2-2gs)^{3/2}},$$

where $g \in (-1,1)$ is a measure of anisotropy with $g = 0$ representing isotropic scattering and $g = 1$ representing single-directional scattering.

Inverse Problem: Recover the parameters μ_a and μ_s from measurements of the outgoing flux g,

$$g(x,t) = \frac{1}{4\pi}\int_{S^2} \nu(x) \cdot \theta u(x,\theta,t)d\theta, \quad x \in \partial\Omega, \quad t \geq 0.$$

Note that μ_a and μ_s are very small compared to a typical object size. The parameter μ_a is between 0.1 and 1.0 mm^{-1} and μ_s is between 100 and 200 mm^{-1}, so the mean free path is between 0.005 and 0.01 mm.

Diffusion model

A more commonly used model is the diffusion equation, which we will derive here. Consider the first few moments of the particle flux u:

$$u_0(x,t) = \frac{1}{4\pi}\int_{S^2} u(x,\theta,t)d\theta,$$
$$u_1(x,t) = \frac{1}{4\pi}\int_{S^2} \theta u(x,\theta,t)d\theta \quad \text{(a vector)},$$
$$u_2(x,t) = \frac{1}{4\pi}\int_{S^2} \theta\theta^T u(x,\theta,t)d\theta \quad \text{(a matrix)}.$$

Integrating the transport equation over S^2 and using $\int_{S^2} \eta(\theta \cdot \theta')d\theta' = 1$, we have

$$\frac{1}{c}\frac{\partial u_0(x,t)}{\partial t} + \nabla \cdot u_1(x,t) + \mu_a(x)u_0(x,t) = 0. \quad (11.21)$$

11.4. Examples of nonlinear inverse problems

Multiplying the transport equation by θ and integrating over S^2 gives

$$\frac{1}{c}\frac{\partial u_1(x,t)}{\partial t} + \nabla \cdot u_2(x,t) + \mu(x)u_1(x,t) = \bar{\eta}\mu_s(x)u_1(x,t), \tag{11.22}$$

where $\bar{\eta} = \frac{1}{4\pi}\int_{S^2} \theta' \cdot \theta \eta(\theta \cdot \theta')d\theta$. Let

$$\mu_s' = (1-\bar{\eta})\mu_s.$$

Then

$$\frac{1}{c}\frac{\partial u_0}{\partial t} + \nabla \cdot u_1 + \mu_a u_0 = 0, \tag{11.23}$$

$$\frac{1}{c}\frac{\partial u_1}{\partial t} + \nabla \cdot u_2 + (\mu_a + \mu_s')u_1 = 0. \tag{11.24}$$

Assume u depends linearly on θ. Then

$$u(x,\theta,t) = \alpha u_0(x,t) + \beta\theta \cdot u_1(x,t). \tag{11.25}$$

Computing moments of order 0 and order 1 of u implies $\alpha = 1$ and $\beta = 3$. Thus,

$$u_2(x,t) = \frac{1}{4\pi}\int_S^2 \theta\theta^T u(x,\theta,t)d\theta,$$

which implies

$$\nabla \cdot u_2 = \frac{1}{3}\nabla u_0.$$

Now (11.24) becomes

$$\frac{1}{c}\frac{\partial u_1}{\partial t} + \frac{1}{3}\nabla u_0 + (\mu_a + \mu_s')u_1 = 0. \tag{11.26}$$

If we assume that $\frac{\partial u_1}{\partial t}$ is negligible, then

$$\frac{1}{3}\nabla u_0 = -(\mu_1 + \mu_s')u_1$$

and

$$u_1 = -\frac{1}{3(\mu_a + \mu_s')}\nabla u_0.$$

Consistent with Fick's law, we can let

$$D = -\frac{1}{3(\mu_a + \mu_s')}.$$

Then (11.23) becomes

$$\frac{1}{c}\frac{\partial u_0}{\partial t} - \nabla \cdot (D\nabla u_0) + \mu_a u_0 = 0. \tag{11.27}$$

This is the diffusion equation for optical tomography. The initial and boundary conditions are

$$u_0(x,0) = 0, \quad x \in \Omega,$$

$$u_0 + 2D\frac{\partial u_0}{\partial \nu} = g^- \quad \text{on } \partial\Omega.$$

The measurements are

$$-D\frac{\partial u_0}{\partial \nu} = g \quad \text{on } \partial\Omega.$$

Inverse Problem: Determine D and μ_a from all values of g from incoming isotropic light distributions g^-.

Suppose $g^-(x,t) = g^-(x)e^{i\omega t}$ and that for large t, $u_0(x,t) = v(x)e^{i\omega t}$, where v satisfies

$$-\nabla \cdot (D\nabla v) + \left(\mu_a + i\frac{\omega}{c}\right)v = 0 \quad \text{in } \Omega,$$

$$v + 2D\frac{\partial v}{\partial \nu} = g^- \quad \text{on } \partial\Omega,$$

$$g = -D\frac{\partial v}{\partial \nu} \quad \text{on } \partial\Omega.$$

Inverse Problem: Determine D and μ_a from knowledge of g for all g^- at one or several frequencies ω.

Chapter 12
Electrical impedance tomography

The purpose of this chapter is to provide an explanation of some of the key mathematical and physical considerations for the forward and inverse problem of electrical impedance tomography (EIT), the guiding example for nonlinear inversion in this book.

EIT is an imaging technique in which electrodes are placed on the surface of the body, and low-frequency current is applied on the electrodes below the threshold of human perception. This results in a voltage distribution on the electrodes which can then be measured. The measurement is repeated for a specified set of current patterns, or choices of current amplitudes at each electrode. The resulting current-to-voltage map serves as data for the inverse problem.

Medical EIT is based on the fact that the conductivity and permittivity of the tissues in the body vary significantly (see Table 12.1), and reconstruction of these spatially and temporally dependent properties allows one to form an image. An illustration of EIT chest imaging is found in Figure 12.1.

The mathematical model of EIT is called the inverse conductivity problem: recover the conductivity distribution inside the body given electric boundary measurements performed on the surface of the body. It is a nonlinear and severely ill-posed problem.

Two questions were posed by Calderón in a classic paper [64] from 1980 which is often pointed to as the mathematical beginnings of the inverse conductivity problem.

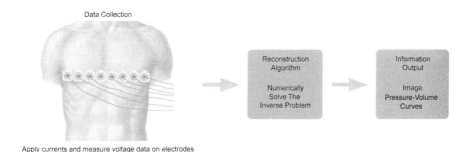

Figure 12.1. Schematic of EIT imaging of a human chest.

Table 12.1. Conductivity values and permittivity values for tissues and organs in the human chest at 100 kHz. (See, for example, [27, 155].)

Tissue	Conductivity (mS/cm)	Permittivity (μ F/m)
Blood	6.70	0.05
Liver	2.80	0.49
Bone	0.06	0.0027
Cardiac Muscle	6.3 (longitudinal)	0.88
	2.3 (transversal)	0.36
Lung	1.0 (expiration)	0.44
	0.4 (inspiration)	0.22
Fat	0.36	0.18
Skin	0.0012	0.0144

The first question is, Is it possible to uniquely determine the conductivity of an unknown object from boundary measurements? The other question is, How can this conductivity be reconstructed? Calderón shows that the linearized problem has an affirmative answer to the uniqueness question, and he proposed a linearized reconstruction scheme. His methods have inspired a multitude of research on the problem, including the use of CGOs solutions for answering both of his questions and for designing a regularized inversion method for practical EIT.

See the articles [53, 80, 192] for reviews of EIT from a mathematical perspective.

12.1 Applications of EIT

EIT is an attractive option for many imaging tasks since the equipment is cost-effective, electric currents can penetrate a variety of materials without damaging them, and fast electronics are available for real-time applications.

However, the underlying inverse conductivity problem is very sensitive to modeling errors and measurement noise, so EIT imaging algorithms need to be properly discretized and regularized. Moreover, the computational methods must be fast enough to meet the specifications coming from the application. Designing the instrumentation requires nontrivial electrical engineering as the noise level needs to be very low. The application of optimal current patterns (necessary for achieving maximum resolution) requires the ability to inject currents through all the electrodes while simultaneously measuring voltages from the same electrodes. Also, interpreting the reconstructed EIT images can be tricky due to the characteristically low spatial resolution. Especially in medical applications, the access to detailed CT and MRI images has created expectations of high spatial resolution, while the strength of EIT lies in other aspects, such as time and contrast resolution. This is the case, for example, in early detection of breast cancer and in lung imaging.

Because of the above, the application of EIT in a given imaging task is never straightforward. All aspects of EIT are delicate: mathematical, physical, computational, instrumentational, and interpretational. Both algorithms and hardware must be top-notch, and considerable application-dependent tuning is to be expected before EIT delivers useful results.

It would take too much space to list all the applications where EIT is useful; we will describe some of the most exciting ones and point the reader to comprehensive review articles.

Medical imaging

Perhaps the earliest motivation for medical EIT is electrocardiography (ECG) on one hand and defibrillation on the other hand. In ECG one measures electrical potentials on the skin and aims to recover electrical activity on the surface of the heart. Defibrillation aims to deliver strong electrical field through the heart, causing the muscles to contract and restore normal rhythm afterwards. Knowledge of the electric conductivity inside the patient would help the interpretation of ECG signals and enable more precise targeting of electrical power to the heart when defibrillating.

The above applies to electroencephalography (EEG) as well: interpretation of EEG signals is more precise when the conductivity distribution between the scalp and the cortex is known. Also, controlled targeting of applied electric current inside the brain may enhance electroconvulsive therapy used, for example, in the treatment of depression.

One of the most promising medical applications of EIT is in pulmonary imaging. Since air has a lower conductivity than any of the tissues in the chest (the conductivity of air is zero) and since blood is highly conductive (see Table 12.1), EIT is very well-suited for imaging ventilation and perfusion. EIT has already been established as a technique that shows promise for neonatal pulmonary measurement [378], monitoring pulmonary perfusion [58, 466, 418, 342, 343, 357, 233], detecting extravascular lung water [294], and evaluating shifts in lung fluid in congestive heart failure patients [149]. It has been shown to provide reproducible regional ventilation information [153, 460, 152, 151]. Regional ventilation results have been validated with CT images on animal studies [102, 153, 182, 460] and humans [418] in the presence of pathologies such as atelectasis, pleural effusion, and pneumothorax. The review articles [103, 150] provide a further overview of EIT for pulmonary imaging. EIT also holds promise for cardiac imaging [464, 465], and for further discussion of chest imaging applications, see [215, Chapter 3].

EIT has been studied as a way to monitor brain function [438, 317], in particular locating epileptic loci [52, 24, 137]. See [215, Chapter 4] for more details.

Early detection of breast cancer is a promising application of EIT. Often malignant breast tumors have roughly the same X-ray attenuation coefficient as healthy breast tissue, so cancer is not clearly visible as light or dark spots in X-ray mammograms, but the difference in electrical conductivity can be as high as fourfold [242, 243, 327]. This fact was the basis of the T-Scan product [13] that was shown to improve breast cancer diagnosis when combined with traditional X-ray mammography. However, the original T-Scan did not perform any EIT reconstruction; it simply displayed the raw data in a suitable form. Later, other researchers developed reconstruction methods for T-Scan-type geometry and spectral (multifrequency) measurements; see [10, 84, 83, 268, 267] and the review article [498]. In particular, the team at Rensselaer Polytechnic Institute [268, 255] has combined X-ray mammography with simultaneously measured EIT using radiolucent electrodes attached to the compression plates of the mammography device [256]. The sensitivity and specificity of the detection of breast cancer can be enhanced significantly with this setup.

The biomedical engineering handbook [215] gives an overview of medical EIT, and the articles [57, 103, 214] give reviews from various clinical perspectives.

Industrial applications

Nondestructive testing and process tomography are perhaps the main applications of EIT in industrial environments. EIT-based nondestructive testing can be used, for example, for concrete structures [257]. For process tomography, see [482, 481] for early examples, and [408, 42, 487, 401, 204] for more recent results. The review article [486] lists many industrial applications and gives references.

Geophysical applications

EIT can be used, for example, for subsurface flow monitoring and remediation [106, 108, 387, 426, 265, 497] and underground contaminant detection [107, 388, 264, 241]. See the review article [3].

12.2 Derivation from Maxwell's equations

A mathematical model for the propagation of electromagnetic fields in the body can be obtained from Maxwell's equations. The derivation presented here generally follows the derivation in [231]. Another useful reference for Maxwell's equations is [339]. The treatment in this section is necessarily three-dimensional, and we use the notation $x \in \mathbb{R}^3$ here (but not elsewhere in the book).

Maxwell's equations are given by the following set of four equations:

$$\nabla \times E = -\partial_t B \quad \text{(Faraday's law)}, \tag{12.1}$$

$$\nabla \times H = J + \partial_t D \quad \text{(Ampere's law)}, \tag{12.2}$$

$$\nabla \cdot B = 0 \quad \text{(Gauss' law for magnetism)}, \tag{12.3}$$

$$\nabla \cdot D = \rho \quad \text{(Gauss' law)}. \tag{12.4}$$

For a point $x \in \mathbb{R}^3$, $E(x,t)$ is the electric field, $B(x,t)$ the magnetic flux, $H(x,t)$ the magnetic strength, $J(x,t)$ the current density, $D(x,t)$ the electric displacement field, and $\rho(x,t)$ the free electric charge density.

In EIT current densities are applied on the surface of the body of the form $J = \text{Re}(J(x)e^{i\omega t})$, where ω is the temporal angular frequency of the applied current (measured in 2π Hz), and t is time. These time-harmonic currents J result in fields of the same form, and so

$$E(x,t) = E(x)e^{i\omega t}, \quad D(x,t) = D(x)e^{i\omega t},$$
$$B(x,t) = B(x)e^{i\omega t}, \quad H(x,t) = H(x)e^{i\omega t},$$
$$J(x,t) = J(x)e^{i\omega t}, \quad \rho(x,t) = \rho(x)e^{i\omega t}.$$

Equations (12.1) and (12.2) then become

$$\nabla \times E(x) = -\partial_t(B(x)e^{i\omega t}) = -i\omega B(x), \tag{12.5}$$

$$\nabla \times H(x) = J(x) + i\omega D(x). \tag{12.6}$$

In order to solve these equations, one must know how D and H depend on E and B. Assuming the electric and magnetic response of the body is linear and isotropic, one has the linear constitutive relationships

$$D = \epsilon(x,\omega)E, \tag{12.7}$$

$$B = \mu(x,\omega)H, \tag{12.8}$$

where $\epsilon(x)$ is the electric permittivity, $\mu(x)$ the magnetic permeability, and $\sigma(x)$ the conductivity. Ohm's law implies

$$J = \sigma(x,\omega)E. \tag{12.9}$$

Since the magnetic permeability in the human body is very small, E and B can be linearized about $\mu = 0$ to obtain

$$E(x,\omega;\mu) = E(x,\omega;0) + \partial_\mu E(x,\omega;0)\mu + \frac{1}{2!}\partial_\mu^2 E(x,\omega;0)\mu^2 + \cdots$$
$$\equiv E_0 + \mu E_1 + \mu^2 E_2 + \cdots,$$
$$B(x,\omega;\mu) = B_0 + \mu B_1 + \mu^2 B_2 + \cdots.$$

To leading order (the $\mu = 0$ terms) equation (12.5) then becomes

$$\nabla \times E_0(x,\omega) = -i\omega B_0(x,\omega). \tag{12.10}$$

Taking the approximation $\mu = 0$ in the linear constitutive relation (12.8), it follows that $B_0(x,\omega) = 0$, and so

$$\nabla \times E_0 = 0.$$

Thus, the zero-order electric field E_0 is a potential field and can be expressed as the gradient of a function u called the electric potential:

$$E_0 = -\nabla u. \tag{12.11}$$

From (12.6), (12.7), and (12.9),

$$\nabla \times H = (\sigma + i\omega\epsilon)E.$$

Since the divergence of a curl is zero, this results in

$$\nabla \cdot (\nabla \times H) = \nabla \cdot [(\sigma + i\omega\epsilon)E] = 0,$$

and to zero-order this becomes

$$\nabla \cdot [(\sigma + i\omega\epsilon)E_0] = 0. \tag{12.12}$$

By (12.11) we arrive at the generalized Laplace equation

$$\nabla \cdot [(\sigma(x) + i\omega\epsilon(x))\nabla u(x)] = 0, \quad x \in \Omega. \tag{12.13}$$

12.3 Continuum model boundary measurements

EIT is based on determining the unknown conductivity $\sigma(z)$ inside $\Omega \subset \mathbb{R}^2$ from electrical boundary measurements. In this section we discuss ideal voltage-to-current density measurements (the Dirichlet-to-Neumann map Λ_σ) and ideal current density-to-voltage measurements (the Neumann-to-Dirichlet map $\mathcal{R}_\sigma$) in the context of the so-called continuum model of EIT.

12.3.1 The Dirichlet-to-Neumann map

We consider the two-dimensional case of the quasi-static ($\omega = 0$) version of (12.13):

$$\nabla \cdot (\sigma \nabla u) = 0 \quad \text{in } \Omega \subset \mathbb{R}^2. \tag{12.14}$$

The application of a voltage distribution f on the boundary of the domain corresponds to the Dirichlet boundary condition

$$u(z) = f(z), \quad z \in \partial\Omega. \tag{12.15}$$

We measure the resulting current density distribution J on the boundary:

$$J(z) = \sigma(z)\frac{\partial u}{\partial \nu}(z), \quad z \in \partial\Omega,$$

where ν is the outward unit normal on the boundary. Note that Kirchhoff's law (conservation of charge) dictates that the current density is conserved over the boundary and so we have

$$\int_{\partial\Omega} J\, ds = \int_{\partial\Omega} \sigma \frac{\partial u}{\partial \nu} ds = \int_{\Omega} \nabla \cdot \sigma \nabla u\, dz = 0, \tag{12.16}$$

where we used Green's formula in (12.16).

The *Dirichlet-to-Neumann (DN) map* takes the given voltage distribution on the boundary to the current-density distribution. This mapping is also called the voltage-to-current-density map and is denoted as Λ_σ.

$$\Lambda_\sigma : u|_{\partial\Omega} \longrightarrow \sigma \frac{\partial u}{\partial \nu}\bigg|_{\partial\Omega}. \tag{12.17}$$

If $\sigma \in L^\infty(\Omega)$ is strictly positive almost everywhere in Ω, then the linear operator Λ_σ is bounded between the following Sobolev spaces:

$$\Lambda_\sigma : H^{1/2}(\partial\Omega) \to H^{-1/2}(\partial\Omega).$$

This can be seen using the standard theory of elliptic PDEs [158].

Knowledge of Λ_σ is equivalent to knowing the current-density distribution arising from any given voltage distribution on the boundary.

12.3.2 The Neumann-to-Dirichlet map

Let u be the unique $H^1(\Omega)$ solution of equation (12.14) with the Neumann boundary condition

$$\sigma \frac{\partial u}{\partial \nu} = g \text{ on } \partial\Omega, \tag{12.18}$$

satisfying $\int_{\partial\Omega} u\, ds = 0$. Then we define the *Neumann-to-Dirichlet (ND) map* by

$$\mathcal{R}_\sigma g = u|_{\partial\Omega}. \tag{12.19}$$

It follows from standard theory of elliptic PDEs [158] that $\mathcal{R}_\sigma : \widetilde{H}^{-1/2}(\partial\Omega) \to \widetilde{H}^{1/2}(\partial\Omega)$ is bounded, where $\widetilde{H}^s$ spaces consist of H^s functions with mean value zero.

The ND map $\mathcal{R}_\sigma$ is self-adjoint, smoothing, and compact.

We note two key equalities concerning Λ_σ and $\mathcal{R}_\sigma$. Define a projection operator $P\phi := |\partial\Omega|^{-1} \int_{\partial\Omega} \phi$. Then for any $f \in H^{1/2}(\partial\Omega)$ we have

$$P\Lambda_\sigma f = |\partial\Omega|^{-1} \int_{\partial\Omega} \sigma \frac{\partial u}{\partial \nu} ds = |\partial\Omega|^{-1} \int_\Omega \nabla \cdot \sigma \nabla u \, dz = 0,$$

so actually $\Lambda_\sigma : H^{1/2}(\partial\Omega) \to \widetilde{H}^{-1/2}(\partial\Omega)$. From the definitions of Λ_σ and $\mathcal{R}_\sigma$ we now have

$$\Lambda_\sigma \mathcal{R}_\sigma = I \qquad : \widetilde{H}^{-1/2}(\partial\Omega) \to \widetilde{H}^{-1/2}(\partial\Omega), \qquad (12.20)$$

$$\mathcal{R}_\sigma \Lambda_\sigma = I - P \qquad : H^{1/2}(\partial\Omega) \to \widetilde{H}^{1/2}(\partial\Omega). \qquad (12.21)$$

12.4 Nonlinearity of EIT

While the DN map is a linear operator, the actual reconstruction problem of EIT is nonlinear. This is because the forward map $\sigma \mapsto \Lambda_\sigma$ is nonlinear, as we will see below.

Using (12.15), the DN map can also be expressed in weak form as follows:

$$\langle \Lambda_\sigma f, u \rangle = \int_{\partial\Omega} u|_{\partial\Omega} \sigma \frac{\partial u}{\partial \nu} ds \qquad (12.22)$$

$$= \int_{\partial\Omega} u\sigma \nabla u \cdot \nu \, ds - \int_\Omega (\nabla \cdot \sigma \nabla u) u \, dz$$

$$= \int_\Omega \sigma |\nabla u|^2 dz. \qquad (12.23)$$

Or, more generally, the weak-form definition of the DN map is given by

$$\Lambda_\sigma : H^{1/2}(\partial\Omega) \to H^{-1/2}(\partial\Omega), \qquad \langle \Lambda_\sigma f, g \rangle = \int_\Omega \sigma \nabla u \cdot \nabla v \, dz. \qquad (12.24)$$

where v is any $H^1(\Omega)$ function with trace g on the boundary and u is the unique $H^1(\Omega)$ solution of the Dirichlet problem (12.15). Physically, the integral in (12.24) represents the power needed to maintain a voltage g on the boundary, and has units of power.

Since u itself is a function of σ, we see from (12.24) that the DN map Λ_σ is a nonlinear function of σ due to the product of σ and ∇u in the integrand. Thus, the current densities on the electrodes are nonlinear functions of the conductivity σ. The nonlinearity of the dependence of the data on the conductivity is in sharp contrast to the case of X-ray tomography, where the measured data is a linear function of the density of the medium. This is illustrated in Figure 12.2.

12.5 Ill-posedness of EIT

The inverse problem of EIT is a very ill-posed one, and the situation is compounded in the practical case by the fact that the measurements are discrete and of finite precision and often only partial boundary data is available. Even in the ideal case, the solution fails to depend continuously on the data.

Ill-posedness of EIT is illuminated in this section by Alessandrini's classical example [5] and by a numerical simulation.

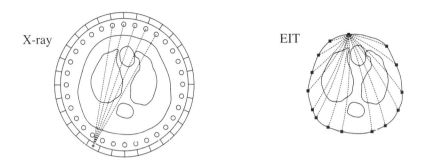

Figure 12.2. Illustration comparing the linearity of X-ray tomography (left) with the nonlinearity of EIT (right). On the left, a fan-bean configuration of X-ray photons passes through the medium in straight lines. Some photons are attenuated through scattering and interaction, but the measured beam has traveled linearly through the medium. In contrast, in EIT, the propagation of the electromagnetic wave depends on the medium, and the path is unknown when the composition of the medium is unknown.

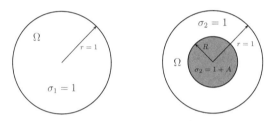

Figure 12.3. The conductivity distributions σ_1 (left) and σ_2 (right).

12.5.1 Alessandrini's example

Discontinuous dependence is established by showing that given any $\epsilon > 0$ and any $\delta > 0$, there exist conductivity distributions $\sigma_1(z)$ and $\sigma_2(z)$ such that

$$\|\Lambda_{\sigma_1} - \Lambda_{\sigma_2}\|_{H^{1/2}(\partial\Omega) \to H^{-1/2}(\partial\Omega)} < \delta$$

but

$$\|\sigma_1 - \sigma_2\|_{L^\infty(\Omega)} > \epsilon.$$

Note that we are specifically considering the $L^\infty(\Omega)$-norm for the space of conductivity distributions.

Consider the two conductivity distributions on the unit disc Ω (see Figure 12.3),

$$\sigma_1(r,\theta) = 1, \qquad \sigma_2(r,\theta) = \begin{cases} 1+A, & 0 \leq r \leq R, \\ 1, & R < r \leq 1, \end{cases} \qquad (12.25)$$

satisfying the generalized Laplace equation with the same Dirichlet boundary condition

$$\nabla \cdot (\sigma_1 \nabla u_1) = 0, \quad \text{in } \Omega, \qquad \nabla \cdot (\sigma_2 \nabla u_2) = 0, \quad \text{in } \Omega,$$
$$u_1|_{\partial\Omega} = \phi, \qquad u_2|_{\partial\Omega} = \phi.$$

Note that this implies that the voltage on the boundary is the same in both cases. The

12.5. Ill-posedness of EIT

solution of the forward problem for σ_1 can be found by separation of variables to be

$$u_1(r,\theta) = \sum_{n=-\infty}^{\infty} \phi_n r^{|n|} e^{in\theta},$$

where ϕ_n are the Fourier coefficients of ϕ. Thus,

$$\Lambda_{\sigma_1}\phi(\theta) = \frac{\partial u_1}{\partial r}\bigg|_{\partial\Omega} = \sum_{n=-\infty}^{\infty} \phi_n |n| e^{in\theta}, \quad (12.26)$$

The solution of the forward problem with conductivity σ_2 can also be found by separation of variables, as outlined below. Denoting the solution inside $r = R$ by u_2^{int} and the solution in the annulus $R < r < 1$ by u_2^{ext}, one finds that

$$u_2^{\text{int}}(r,\theta) = \sum_{n=-\infty}^{\infty} u_n r^{|n|} e^{in\theta}, \quad (12.27)$$

while

$$u_2^{\text{ext}}(r,\theta) = \sum_{n=-\infty}^{\infty} v_n r^{|n|} e^{in\theta} + w_n r^{-|n|} e^{in\theta}. \quad (12.28)$$

The boundary condition $u_2|_{\partial\Omega} = \phi$ implies

$$v_n + w_n = \phi_n.$$

There are two matching conditions at $r = R$:

$$u_2^{\text{int}}(R,\theta) = u_2^{\text{ext}}(R,\theta) \quad \text{and} \quad (1+A)\frac{\partial u_2^{\text{int}}}{\partial r}(R,\theta) = \frac{\partial u_2^{\text{ext}}}{\partial r}(R,\theta).$$

The first condition implies $u_n R^{|n|} = v_n R^{|n|} + w_n R^{-|n|}$, while the second condition implies $(1+A)|n|u_n R^{|n|-1} = |n|v_n R^{|n|-1} - |n|w_n R^{-|n|-1}$. These two conditions together with the boundary condition give

$$u_n = \frac{2}{2+A(1-R^{2|n|})}\phi_n,$$

$$v_n = \frac{2+A}{2+A(1-R^{2|n|})}\phi_n, \quad \text{and} \quad w_n = \frac{-AR^{2|n|}}{2+A(1-R^{2|n|})}\phi_n.$$

Thus,

$$\Lambda_{\sigma_2}\phi = \frac{\partial u_2^{\text{ext}}}{\partial r}(1,\theta) = \sum_{n=-\infty}^{\infty} (v_n - w_n)|n|e^{in\theta}$$

$$= \sum_{n=-\infty}^{\infty} |n|\left(\frac{2+A(1+R^{2|n|})}{2+A(1-R^{2|n|})}\right)\phi_n e^{in\theta}. \quad (12.29)$$

As a result, the difference of the DN maps applied to ϕ is

$$(\Lambda_{\sigma_1} - \Lambda_{\sigma_2})\phi = \sum_{n=-\infty}^{\infty} |n| \left(\frac{-2AR^{2|n|}}{2 + A(1 - R^{2|n|})} \right) \phi_n e^{in\theta}. \tag{12.30}$$

Since

$$\left| \frac{-2AR^{2|n|}}{2 + A(1 - R^{2|n|})} \right| \leq AR,$$

we have the bound

$$\|\Lambda_{\sigma_1} - \Lambda_{\sigma_2}\|_{H^{1/2}(\partial\Omega) \to H^{-1/2}(\partial\Omega)} \leq AR,$$

which can be made arbitrarily small, depending on the choice of R. However, independent of R,

$$\|\sigma_1 - \sigma_2\|_{L^{\infty}(\Omega)} = A,$$

which can be chosen greater than ϵ.

From (12.26) we see that if the applied voltages $\phi(\theta)$ are, for example, the trigonometric functions $\cos(n\theta)$, then the solution u_1 to the homogeneous problem with constant conductivity $\sigma_1 = 1$ is $\Lambda_{\sigma_1}\cos(n\theta) = |n|\cos(n\theta)$, while from (12.27) and (12.28) the solution u_2 to the nonhomogeneous problem with conductivity σ_2 is $\Lambda_{\sigma_2}\cos(n\theta) = |n|\left(\frac{2+A(1+R^{2|n|})}{2+A(1-R^{2|n|})}\right)\cos(n\theta)$. As the spatial frequency of the applied currents increases (i.e., as $n \to \infty$), the difference in the current densities goes to zero. For example, if $R = 0.5$ and $A = 8$, the difference in the current densities for $n = 1$ is $0.5\cos\theta$, while for $n = 8$ the difference is only $2.4414 \times 10^{-5}\cos\theta$. This example illuminates the ill-posedness of the inverse conductivity problem and has significant implications in the design of an EIT system. See also [159].

The results of the computations above can also be used to illustrate that one measurement is not sufficient to uniquely identify $\sigma(z)$. This is addressed in Exercise 12.5.2.

Exercise 12.5.1. *Repeat the above computations using the ND map.*

Exercise 12.5.2. *Consider two conductivity distributions σ_1 and σ_2 of the form of σ_2 in (12.25) with contrasts A_1 and A_2, respectively, and suppose the boundary data $\phi = \cos\theta$. Using formula (12.29), find the difference in the resulting current densities on the boundary. Letting $\mu_1 = \frac{A_1}{2+A_1}$ and $\mu_2 = \frac{A_2}{2+A_2}$, show that the difference in the current densities is 0 when*

$$\frac{\mu_1}{\mu_2} = \frac{R_2^2}{R_1^2}.$$

Use this to find several conductivity distributions that result in the same current density on the boundary. Do your chosen σ_1 and σ_2 give rise to different current densities on the boundary if $\phi = \cos 2\theta$?

12.5.2 Ill-posedness of EIT: A simulation study

We construct two computer-simulated models, σ_1 and σ_2, of the cross-section of a human chest. The difference between them is that in σ_2 a piece of the lung is "missing" or has the background conductivity. See the top plots in Figure 12.4 for σ_1 and σ_2. As an anatomical

12.5. Ill-posedness of EIT

Conductivity σ_1 with lungs intact

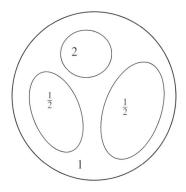

Conductivity σ_2 with anomaly

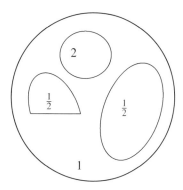

$\mathrm{Re}(u_1)$ with $\nabla \cdot \sigma_1 \nabla u_1 = 0$ in Ω, $\frac{\partial u_1}{\partial \nu}|_{\partial \Omega} = \cos\theta$

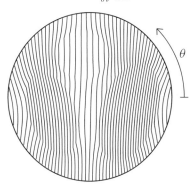

$\mathrm{Re}(u_2)$ with $\nabla \cdot \sigma_2 \nabla u_2 = 0$ in Ω, $\frac{\partial u_2}{\partial \nu}|_{\partial \Omega} = \cos\theta$

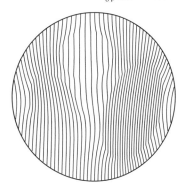

Real part of voltage $u_1|_{\partial \Omega}$

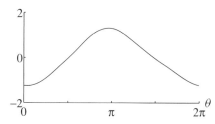

Real part of voltage $u_2|_{\partial \Omega}$

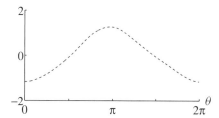

Figure 12.4. Left column: EIT measurement corresponding to heart-and-lungs conductivity phantom σ_1. The heart has conductivity 2, the lungs conductivity $\frac{1}{2}$, and the background conductivity is 1. Right column: EIT measurement corresponding to the anomalous phantom σ_2. See also Figure 12.5.

model this is quite crude, but it is good enough for demonstrating that big differences in the conductivity may result in only small differences in measured EIT data.

In Figure 12.4 we show what happens when we feed the current pattern $\cos\theta$ through the boundary. This is an idealized measurement according to the continuum model; electrodes are not modeled. Here the unit circle is parameterized as $(\cos\theta,\sin\theta)$ with the angle $0 \leq \theta < 2\pi$. We use the FEM to solve the Neumann problems

$$\nabla \cdot \sigma_1 \nabla u_1 = 0 \text{ in } \Omega, \qquad \left.\frac{\partial u_1}{\partial \nu}\right|_{\partial\Omega} = \cos\theta,$$

and

$$\nabla \cdot \sigma_2 \nabla u_2 = 0 \text{ in } \Omega, \qquad \left.\frac{\partial u_2}{\partial \nu}\right|_{\partial\Omega} = \cos\theta.$$

Uniqueness of the above solutions is ensured by requiring $\int_{\partial\Omega} u_j \, dS = 0$. The resulting potential distributions u_1 and u_2 are shown inside Ω in the middle row of Figure 12.4.

In the continuum model of EIT we are given as measurements the voltage distributions $u_1|_{\partial\Omega}$ and $u_2|_{\partial\Omega}$ at the boundary. The bottom row of plots in Figure 12.4 shows these voltage distributions as a function of the angular parameter θ. Furthermore, we repeat the calculation depicted in Figure 12.4 with three more Neumann data: $\cos(n\theta)$ with $n = 2, 3, 4$. As seen in the plots of Figure 12.5, the quite substantial difference in the conductivities σ_1 and σ_2 shows up as a small difference in measured voltage data. This is ill-posedness in the sense of Hadamard's third condition.

12.6 Electrode models

The model given by equations (12.18) and (12.17) is an idealization and is referred to as the "continuum model" or the "continuous model" in the EIT literature. This is because (12.17) does not take into account the following fact. In practice, current is applied through a finite number of electrodes on the surface of the body as shown in Figure 12.6, and not current-density continuously along the boundary. Various models have been proposed to better simulate the experimental situation, and we will briefly survey them here.

12.6.1 The gap model

To account for the discrete nature of the electrodes, the "gap model" sets the current density J equal to 0 off the electrodes and approximates the current density on the ℓth electrode by the current on electrode e_ℓ divided by the area of the ℓth electrode:

$$J(z) = \begin{cases} \frac{I_\ell}{A_\ell} & \text{if } z \text{ lies on electrode } e_\ell, \ell = 1, \ldots, L, \\ 0 & \text{off } \cup_{\ell=1}^{L} e_\ell. \end{cases} \qquad (12.31)$$

Here L denotes the number of electrodes, e_ℓ the ℓth electrode, I_ℓ the current applied on the ℓth electrode, and A_ℓ the area of the ℓth electrode. The conservation of charge condition (12.16) is replaced by the corresponding condition for conservation of currents,

$$\sum_{\ell=1}^{L} I_\ell = 0. \qquad (12.32)$$

12.6. Electrode models

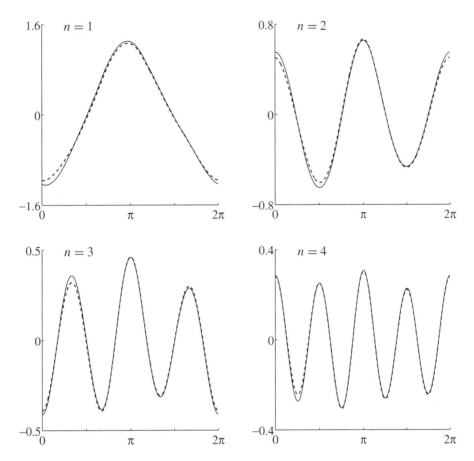

Figure 12.5. Continuum model EIT data (voltages at the boundary) corresponding to four different current distributions applied at the boundary: $\cos(n\theta)$ for $n = 1, 2, 3, 4$. Here $u_1|_{\partial\Omega}$ is shown with a solid line, and $u_2|_{\partial\Omega}$ is shown with a dashed line. The tiny differences between the solid and dashed lines are how the big difference between the two conductivities σ_1 and σ_2 (see the top of Figure 12.4) shows up in EIT data. Note that the top left plot here shows the same two functions than the bottom row in Figure 12.4.

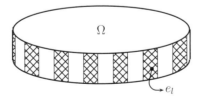

Figure 12.6. A cross-sectional slab of material in the plane of the electrodes represented by Ω. This figure serves primarily as a two-dimensional model in spite of the fact that the electrodes have both height and width. The ℓth electrode is denoted by e_ℓ and has area A_ℓ.

Measured voltages are assumed to have the values of the potential at the center of each electrode,
$$u(\text{center of } e_\ell) = V_\ell. \tag{12.33}$$

The reference potential is chosen so that
$$\sum_{\ell=1}^{L} V_\ell = 0. \tag{12.34}$$

In the gap model, we consider the two-dimensional conductivity equation (12.14) equipped with the boundary condition (12.31). The three-dimensional shape of the body in Figure 12.6 is thus otherwise ignored except for the area of the two-dimensional electrodes that appears in (12.31).

In the gap model formulation, the inverse conductivity problem is as follows: *"Given all possible current patterns* $\mathbf{I} = (I_1, \ldots, I_L)$ *and their corresponding voltage patterns* $\mathbf{V} = (V_1, \ldots, V_L)$, *find the conductivity* σ *inside the body"* [81].

12.6.2 The shunt model

This simpler model does not, however, take into account the shunting effect of the electrodes when they come into contact with the surface of the medium being imaged. The electrodes themselves provide a highly conductive path for the current. By assuming the electrode is a perfect conductor, the voltage is then assumed to be constant over the electrode
$$u(z) = V_\ell \quad \text{for } z \text{ on } e_\ell, \quad \ell = 1, 2, \ldots, L, \tag{12.35}$$

and
$$u(z) = 0 \quad \text{off} \quad \cup_{\ell=1}^{L} e_\ell. \tag{12.36}$$

To avoid an overdetermined forward problem, the condition on the applied current is weakened to equal the integral of the current density over the electrode
$$\int_{e_\ell} \sigma \frac{\partial u}{\partial \nu} ds = I_\ell, \quad \ell = 1, 2, \ldots, L, \tag{12.37}$$

and
$$\sigma \frac{\partial u}{\partial \nu} = 0 \quad \text{off} \quad \cup_{\ell=1}^{L} e_\ell. \tag{12.38}$$

Together with (12.14), equations (12.35)–(12.38) constitute the "shunt model" and, with the conditions (12.32) and (12.34), form a well-posed forward problem [422].

12.6.3 The complete electrode model

When imaging human subjects, the shunt model does not agree closely with experimentally measured currents and voltages. An electrochemical effect between the electrode and the skin causes a thin, highly resistive layer to form at the electrode-skin interface with an impedance ζ_ℓ, which is referred to as the contact impedance. In tank experiments, a small contact impedance is needed to take into account the electrochemical effect between the

12.7. Current patterns and distinguishability

electrode and the electrolyte in the tank. In the "complete model" this arises as a Robin boundary condition,

$$u + \zeta_\ell \sigma \frac{\partial u}{\partial \nu} = V_\ell \quad \text{on} \quad e_\ell, \quad \ell = 1, 2, \ldots, L. \tag{12.39}$$

Replacing equation (12.35) by (12.39) results in a forward problem with a unique solution that is capable of closely reproducing measured voltages [82, 422]. The existence and uniqueness for the model comprised of equations (12.14), (12.39), (12.37), (12.38) is established in [422].

12.7 Current patterns and distinguishability

The application of "all possible current patterns" as stated in the inverse problem is impossible in practice since on L electrodes, there are $L-1$ linearly independent current patterns, and so any other current pattern will be a linear combination of these. Furthermore, some commonly used choices of current patterns do not result in a complete linearly independent set. If K linearly independent current patterns are applied, there will be K linearly independent voltage measurements. Thus, our voltage to current-density data will be a finite-dimensional approximation to the DN map.

12.7.1 Popular choices of current patterns

The choice of applied current patterns in EIT is an important one that is discussed further in Section 12.7.2. Here we describe some commonly-used current patterns to give the reader a greater sense of the physical setup. We will restrict our discussion to two dimensions, and the ideas are easily generalized to three-dimensional electrode configurations. The paper [79] provides an excellent discussion of current patterns and distinguishability, which will be discussed in Section 12.7.2.

The Fourier basis functions $\left\{ \frac{1}{\sqrt{2\pi}} e^{in\theta} \right\}$ of $L^2(\partial \Omega)$ afford a natural choice for a basis of current patterns. These are discussed in the context of data simulation in Chapter 13. Since the applied currents are real-valued functions, we consider sines and cosines in light of Euler's formula. The low spatial frequency patterns give rise to the maximum voltage signal, and so we consider

$$I_\ell^n = \begin{cases} M \cos(n\theta_\ell), & n = 1, \ldots, \frac{L}{2} - 1, \\ M \cos(\pi l), & n = L/2, \\ M \sin((n - L/2)\theta_\ell), & n = \frac{L}{2} + 1, \ldots, L - 1, \end{cases} \tag{12.40}$$

where M is the (maximum) current amplitude. The vector I^n is a discrete approximation to $M \cos n\theta$ for $1 \leq n \leq L/2$ or $M \sin n\theta$ for $L/2 - 1 \leq n \leq L - 1$. There are several ways of visualizing the trigonometric current patterns. Figure 12.7 provides a depiction of the patterns for $n = 1, 2,$ and 8 on a 16 electrode system. Figure 12.8 provides another kind of illustration for visualizing the current pattern I^5. The current pattern I^1 is depicted in Figure 12.11. The vectors defined by (12.40) form a set of $L-1$ linearly independent vectors, and the size of the DN map computed from (13.40) will be $L-1 \times L-1$. One can show that for the unit disc with a circular inclusion, the functions $\phi^n(z) = e^{in\theta}$ are the eigenfunctions of Λ_σ [230]. See Section 13.1.

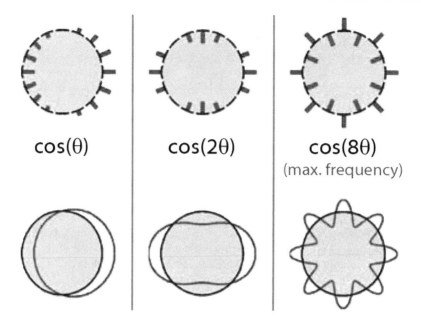

Figure 12.7. An illustration of trigonometric current patterns $n = 1, 2, 8$ approximating $\cos n\theta$ on a 16 electrode system. The short thick lines represent the 16 electrodes with length illustrating the magnitude of the applied current and the direction they are pointing the sign. On a 16 electrode system, $\cos(8\theta)$ represents the maximum frequency in (12.40).

From an engineering standpoint, an EIT system that applies the trigonometric patterns requires L current sources, one for each electrode. With such a system, one has the freedom to apply any desired current pattern. Another pattern that makes use of all L electrodes simultaneously as current sources is the Walsh pattern (see Figure 12.11):

$$I_\ell^n = \begin{cases} M, & l = n \bmod L, \ldots, (n + \tfrac{L}{2}) \bmod L, \\ -M, & \ell = (n + \tfrac{L}{2} + 1) \bmod L, \ldots, (n + L - 1) \bmod L. \end{cases} \tag{12.41}$$

For a system with a single current source, current patterns that apply equal and opposite current on pairs of electrode must be considered to preserve Kirchhoff's law. We will refer to this configuration as a pairwise current injection system. Two examples of pairwise current patterns include the adjacent pattern depicted in Figures 12.9 and 12.11 given by

$$I_\ell^k = \begin{cases} M, & \ell = k, k = 1, \ldots, L, \\ -M, & \ell = k+1, k = 1, \ldots, L-1, \\ -M, & \ell = 1, k = L, \\ 0 & \text{otherwise} \end{cases} \tag{12.42}$$

12.7. Current patterns and distinguishability

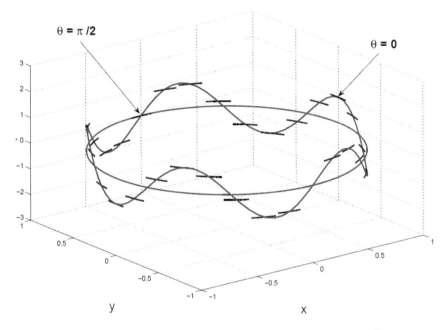

Figure 12.8. An illustration of a trigonometric current pattern I^5 which approximates $\cos(5\theta)$. The circle represents the boundary of a unit disc Ω. The cosine function is shown in gray with amplitude of $M = 1\text{mA}$. The short thick black lines represent the 32 electrodes, and while the electrodes lie physically along the boundary of Ω, they are drawn to intersect the cosine function to illustrate the fact that the current sent to the ℓth electrode has amplitude equal to the intersection of cosine function with the short black line representing the ℓth electrode.

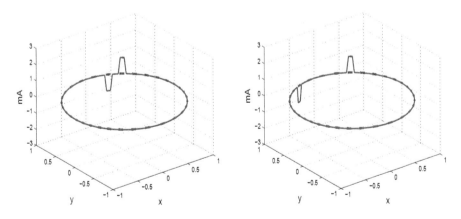

Figure 12.9. An illustration of an adjacent current pattern defined by (12.42) on the left, and a "skip 3" pattern defined by (12.43) on the right.

and a "skip 3" pattern in which three electrodes are skipped between the two injection electrodes given by

$$I_\ell^k = \begin{cases} M, & \ell = k, k = 1, \ldots, L, \\ -M, & \ell = k+4, k = 1, \ldots, L-2, \\ -M, & \ell = m, k = L-m, m = 0, 1, 2, 3, \\ 0 & \text{otherwise,} \end{cases} \quad (12.43)$$

also depicted in Figure 12.9. Another way to visualize the "skip 3" pattern is provided in Figure 12.10.

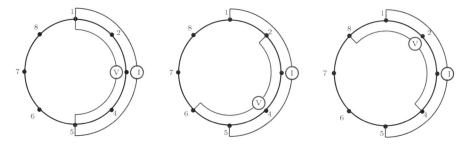

Figure 12.10. An illustration of the "skip 3" current pattern.

The size of the smallest object a system can detect is dependent upon the set of applied current patterns. This will be discussed further in Section 12.7.2. A study by Cheney and Isaacson [78] provides theoretical estimates of the smallest disc of contrast 2 in a homogeneous disc of radius 15 cm that a system can detect whose precision is such that it can reliably measure 1 mV in response to an applied current of 1 mA. The results are given in Table 12.2. The optimal current patterns are discussed in Section 12.7.2. In this case, they correspond to the trigonometric patterns (12.40).

Table 12.2. Theoretical estimates [78] of the smallest detectable disc of contrast 2 in a homogeneous disc of radius 15 cm.

Current Pattern	Currents with max amp 1 mA		Currents with power 7.9 mW	
	Power (μW)	Radius (cm)	max amp $\left(\frac{\text{mA}}{\text{cm}^2}\right)$	Radius (cm)
Optimal	7922	0.95	0.094	0.953
Walsh	13540	0.84	0.072	0.958
Opposite	238	2.7	0.545	1.123
Adjacent	85	8.6	0.909	2.576

12.7.2 Distinguishability and optimal current patterns

It is very important to understand the nature of the measurements and hence the data that one has in an inverse problem. The DN map Λ_σ is a mapping from the Sobolev space $H^{1/2}(\partial\Omega)$ to $H^{-1/2}(\partial\Omega)$, whereas the ND map $\mathcal{R}_\sigma$ is a smoothing map; it takes functions from $\widetilde{H}^{-1/2}(\partial\Omega)$ to $\widetilde{H}^{1/2}(\partial\Omega)$. Thus, a function with noise added to it, such as a superimposed highly oscillatory function, will map to a function with noise of even higher

12.7. Current patterns and distinguishability

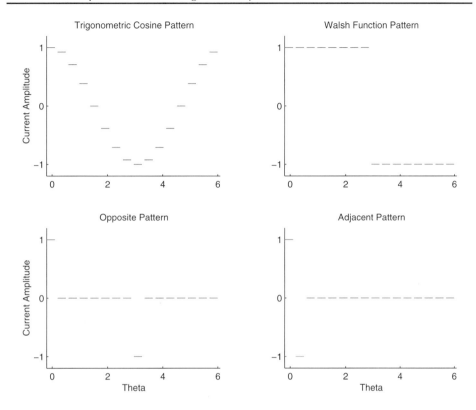

Figure 12.11. An illustration of the four current patterns in Table 12.2 corresponding to a 16 electrode system. This example assumes homogeneous conductivity $\sigma \equiv 1$, so the optimal current patterns are trigonometric; here $n = 1$.

amplitude under Λ_σ, which is undesirable in terms of the signal-to-noise ratio. However, applying $\mathcal{R}_\sigma$ to such a function will result in a smoother function than the original one. It is for this reason that in practical EIT, currents are applied and voltages are measured. Thus, in practice the data are the approximate and noisy current-density-to-voltage, or ND map.

David Isaacson quantified a notion of distinguishability in [230]. The distinguishability, or ability to detect an object, depends on the properties and location of the object as well as the applied currents. The *distinguishability* $\delta(j)$ is defined to be the norm of the voltage difference divided by the norm of the current density j.

$$\delta(\sigma_1, \sigma_2, j) = \frac{\|V(\cdot, \sigma_1, j) - V(\cdot, \sigma_2, j)\|}{\|j\|}. \tag{12.44}$$

This definition can be used to answer such questions as

- "To what precision must the EIT system measure to detect an object from the background with a certain size, location, and contrast?"
- "What is the smallest inhomogeneity that my system can detect?"
- "What are the best current patterns to apply to detect this object?"

The answers to these questions can depend on several choices, such as what norm is chosen for defining distinguishability. The norm in definition (12.44) could be the norm on a function space or a vector norm when considering discrete measurements. The choice of norm determines the units on δ and the permissible current densities. Thus, one definition may result in a smaller distinguishability than another definition [78]. If the distinguishability is less than the measurement precision of the system, the two conductivity distributions are indistinguishable using the measured data. Two excellent studies of distinguishability-related questions are [159] and [78]. We will include some of the results from those works here.

It is proved in [159] that if one applies a pairwise current pattern j in which equal and opposite currents are applied on two electrodes and the voltage is measured on all electrodes, as the number of electrodes L increases, the distinguishability goes to zero. This result can be generalized to show that if the number of current sources is held fixed at a finite number, and the number of electrodes is allowed to go to infinity, the distinguishability will go to zero. We include essentially the proof from [231] for a pairwise current pattern here.

Theorem 12.1. *Suppose the boundary of Ω is smooth and j denotes a pairwise current injection pattern with current applied on electrodes e_a and e_b. Denoting the number of electrodes by L and the (uniform) electrode area by A,*

$$\lim_{L \to \infty, A \to 0} \delta(\sigma_1, \sigma_2, j) = 0. \tag{12.45}$$

Proof. Let σ_1 and σ_2 be two nonzero finite conductivity distributions, and $\mathcal{R}_{\sigma_i}$, $i = 1, 2$, the corresponding ND map. We will choose the L^2-norm in our definition of distinguishability so that

$$\delta(\sigma_1, \sigma_2, j) = \frac{\|\mathcal{R}_{\sigma_1} j - \mathcal{R}_{\sigma_2} j\|_2}{\|j\|_2}. \tag{12.46}$$

Let $\delta \mathcal{R} \equiv \mathcal{R}_{\sigma_1} - \mathcal{R}_{\sigma_2}$. Since the maps $\mathcal{R}_{\sigma_i}$ are compact, their difference is compact, and hence $\delta \mathcal{R}$ has eigenvalues $\{\lambda_n\}$ such that $|\lambda_1| \geq |\lambda_2| \geq \cdots$ and such that $\lim_{n \to \infty} \lambda_n = 0$. Since the $\mathcal{R}_{\sigma_i}$ and thus $\delta \mathcal{R}$ are self-adjoint, there exists a complete orthonormal set of eigenfunctions for $\delta \mathcal{R}$, which we will denote by ϕ_n. Expanding $j(z)$ in terms of these eigenfunctions,

$$j(z) = \sum_{n=1}^{\infty} \langle j, \phi_n \rangle \phi_n(z), \tag{12.47}$$

we have

$$\delta \mathcal{R} j(z) = \sum_{n=1}^{\infty} \lambda_n \langle j, \phi_n \rangle \phi_n(z). \tag{12.48}$$

We can then write the distinguishability function in terms of the orthonormal basis functions ϕ_n as follows

$$(\delta(\sigma_1, \sigma_2, j))^2 = \frac{\sum_{n=1}^{\infty} |\lambda_n|^2 |\langle j, \phi_n \rangle|^2}{\sum_{n=1}^{\infty} |\langle j, \phi_n \rangle|^2}.$$

12.7. Current patterns and distinguishability

Let $\epsilon > 0$ be arbitrary and small. Choose K sufficiently large that $|\lambda_n| < \epsilon$ for $n > K$. Then

$$(\delta(\sigma_1,\sigma_2,j))^2 \leq \frac{\sum_{n=1}^{K}|\lambda_n|^2|\langle j,\phi_n\rangle|^2}{\|j\|_2^2} + \frac{\epsilon^2 \sum_{n=K+1}^{\infty}|\langle j,\phi_n\rangle|^2}{\sum_{n=1}^{\infty}|\langle j,\phi_n\rangle|^2}$$

$$\leq \frac{K|\lambda_1|^2}{\|j\|_2^2} \max_{1\leq n\leq K} |\langle j,\phi_n\rangle|^2 + \epsilon^2.$$

Since current is applied on electrodes e_a and e_b, by Hölder's inequality

$$|\langle j,\phi_n\rangle| \leq \left|\int_{e_a} j(z)\phi_n(z)ds(z)\right| + \left|\int_{e_b} j(z)\phi_n(z)ds(z)\right|$$
$$\leq \|j\|_{L^2(e_a)}\|\phi_n\|_{L^2(e_a)} + \|j\|_{L^2(e_b)}\|\phi_n\|_{L^2(e_b)}$$
$$\leq 2\|j\|_2\sqrt{A}\|\phi_n\|_\infty.$$

Thus,

$$(\delta(\sigma_1,\sigma_2,j))^2 \leq 4AK|\lambda_1|^2 \max_{1\leq n\leq K}\|\phi_n\|_\infty^2 + \epsilon^2,$$

which goes to zero as $A \to 0$ since ϵ can be made arbitrarily small. $\square$

As remarked in [159], a technical aspect not addressed here is that if σ_1 and σ_2 actually differ *on* the boundary, the distinguishability may not approach zero as $A \to 0$. For a study of the effects of measurement precision and finite numbers of electrodes on linear impedance imaging algorithms, see [232].

Another classical result that arises from the concept of distinguishability is that of optimal current patterns [230], where the optimal current patterns are defined to be those that maximize the distinguishability. Of course, one must first decide which conductivities one wants to distinguish. Typically, one has an inhomogeneity of interest that one wishes to distinguish from a known background, often constant. Note that the choice of a constant background corresponds to maximizing the voltage signal.

Expressing the voltages, $V(z,\sigma_i,j)$, $i=1,2$, as $V(z,\sigma_i,j) = \mathcal{R}_{\sigma_i} j(z)$, the optimal currents j^{opt} satisfy

$$\delta(\sigma_1,\sigma_2,j^{\text{opt}}) = \max_{\|j\|=1} \|\mathcal{R}_{\sigma_1} j - \mathcal{R}_{\sigma_2} j\| = \max_{\|j\|=1} \|\delta\mathcal{R} j\|. \tag{12.49}$$

We use again the fact that since $\delta\mathcal{R}$ is compact and self-adjoint, it has a complete set of orthonormal eigenfunctions whose eigenvalues $\lambda_1 \geq \lambda_2 \geq \cdots$ satisfy $\lim_{n\to\infty}\lambda_n = 0$. In the L^2-norm, the maximum of (12.49) corresponds to the largest eigenvalue of $\delta\mathcal{R}$, namely, $|\lambda_1|$. Thus, the optimal currents are the eigenfunctions of $\delta\mathcal{R}$. An algorithm for computing these can be found in [159], and it is essentially the power method for computing the largest eigenvalue of a matrix. The beauty of this approach is that the unknown medium itself tells the most efficient current patterns for the imaging of this particular target.

See also the current pattern study [125]. Optimal current patterns can be defined in the framework of statistical (Bayesian) inversion as well; see [245, 246].

Exercise 12.7.1. *Find the radius of the smallest circular inclusion of the form* (12.25) *with contrast $A = 2$ detectable by a system with measurement precision 0.01 and current pattern $\phi = \sqrt{\pi}^{-1} \cos\theta$.*

Exercise 12.7.2. *Find the first* 8 *eigenvalues of the DN map for a conductivity* σ_2 *of the form* (12.25) *with* $A = 4$ *and* $R = 0.25$.

Exercise 12.7.3. *Verify that the Fourier basis functions define optimal current patterns for the case of a single circular inclusion in a disc. Explain why these patterns cannot be applied in practice. What is the closest practical approximation?*

Exercise 12.7.4. *Are the Fourier basis functions optimal current patterns an arbitrary radially symmetric conductivity defined in a disc? Why or why not?*

12.8 Further reading

For simplicity, we restrict the EIT study in this book to a particular case defined by the following assumptions. The imaging domain is the unit disc $\Omega = D(0, 1) \subset \mathbb{R}^2$. The conductivity is real-valued and isotropic (scalar-valued) and equals one near the boundary: $\sigma(z) \equiv 1$ for all z in a neighborhood of $\partial \Omega$. The conductivity is strictly bounded from below and above: $0 < c \leq \sigma(z) \leq C < \infty$ for all $z \in \Omega$. Also, we assume that we have available the "dummy load" measurement Λ_1 corresponding to the homogeneous conductivity $\sigma(z) \equiv 1$.

In the following, we indicate further reading about EIT imaging under assumptions other than those mentioned above.

12.8.1 Three-dimensional EIT

Most objects imaged in real life using EIT are three-dimensional, including patients. It is often reasonable to consider two-dimensional approximations to three-dimensional EIT configurations, for example, when imaging heart and lungs using a belt of electrodes around the chest as depicted in Figure 12.1. However, electric current will not be confined to a two-dimensional slice even in that case.

Computational inversion for three-dimensional EIT is demanding. Published studies include the following. Calderón's original method for the linearized problem was presented in [64] and implemented in dimension three in [54, 117]. First theoretical reconstructions were described, independently, in [366] and [348]. Numerical reconstruction method based on [348] was discussed in [101] and implemented in [47, 278, 116]. Practical reconstruction methods based on least-squares and regularization are reported in [50, 162, 332, 333, 334, 341, 454, 455, 453, 476].

12.8.2 Reconstruction at the boundary

If the conductivity is not constant near the boundary, it is advisable to recover the trace (and possibly the normal derivative) of σ at $\partial \Omega$. Numerical methods for such recovery have been studied in dimension two [352] and in dimension three [351].

12.8.3 Noncircular boundary

Circular boundaries do appear in some applications of EIT, most notably in process tomography where the insides of pipelines are imaged. But in medical imaging this is never the

12.8. Further reading

case in practice; rather the boundary shape is unknown and its recovery part of the inverse problem.

Statistical methods for recovering the shape of the domain from measured EIT data have been published in [361, 362, 363, 364]. The boundary shape can also be recovered using Teichmüller space techniques and quasi-conformal mappings; see [284, 285, 286, 287] and Figure 12.12.

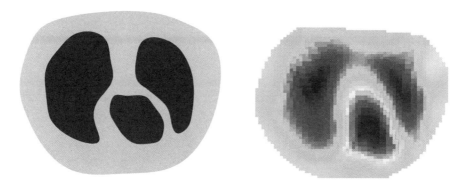

Figure 12.12. Left: original synthetic conductivity. Right: reconstruction, *without knowing the shape of the boundary*, from EIT data simulated with the complete electrode model using 16 electrodes and 42 dB noise level. See [287] for the details of computation.

See [346, 345, 347] for studies of EIT on noncircular (but known) boundaries.

12.8.4 Complex-valued conductivity

In medical EIT, alternating currents are always used to ensure the safety of the subject. The temporally sinusoidal measurement causes an offset, or delay, in the measured voltage, resulting in a complex-valued voltage measurement. The measurement of this offset to high precision is challenging from an engineering standpoint, and many EIT systems do not record it in the voltage measurement. Furthermore, the mathematical theory is more advanced for the real case than the case of complex conductivities. Recall equation (12.13) in which the complex conductivity has the form $\sigma + i\omega\epsilon$, where ω is the frequency of the alternating current and ϵ is the electric permittivity.

An important theoretical uniqueness study concerning complex admittivities is [148], in which it is proved that $\Omega \in \mathbb{R}^2$ is a bounded open domain with Lipschitz boundary, σ is bounded away from zero by a constant σ_0, and there exists a positive constant β such that $\|\sigma\|_{W^{2,\infty}(\Omega)}, \|\sigma\|_{W^{2,\infty}(\Omega)} \leq \beta$, and there exists a frequency ω_0 dependent on β, σ_0, and Ω such that for $\omega < \omega_0$ the complex admittivity is uniquely determined by the DN map. The proof in [148] is nearly constructive, but equations relating the DN map to the scattering transform and CGO solutions are not present in that work and are derived in [189], completing a D-bar method for the reconstruction of complex admittivities. An implementation is also found in [189]. A reconstruction on a chest-shaped domain from simulated data with 0.01% noise is found in Figure 12.13.

Complex admittivities hold promise in lung imaging, for cryoablation, and for the early detection of breast cancer, particularly when augmenting X-ray mammography with

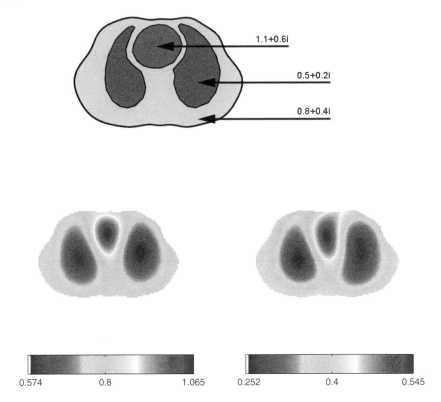

Figure 12.13. Top: an illustration of the phantom chest domain with the simulated heart and lungs. The simulated voltage data included 0.01% added noise. Lower left: reconstruction of the real part of the conductivity. Lower right: reconstruction of the imaginary part of the conductivity. Reconstructions were computed by Sarah Hamilton. The dynamic range is 81.9% for the conductivity, and 73.5% for the permittivity.

spectroscopic EIT imaging. Namely, sweeping over a variety of frequencies of alternating current differentiates between healthy and cancerous cells quite effectively, as is shown in [268, 255, 267].

12.8.5 Lack of the dummy load measurement

Many inversion methods for EIT assume that two measurements are available: Λ_σ and Λ_1, where the latter corresponds to the homogeneous target. Such a target is rarely available (imagine having in addition to a real patient a lump of homogeneous material shaped exactly the same as the patient), so it is often thought that Λ_1 is simulated computationally. However, the exact shape of the target is not always known, so there will be some modeling error arising from the simulation of Λ_1. Note that the Calderón method does not require the use of a dummy load.

The use of different frequencies, as discussed in Section 12.8.4, can sometimes be used to overcome the lack of reference data. See [199] for a study of detecting inclusions using EIT measurements with two different frequencies.

12.8.6 Counterexamples and invisibility

Relaxing the boundedness and strict positivity assumptions on the conductivity leads to interesting results on uniqueness and nonuniqueness. These results in turn are closely connected to studies of invisibility.

Before 2003, there had appeared some occasional works on invisibility; see, for example, [266, 43]. However, the serious theoretical study was launched in 2003 in the articles [175, 177, 305] and continued by the later works [8, 336, 311, 376]. A widely reported experiment [406] applied the theory to actual cloaking devices, or structures that would not only render an object invisible but also undetectable to electromagnetic waves. The most popular cloaking technique is *transformation optics* [470], where wave propagation is controlled by modifying the transformation rules for the material properties.

The aim of transformation optics is to design objects that appear in all external measurements as another object. As an example, consider an anisotropic conductivity on a domain $\Omega \subset \mathbb{R}^n$ given by a symmetric, positive semidefinite matrix-valued function $\sigma = [\sigma^{ij}(x)]_{i,j=1}^n$. In the absence of sources or sinks of electric current, an electrical potential u satisfies

$$\nabla \cdot \sigma \nabla u(x) = \sum_{j,k=1}^n \frac{\partial}{\partial x^j}\left(\sigma^{jk}(x)\frac{\partial u}{\partial x^k}(x)\right) = 0 \quad \text{in } \Omega, \tag{12.50}$$

where $u|_{\partial\Omega} = f$ with a prescribed voltage f on the boundary. The resulting DN map (or voltage-to-current map) is then defined by

$$\Lambda_\sigma(f) = Bu|_{\partial\Omega}, \tag{12.51}$$

where $Bu = \sum_{j=1}^n \nu_j \sigma^{jk} \partial_k u$, with u the solution of (12.50) and ν the unit normal vector of $\partial\Omega$. Applying the divergence theorem, we have

$$Q_\sigma(f) = \sum_{j,k=1}^n \int_\Omega \sigma^{jk}(x) \frac{\partial u}{\partial x^j} \frac{\partial u}{\partial x^k} dx = \int_{\partial\Omega} \Lambda_\sigma(f) f \, dS, \tag{12.52}$$

where u solves (12.50) and dS denotes surface measure on $\partial\Omega$. $Q_\sigma(f)$ represents the power needed to maintain the potential f on $\partial\Omega$. By (12.52), knowing Q_σ is equivalent to knowing Λ_σ. If $F: \Omega \to \Omega$, $F = (F^1, F^2, \ldots, F^n)$, is a diffeomorphism with $F|_{\partial\Omega} =$ Identity, then by making the change of variables $y = F(x)$ and setting $u = v \circ F^{-1}$ in the first integral in (12.52), we obtain the identity $Q_{F_*\sigma}(f) = Q_\sigma(f)$ for all f on $\partial\Omega$. Here

$$(F_*\sigma)^{jk}(y) = \left. \frac{1}{\det\left[\frac{\partial F^j}{\partial x^k}(x)\right]} \sum_{p,q=1}^n \frac{\partial F^j}{\partial x^p}(x) \frac{\partial F^k}{\partial x^q}(x) \sigma^{pq}(x) \right|_{x=F^{-1}(y)} \tag{12.53}$$

is the *push-forward* of the conductivity σ by F.

It follows from $Q_{F_*\sigma}(f) = Q_\sigma(f)$ that

$$\Lambda_{F_*\sigma} = \Lambda_\sigma, \tag{12.54}$$

that is, all boundary measurements for the conductivity $F_*\sigma$ and σ are the same. Thus, there is an infinite-dimensional class of anisotropic conductivities which give rise to the same electrical measurements at the boundary! This was first observed in [282] following a remark by Luc Tartar. Consequently, Calderón's uniqueness question for anisotropic conductivities is whether two conductivities with the same DN map must be push-forwards of each other. This question has been studied in [430, 429, 310, 306, 20, 15].

The counterexamples to uniqueness in Calderón's problem in [175, 177] were based on using formula (12.54) with a singular transformation

$$F_1 : B(2) \setminus \{0\} \to B(2), \quad F_1(x) = \left(\frac{|x|}{2} + 1\right)\frac{x}{|x|}, \quad 0 < |x| \leq 2, \qquad (12.55)$$

where $B(R) \subset \mathbb{R}^n$ is a ball of radius R and including inside a closed ball $\overline{B}(1)$ an arbitrary conductivity $\sigma_c(x)$ that corresponds to a hidden object. This produced a conductivity

$$\tilde{\sigma} = \begin{cases} 2(I - P(x)) + 2|x|^{-2}(|x|-1)^2 P(x) & \text{for } x \in B(2) \setminus \overline{B}(1), \\ \sigma_c(x) & \text{for } x \in \overline{B}(1), \end{cases} \qquad (12.56)$$

where I is the identity matrix and $P(x) = |x|^{-2} x x^t$ is the projection to the radial direction $x/|x|$. With the DN map $\Lambda_{\tilde{\sigma}}$ suitably defined as in [175, 177], the conductivity $\tilde{\sigma}$ is seen to appear in all boundary measurements as the homogeneous conductivity 1:

$$\Lambda_{\tilde{\sigma}} = \Lambda_1.$$

This is invisibility cloaking for zero-frequency measurements: take an arbitrary object having conductivity $\sigma_c(x)$ in $\overline{B}(1)$, surround it with an exotic layer corresponding to $(F_1)_\sigma I = \tilde{\sigma}|_{B(2)\setminus \overline{B}(1)}$, and the coated body $\tilde{\sigma}$ appears in boundary measurements as a homogeneous and isotropic body.

The above counterexamples were motivated by studies of degenerate families of Riemannian metrics corresponding to singular conductivities. A related example of a complete but noncompact two-dimensional Riemannian manifold with boundary having the same DN map as a compact one was given in [305]; see also [176]. The techniques of Greenleaf, Lassas, and Uhlmann [175, 177] are valid in dimensions three and higher, and a similar construction has been shown to work in two dimensions [281].

The breakthrough in the physical study of invisibility happened in 2006 when the two papers [311, 376] appeared in the same issue of *Science* with transformation optics-based proposals for cloaking based on metamaterials, inspired by Veselago [458]. Leonhardt [311] gave a description, based on conformal mapping, of inhomogeneous indices of refraction n in two dimensions that would cause light rays to go around a region and emerge on the other side as if they had passed through empty space. On the other hand, Pendry, Schurig, and Smith [376] gave a prescription for values of ϵ and μ in (12.7) and (12.8) giving a cloaking device for electromagnetic waves, based on the fact that ϵ and μ transform in the same way (12.53) as the conductivity tensor in electrostatics. They used the singular transformation (12.55) resulting in

$$\epsilon(x) = c_1 \tilde{\sigma}(x), \quad \mu(x) = c_2 \tilde{\sigma}(x), \qquad (12.57)$$

where c_1 and c_2 are constants having the appropriate physical units and $\tilde{\sigma}$ is the conductivity (12.56). Rigorous theory on cloaking with singular permittivity and permeability (12.57) is studied in [172] and [475]; see also [173, 174].

Chapter 13
Simulation of noisy EIT data

At least two kinds of data simulation approaches are useful for developing reconstruction methods for electrical impedance tomography (EIT), based on either the continuum model or some of the electrode models of Section 12.6. The former involve matrix approximations or linear operators acting on infinite-dimensional function spaces, whereas the latter approach leads naturally to data matrices whose size is determined by the number of electrodes.

Why do we need continuum data at all, given the facts that the complete electrode model can reproduce practical data very accurately [422] and that electrodes are used in every practical EIT experiment? There are at least two reasons for that. EIT is such a complicated inverse problem that the development of reconstruction methods is best done in small steps, starting with the continuum model for simplicity and then adding one complication at a time. Then it is easier to analyze where possible problems arise: Are they perhaps due to the inherent ill-posedness of EIT, or due to some bug in electrode modeling? Furthermore, the DN map is the mathematically idealized EIT measurement that arises in other applications as well, such as diffuse optical tomography, groundwater modeling, and elastic probing.

The DN map is completely determined by its spectral data [159], and so one point of view is to approximate the DN map by computing approximations to its eigenvalues. In cases in which one wants to study a known and relatively simple conductivity, this approach can be very practical and was adopted in [415, 344, 47], for example. For more complicated conductivities and domains, the measurements can be simulated by computing the measured voltages by solving the forward problem using, for example, the FEM [427] or boundary element method (BEM) [22, 23]. Both the BEM and FEM methods are capable of including the electrode models. The computational tool we use in this text is the FEM.

13.1 Eigenvalue data for symmetric σ

Consider the case of a rotationally symmetric conductivity $\sigma(z) = \sigma(|z|)$ defined in the unit disc Ω. Parameterize the boundary (unit circle) as

$$\partial \Omega = \{(\cos\theta, \sin\theta) \mid 0 \leq \theta < 2\pi\}.$$

Then the *Fourier basis functions*

$$\varphi_n(\theta) = (2\pi)^{-1/2} e^{in\theta}, \quad n \in \mathbb{Z}, \tag{13.1}$$

are eigenfunctions of the DN map: $\Lambda_\sigma \varphi_n = \lambda_n \varphi_n$, where the eigenvalues are nonnegative: $\lambda_n \geq 0$. The proof of these facts is left as an exercise.

The eigenvalues of a rotationally symmetric piecewise constant conductivity with N jump discontinuities are given by the following theorem [415].

Theorem 13.1. *Let $\Omega \in \mathbb{R}^2$ be the unit disc and $0 = r_0 < r_1 < \cdots < r_{N-1} = 1$, where $n \geq 2$. For $j = 1, \ldots, N$, let σ_j be positive real numbers such that $\sigma_j \neq \sigma_{j+1}$ and $\sigma_N = 1$. Define $\sigma(r) = \sigma_j$ for $r_{j-1} \leq r < r_j$, $j = 1, \ldots, N$. Then the eigenvalues of Λ_σ are*

$$\lambda_n = |n| - 2|n|(1 + C_{N-1})^{-1}, \tag{13.2}$$

where the numbers C_j are given recursively by $C_1 = \rho_1 r_1^{-2|n|}$ and $C_j = (\rho_j C_{j-1} + r_j^{-2|n|})/(\rho_j + C_{j-1} r_j^{2|n|})$ for $j = 2, \ldots, N-1$, where $\rho_j = (\sigma_{j+1} + \sigma_j)/(\sigma_{j+1} - \sigma_j)$.

Proof. Since $\lambda_0 = 0$, we take $n \neq 0$. We will construct an $H^1(\Omega)$ function u_n that solves $\nabla \cdot \sigma \nabla u_n = 0$ in Ω, with $u_n|_{r=1} = \varphi_n$. Write $u_n = v_n(r) e^{in\theta}$ and $v_n(r) = a_j r^{|n|} + b_j r^{-|n|}$ for $r_{j-1} \leq r < r_j$, $j = 1, \ldots, N$. Set $b_1 = 0$ so that u_n is harmonic in the innermost disc. Matching the limits

$$v_n(r_j^-) = v_n(r_j^+) \quad \text{and} \quad \frac{\sigma_j \partial v_n(r_j^-)}{\partial r} = \frac{\sigma_{j+1} \partial v_n(r_j^+)}{\partial r}$$

results in $a_j = C_{j-1} b_j$ for $j = 2, \ldots, N$. On the outermost radius the boundary condition implies $a_N + b_N = 1/\sqrt{2\pi}$.

It is left to the reader to verify that the function u_n constructed in this manner is in $H^1(\Omega)$ and satisfies $\nabla \cdot \sigma \nabla u_n = 0$ in the weak sense. Finally,

$$\frac{\partial u_n}{\partial r}\Big|_{r=1} = |n|(a_N - b_N)\varphi_n,$$

so $\lambda_n = |n|(1 - b_N)$ and $b_N = (1 + C_{N-1})^{-1}$, which proves the theorem. $\square$

A method of computing the approximate eigenvalues of a continuous rotationally symmetric conductivity was also given in [415], based on Theorem 13.1 and [159]. The continuous conductivity σ is approximated above and below pointwise by piecewise constant conductivities σ_L and σ_U so that $\sigma_L(z) \leq \sigma(z) \leq \sigma_U(z)$ for all $z \in \Omega$. Then the eigenvalues of the corresponding DN maps satisfy the same monotonicity: $\lambda_n^L \leq \lambda_n \leq \lambda_n^U$; these inequalities can be proved by slightly modifying the ND map-based analysis in [159].

Note that the eigenvalues are of the form $\lambda_n = |n| + \epsilon$, where ϵ is exponentially small [159] (see also Exercise 13.1.2.) Thus, it is difficult to accurately compute eigenvalues of the DN map. However, if the reconstruction method is based on the difference $\Lambda_\sigma - \Lambda_1$, then Theorem 13.1 becomes incredibly efficient. Namely, the nth eigenvalue of the operator Λ_1 is $|n|$, so the nth eigenvalue of $\Lambda_\sigma - \Lambda_1$ is, by formula (13.2), given exactly by $2|n|(1+C_{N-1})^{-1}$. This provides a way to compute DN maps with unrealistically high accuracy, enabling numerical testing of EIT algorithms with extremely low noise levels.

Exercise 13.1.1. *Assume that $\sigma(z) = \sigma(|z|)$ holds for all z in the unit disc for a strictly positive conductivity $\sigma \in L^\infty(\Omega)$. Let the functions $\varphi_n : \partial\Omega \to \mathbb{C}$ be defined by (13.1) for all $n \in \mathbb{Z}$. Show that $\Lambda_\sigma \varphi_n = \lambda_n \varphi_n$ and that $\lambda_n \geq 0$.*

Exercise 13.1.2. *Compute the eigenvalues of Λ_1, the DN map corresponding to the homogeneous conductivity $\sigma = 1$ on the unit disc.*

Exercise 13.1.3. *Compute the eigenvalues for the example in Section 12.5 with $A = 2$ and $A = 4$ and $R = 0.5$. Observe the exponential decay for $\Lambda_\sigma - \Lambda_1$.*

Exercise 13.1.4. *Show that the function u_n constructed in Theorem 13.1 is in $H^1(\Omega)$ and satisfies $\nabla \cdot \sigma \nabla u_n = 0$ in the weak sense.*

Exercise 13.1.5. *Use Theorem 13.1 to approximate the eigenvalues of the DN map by finding a set of upper bounds for the eigenvalues and a set of lower bounds for the eigenvalues for the conductivity defined as follows: Let σ be a radially symmetric conductivity defined by*

$$\sigma(z) := (\alpha F_\rho(|z|) + 1)^2, \tag{13.3}$$

where

$$F_\rho(z) := (z^2 - \rho^2)^4 \cos\frac{3\pi z}{2\rho}, \quad -\rho < z < \rho, \tag{13.4}$$

and $F_\rho(z) = 0$ for $|z| > \rho$. Choose $\alpha = 10$ and $\rho = 3/4$. Plot the eigenvalues of $\Lambda_\sigma - \Lambda_1$ on a logarithmic scale.

13.2 Continuum model data and FEM

We restrict ourselves here to $\Omega \subset \mathbb{R}^2$ being the unit disc and explain how to construct a matrix approximation to a bounded linear operator $\mathcal{A} : H^s(\partial\Omega) \to H^r(\partial\Omega)$, focusing on the DN map Λ_σ and the ND map $\mathcal{R}_\sigma$. See Section C.2 for the definition of the Sobolev spaces $H^s(\partial\Omega)$ and $H^r(\partial\Omega)$.

13.2.1 Computing the DN matrix directly

Choose some $N > 0$ and use the truncated basis (13.1) with $-N \leq n \leq N$. We approximate the DN map Λ_σ by the $(2N+1) \times (2N+1)$ matrix $\mathbf{L}_\sigma = [(\mathbf{L}_\sigma)_{m,n}]$ defined by

$$(\mathbf{L}_\sigma)_{m,n} := \langle \Lambda_\sigma \varphi_n, \varphi_m \rangle = \frac{1}{\sqrt{2\pi}} \int_0^{2\pi} (\Lambda_\sigma \varphi_n) e^{-im\theta} d\theta, \tag{13.5}$$

where we use the following nonstandard but convenient indexing: $m \in \{-N, \ldots, N\}$ for the rows and $n \in \{-N, \ldots, N\}$ for the columns. Note that formula (13.5) is achieved by taking $\mathcal{A} := \Lambda_\sigma$ and $\mathbf{L}_\sigma := A$ in the analysis of Section C.2.1.

The matrix $\mathbf{L}_\sigma$ can be computed numerically using the FEM as follows. Fix $-N \leq n \leq N$ and define the Dirichlet problem

$$\nabla \cdot \sigma \nabla u_n = 0 \text{ in } \Omega, \qquad u_n|_{\partial\Omega} = \varphi_n, \qquad (13.6)$$

in a suitable FEM implementation (such as provided by MATLAB's PDE Toolbox). Solve for u_n in a finite-dimensional function space spanned by appropriate basis functions. Finally, differentiate the approximate solution numerically to evaluate $\Lambda_\sigma \varphi_n = \sigma(\partial u_n)/(\partial \nu)|_{\partial\Omega}$. Then use a quadrature rule to evaluate approximately the integral in (13.5) for all $-N \leq m \leq N$, yielding the column of $\mathbf{L}_\sigma$ corresponding to index n. Repeating the above for all $-N \leq n \leq N$ gives the matrix $\mathbf{L}_\sigma$ column by column.

For the finite element solution of the conductivity equation (13.6) you need to do the following:

(i) Create a very fine finite element mesh consisting of node points, triangles, and edge segments.

(ii) Specify the conductivity σ in the interior of the domain. The values should be calculated at centerpoints of the triangles as shown in Figure 13.1(a).

(iii) Specify the Dirichlet boundary condition $\varphi_n(\theta)$. The Dirichlet values are specified at the boundary nodes as shown in Figure 13.1(b).

The FEM solution results in approximate point values of the function u at the node points.

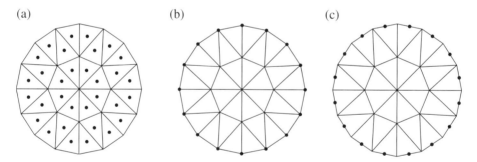

Figure 13.1. Finite element mesh for computing approximate solutions of the conductivity equation. (a) The dots indicate the centers of mass of the triangles; this is where the conductivity σ should be evaluated. (b) The dots indicate nodes at the boundary; this is where Dirichlet data need to be specified (or evaluated). (c) The dots indicate middle points of boundary segments; this is where the normal derivative of the approximate solution is to be evaluated (or specified). The mesh shown here is very coarse for illustrative purposes; meshes used in practice should be much, much finer.

We assume here that the FEM basis functions are piecewise linear. In that case the gradient of the approximate potential u is a constant vector inside each triangle. We need to compute

$$\sigma \frac{\partial u_n}{\partial \nu}\bigg|_{\partial\Omega} = \sigma \nu \cdot \nabla u_n|_{\partial\Omega},$$

where ν is the outward normal vector of $\partial\Omega$. This leads to step (iv):

13.2. Continuum model data and FEM

(iv) Pick out the gradient vectors of u_n at triangles having a side at the boundary. Calculate current through boundary as $\Lambda_\sigma \varphi_n = \sigma \nu \cdot \nabla u|_{\partial\Omega}$, evaluated at centers of edge segments as shown in Figure 13.1(c).

You can find helpful MATLAB routines at the book website, so there is no need to build up all of the above from scratch.

The kind of computation above (but with different Dirichlet data) has been carried out, for example, in [220], where piecewise linear basis functions were used in a square-shaped domain Ω. The finite element triangulation was progressively refined in [220] near the boundary to improve the accuracy of numerical differentiation.

However, it is generally advisable to compute a matrix approximation $\mathbf{R}_\sigma$ to the ND map $\mathcal{R}_\sigma$ instead, and invert $\mathbf{R}_\sigma$ to find $\mathbf{L}_\sigma$ (apart from a few trivial elements related to constant basis functions). Computation of the ND map avoids the unstable step of numerical differentiation and is a better model of EIT measurements as they are typically current-to-voltage rather than voltage-to-current maps. We discuss computing $\mathbf{L}_\sigma$ via $\mathbf{R}_\sigma$ in Section 13.2.2 below.

Exercise 13.2.1. *Use the eigenvalues discussed in Section 13.1 to compute a matrix representation of Λ_1 in the truncated Fourier basis $\{\varphi_n : -N \leq n \leq N\}$.*

Exercise 13.2.2. *Show that Λ_1 is a bounded linear operator between the Sobolev spaces $H^{1/2}(\partial\Omega)$ and $H^{-1/2}(\partial\Omega)$. Use the matrix constructed in Exercise 13.2.1 at the limit $N \to \infty$.*

13.2.2 Computing the ND matrix directly

Consider the ND map $\mathcal{R}_\sigma : \widetilde{H}^{-1/2}(\partial\Omega) \to \widetilde{H}^{1/2}(\partial\Omega)$, where $\widetilde{H}^s$ spaces consist of H^s functions with mean value zero. We wish to compute the matrix approximation $\mathbf{R}_\sigma$ to $\mathcal{R}_\sigma$. To that end, consider the Neumann problem

$$\nabla \cdot \sigma \nabla u_n = 0 \text{ in } \Omega, \qquad \sigma \frac{\partial u_n}{\partial \nu}\bigg|_{\partial\Omega} = \varphi_n, \qquad (13.7)$$

where the solution u_n is only determined up to an additive constant. We make the solution unique by the requirement $\int_{\partial\Omega} u_n \, ds = 0$. Note that the problem (13.7) is well-defined only for $n \neq 0$ because the Neumann data must satisfy $\int_{\partial\Omega} \sigma \frac{\partial u_n}{\partial \nu} \, ds = 0$. Consequently, we define $\mathbf{R}_\sigma$ as a $2N \times 2N$ matrix and use the basis $\{\varphi_{-N}, \ldots, \varphi_{-1}, \varphi_1, \ldots, \varphi_N\}$ that does not include the constant function $\varphi_0(\theta) = (2\pi)^{-1/2}$.

The first step is to solve the Neumann problem (13.7) for

$$n = -N, -N+1, \ldots, -1, 1, \ldots, N$$

using FEM. Neumann data need to be specified at the centers of edge segments as shown in Figure 13.1(c).

One can then evaluate the trace $u_n|_{\partial\Omega}$ by picking out the values of the solution u_n at the boundary nodes as shown in Figure 13.1(b). (Note that this step is stable, as opposed to the numerical differentiation needed in the computation of the DN matrix in Section

13.2.1.) Set $\mathbf{R}_\sigma = [(\mathbf{R}_\sigma)_{m,n}] = [\widehat{u}_n(m)]$ with

$$\widehat{u}_n(m) = \langle u_n|_{\partial\Omega}, \varphi_m \rangle = \frac{1}{\sqrt{2\pi}} \int_0^{2\pi} u_n|_{\partial\Omega}(\theta) e^{-in\theta} d\theta. \qquad (13.8)$$

As usual, here m is the row index and n is the column index. The integration in (13.8) can be implemented with a simple quadrature using the lengths of boundary segments as weights.

Exercise 13.2.3. *Use the eigenvalues discussed in Section* 13.1 *and the identities identities* (12.20) *and* (12.21) *to compute a matrix representation of* $\mathcal{R}_1$ *in the truncated Fourier basis* $\{\varphi_n : n = -N, \ldots, -1, 1, \ldots, N\}$.

13.2.3 Computing the DN matrix using the ND matrix

Now that we have the ND matrix $\mathbf{R}_\sigma$ available, we are ready to discuss the computation of the DN matrix $\mathbf{L}_\sigma$ using the inverse $\mathbf{R}_\sigma^{-1}$. Note that the electric potential $u_n \in H^1(\Omega)$ is the same in both problems (13.6) and (13.7), and therefore the DN and ND maps are inverses of each other in function spaces, as is seen in the identities (12.20) and (12.21).

Define

$$\mathbf{L}'_\sigma := \mathbf{R}_\sigma^{-1};$$

then $\mathbf{L}'_\sigma$ is a matrix of size $2N \times 2N$. However, the DN matrix $\mathbf{L}_\sigma$ acts on the truncated basis (13.1) with $-N \leq n \leq N$, so it needs to have size $(2N+1) \times (2N+1)$ and possess appropriate mapping properties for constant basis functions at the boundary. More precisely, the DN operator satisfies

$$\Lambda_\sigma 1 = 0, \qquad (13.9)$$

$$\int_{\partial\Omega} \Lambda_\sigma f \, ds = 0 \qquad \text{for all } f \in H^{1/2}(\partial\Omega). \qquad (13.10)$$

Here (13.9) follows from the fact that $u \equiv 1$ is the unique solution of the Dirichlet problem $\nabla \cdot \sigma \nabla u = 0$ in Ω and $u_{\partial\Omega} = 1$. In fact, (13.9) holds for any constant function C, since the unique solution of the conductivity equation with $u|_{\partial\Omega} = C$ is $u = C$. Identity (13.10) states that functions of the form $\Lambda_\sigma f$ have zero constant component; this is a direct consequence of Green's formula and represents physically conservation of charge, as the net current through the boundary must be zero.

We enforce the conditions (13.9) and (13.10) in the matrix approximation $\mathbf{L}_\sigma$ of Λ_σ by adding one zero row and one zero column in the middle of the matrix $\mathbf{L}'_\sigma$; this also gives $\mathbf{L}_\sigma$ the correct size $(2N+1) \times (2N+1)$. Divide the matrix $\mathbf{L}'_\sigma$ into four $N \times N$ blocks named like this:

$$\mathbf{L}'_\sigma = \left[\begin{array}{c|c} \mathbf{L}'_\sigma\{1,1\} & \mathbf{L}'_\sigma\{1,2\} \\ \hline \mathbf{L}'_\sigma\{2,1\} & \mathbf{L}'_\sigma\{2,2\} \end{array} \right]. \qquad (13.11)$$

Now construct $\mathbf{L}_\sigma$ by adding a zero row and a zero column:

$$\mathbf{L}_\sigma = \begin{bmatrix} \mathbf{L}'_\sigma\{1,1\} & \mathbf{0} & \mathbf{L}'_\sigma\{1,2\} \\ & {\scriptstyle N \times 1} & \\ \mathbf{0} & \mathbf{0} & \mathbf{0} \\ {\scriptstyle 1 \times N} & {\scriptstyle 1 \times 1} & {\scriptstyle 1 \times N} \\ \mathbf{L}'_\sigma\{2,1\} & \mathbf{0} & \mathbf{L}'_\sigma\{2,2\} \\ & {\scriptstyle N \times 1} & \end{bmatrix}. \qquad (13.12)$$

The effect of the zero column in (13.12) is that multiplying the vector representing the coefficients of the constant function $u = 1$ expressed in the Fourier basis

$$[\underset{1}{0} \quad \cdots \quad \underset{N}{0} \quad \underset{N+1}{1} \quad \underset{N+2}{0} \quad \cdots \quad \underset{2N+1}{0}]^T$$

by the matrix $\mathbf{L}_\sigma$ on the left gives the zero vector; this is the discrete counterpart of the condition (13.9). The effect of the zero row in (13.12) is that for any vector $\mathbf{f} \in \mathbb{R}^{2N+1}$ the vector $\mathbf{L}_\sigma \mathbf{f}$ has the form

$$[\underset{1}{\bullet} \quad \cdots \quad \underset{N}{\bullet} \quad \underset{N+1}{0} \quad \underset{N+2}{\bullet} \quad \cdots \quad \underset{2N+1}{\bullet}]^T.$$

This ensures the coefficient of ϕ_0 is 0; this is the discrete counterpart of the condition (13.10).

Exercise 13.2.4. *Choose a rotationally symmetric, piecewise constant conductivity σ satisfying the assumptions of Theorem 13.1. Choose $N = 16$ and compute the DN matrix $\mathbf{L}_\sigma$ in three ways. First, determine $\mathbf{L}_\sigma$ analytically using the eigenvalues provided by Theorem 13.1; the result will be used as a ground truth. Second, compute $\mathbf{L}_\sigma$ directly using the method described in Section 13.2.1. Third, compute $\mathcal{R}_\sigma$ using FEM and calculate $\mathbf{L}_\sigma$ as explained above. Which method is more accurate (as compared to the ground truth), the second or the third?*

13.3 Complete electrode model and FEM

To simulate actual voltage measurements from EIT hardware to good precision, a FEM model that takes the electrode effects into account is a very good option. It was to this end that the complete electrode model was developed [82], and we describe an implementation here.

13.3.1 Variational formulation

Recall the boundary conditions for the complete electrode model; given in Section 12.6, with z_ℓ denoting the effective contact impedance between the ℓth electrode and the skin,

$$\int_{e_\ell} \sigma \frac{\partial u}{\partial \nu} ds = I_\ell, \quad \ell = 1, 2, \ldots, L, \tag{13.13}$$

$$\sigma \frac{\partial u}{\partial \nu} = 0 \text{ off } \bigcup_{\ell=1}^{L} e_\ell, \tag{13.14}$$

and Robin boundary condition

$$u + z_\ell \sigma \frac{\partial u}{\partial \nu} = U_\ell \text{ on } e_\ell \text{ for } \ell = 1, 2, \ldots, L, \tag{13.15}$$

with the uniqueness condition specifying a choice of ground

$$\sum_{\ell=1}^{L} U_l = 0, \tag{13.16}$$

and Kirchhoff's law

$$\sum_{\ell=1}^{L} I_l = 0. \tag{13.17}$$

Denoting an electric potential inside Ω by a lowercase u or v, and uppercase letters for the vectors of voltages measured on L electrodes, it was shown in [422] that for any (v, V), where $v \in H^1(\Omega)$ and $V \in \mathbb{C}^L$, the *variational form* of the complete electrode model is

$$B_s((u, U), (v, V)) = \sum_{\ell=1}^{L} I_\ell \bar{V}_\ell, \tag{13.18}$$

where $B_s : H \times H \to \mathbb{C}$ is the *sesquilinear* form given by

$$B_s((u, U), (v, V)) = \int_\Omega \sigma \nabla u \cdot \nabla \bar{v} \, dx dy + \sum_{\ell=1}^{L} \frac{1}{z_\ell} \int_{e_\ell} (u - U_\ell)(\bar{v} - \bar{V}_\ell) dS. \tag{13.19}$$

Following [216, 455] we derive a FEM; discretization of the variational problem. We begin by discretizing our domain Ω into very small tetrahedral elements with N nodes in the mesh. Suppose (u, U) is the solution to the complete electrode model, and that trigonometric current patterns are applied. A finite-dimensional approximation to the voltage distribution inside Ω is

$$u^h(z) = \sum_{k=1}^{N} \alpha_k \varphi_k(z) \tag{13.20}$$

and on the electrodes is

$$U^h(z) = \sum_{k=N+1}^{N+(L-1)} \beta_{(k-N)} \vec{n}_{(k-N)}, \tag{13.21}$$

13.3. Complete electrode model and FEM

where the superscript h indicates the discrete approximation, the functions φ_k comprise a basis for the finite-dimensional space $\mathcal{H} \subset H^1(\Omega)$, α_k and $\beta_{(k-N)}$ are coefficients yet to be determined, and

$$\vec{n}_j = (1, 0, \ldots, 0, -1, 0, \ldots, 0)^T \in \mathbb{R}^{L \times 1},$$

where the -1 is in the $(j+1)$st position.

Notice that this choice of $\vec{n}_{(k-N)}$ satisfies the condition for the ground in (13.16) since substituting the $\vec{n}_{(k-N)}$'s into (13.21) results in

$$U^h(z) = \sum_{k=N+1}^{N+(L-1)} \beta_{(k-N)} \vec{n}_{(k-N)} = \sum_{k=1}^{L-1} \beta_k \vec{n}_k$$

$$= \left(\sum_{k=1}^{L-1} \beta_k, -\beta_1, -\beta_2, \ldots, -\beta_{L-1} \right)^T. \tag{13.22}$$

13.3.2 Finite element approximation

In order to implement the FEM computationally in MATLAB, we first need to expand (13.18) using our approximating functions in equations (13.20) and (13.21) with $v = \varphi_j$ for $j = 1, 2, \ldots, N$ and $V = \vec{n}_j$ for $j = N+1, N+2, \ldots, N+(L-1)$ to construct a linear system

$$A\vec{b} = \vec{f}, \tag{13.23}$$

where $\vec{b} = (\vec{\alpha}, \vec{\beta})^T \in \mathbb{C}^{N+L-1}$ with the vector $\vec{\alpha} = (\alpha_1, \alpha_2, \ldots, \alpha_N)$ and the vector $\vec{\beta} = (\beta_1, \beta_2, \ldots, \beta_{L-1})$, and $A \in \mathbb{C}^{(N+L-1) \times (N+L-1)}$ is of the form

$$A = \begin{pmatrix} B & C \\ \tilde{C} & D \end{pmatrix}. \tag{13.24}$$

We will take cases to construct the block matrices in A. The right-hand-side vector is given by

$$\vec{f} = (\mathbf{0}, \tilde{I})^T, \tag{13.25}$$

where $\mathbf{0} \in \mathbb{C}^{1 \times N}$ and $\tilde{I} = (I_1 - I_2, I_1 - I_3, \ldots, I_1 - I_L) \in \mathbb{C}^{1 \times (L-1)}$. Note that the entries of the vector $\vec{\alpha}$ represent the voltages throughout the domain, while those of $\vec{\beta}$ are used to find the voltages on the electrodes by

$$U^h = \mathcal{C}\vec{\beta}, \tag{13.26}$$

where $\mathcal{C}$ is the $L \times (L-1)$ matrix

$$\mathcal{C} = \begin{pmatrix} 1 & 1 & 1 & \ldots & 1 \\ -1 & 0 & 0 & \ldots & 0 \\ 0 & -1 & 0 & \ldots & 0 \\ & & \ddots & & \\ 0 & 0 & 0 & \ldots & -1 \end{pmatrix}. \tag{13.27}$$

The entries of the block matrix B are determined as follows.

Case 1: $1 \leq k, j \leq N$. Here, $u^h \neq 0$, $U^h = 0$, $v \neq 0$, but $V = 0$ and thus $B_s((u^h, U^h), (v, V))$ becomes

$$B_s((u^h, 0), (v, 0)) = \int_\Omega \sigma \nabla u^h \cdot \nabla \bar{v} dx + \sum_{\ell=1}^{L} \frac{1}{z_\ell} \int_{e_\ell} u^h \bar{v} dS$$

$$= \int_\Omega \sigma \nabla \left(\sum_{k=1}^{N} \alpha_k \varphi_k(z) \right) \cdot \nabla \bar{\varphi}_j dx$$

$$+ \sum_{\ell=1}^{L} \frac{1}{z_\ell} \int_{e_\ell} \left(\sum_{k=1}^{N} \alpha_k \varphi_k(z) \right) \bar{\varphi}_j dS$$

$$= 0 \quad \text{by (13.18)}.$$

Thus, for $1 \leq j \leq N$

$$\int_\Omega \sigma \nabla \left(\sum_{k=1}^{N} \alpha_k \varphi_k \right) \cdot \nabla \bar{\varphi}_j dx + \sum_{\ell=1}^{L} \frac{1}{z_\ell} \int_{e_\ell} \left(\sum_{k=1}^{N} \alpha_k \varphi_k \right) \bar{\varphi}_j dS = 0 \quad (13.28)$$

and so the (k, j)th entry of the block matrix B is

$$B_{kj} = \int_\Omega \sigma \nabla \varphi_k \cdot \nabla \bar{\varphi}_j dx + \sum_{\ell=1}^{L} \frac{1}{z_\ell} \int_{e_\ell} \varphi_k \bar{\varphi}_j dS. \quad (13.29)$$

The entries of the block matrix C are determined as follows.

Case 2: $1 \leq k \leq N$ and $N+1 \leq j \leq N+(L-1)$. Here, $u^h \neq 0$, $U^h = 0$, $v = 0$, and $V \neq 0$ and thus $B_s((u^h, U^h), (v, V))$ becomes

$$B_s((u^h, 0), (0, V)) = -\sum_{\ell=1}^{L} \frac{1}{z_\ell} \int_{e_\ell} u^h \bar{V}_\ell ds$$

$$= -\sum_{\ell=1}^{L} \frac{1}{z_\ell} \int_{e_\ell} \sum_{k=1}^{N} \alpha_k \varphi_k (\vec{n}_j)_\ell ds. \quad (13.30)$$

In addition we have

$$(\vec{n}_j)_\ell = \begin{cases} 1, & \ell = 1, \\ -1, & l = j+1, \\ 0 & \text{otherwise}. \end{cases} \quad (13.31)$$

Then using (13.18) we have

$$B_s((u^h, 0), (0, V)) = \sum_{\ell=1}^{L} I_\ell \bar{V}_\ell = I_1 - I_{j+1},$$

and (13.30) becomes

$$-\sum_{\ell=1}^{L} \frac{1}{z_\ell} \int_{e_\ell} \sum_{k=1}^{N} \alpha_k \varphi_k (\vec{n}_j)_\ell ds = I_1 - I_{j+1} \quad \text{for } N+1 \leq j \leq N+(L-1). \quad (13.32)$$

13.3. Complete electrode model and FEM

Thus, the entries of C are given by

$$C_{kj} = -\sum_{\ell=1}^{L} \frac{1}{z_\ell} \int_{e_\ell} \varphi_k \left(\vec{n}_j\right)_\ell ds$$
$$= -\left[\frac{1}{z_1} \int_{e_1} \varphi_k dS - \frac{1}{z_{j+1}} \int_{e_{j+1}} \varphi_k ds\right]. \quad (13.33)$$

The entries of the block matrix $\tilde{C}$ are determined as follows.

Case 3: $N \leq k \leq N+(L-1)$ **and** $1 \leq j \leq N$. Here, $u^h = 0$, $U^h \neq 0$, $v \neq 0$, and $V = 0$ and thus $B_s((u^h, U^h), (v, V))$ becomes

$$B_s((0, U^h), (v, 0)) = -\sum_{\ell=1}^{L} \frac{1}{z_\ell} \int_{e_\ell} U^h \bar{v}_\ell ds$$
$$= -\sum_{\ell=1}^{L} \frac{1}{z_\ell} \int_{e_\ell} \left(\sum_{k=N+1}^{N+(L-1)} \beta_{(k-N)} \vec{n}_{(k-N)}\right) \bar{\varphi}_j ds$$
$$= 0 \quad \text{by (13.18)}.$$

Thus,

$$-\sum_{\ell=1}^{L} \frac{1}{z_\ell} \int_{e_\ell} \left(\sum_{k=N+1}^{N+(L-1)} \beta_{(k-N)} \vec{n}_{(k-N)}\right)_\ell \bar{\varphi}_j ds = 0, \quad 1 \leq j \leq N, \quad (13.34)$$

and the kjth entry of $\tilde{C}$ is

$$\tilde{C}_{kj} = -\sum_{\ell=1}^{L} \frac{1}{z_\ell} \int_{e_\ell} \left(\varphi_{(k-N)}\right)_\ell \bar{\varphi}_j ds$$
$$= -\left[\frac{1}{z_1} \int_{e_1} \bar{\varphi}_j ds - \frac{1}{z_{j+1}} \int_{e_{j+1}} \bar{\varphi}_{j+1} ds\right]. \quad (13.35)$$

Finally, the entries of the block matrix D are determined as follows.

Case 4: $N \leq k, j \leq N+(L-1)$. Here, $u^h = 0$, $U^h \neq 0$, $v = 0$, and $V \neq 0$ and thus $B_s((u^h, U^h), (v, V))$ becomes

$$B_s((0, U^h), (0, V)) = \sum_{\ell=1}^{L} \frac{1}{z_\ell} \int_{e_\ell} U^h \bar{V}_\ell ds$$
$$= \sum_{\ell=1}^{L} \frac{1}{z_\ell} \int_{e_\ell} \left(\sum_{k=N+1}^{N+(L-1)} \beta_{(k-N)} \vec{n}_{(k-N)}\right) \left(\vec{\bar{n}}_j\right)_\ell ds$$
$$= I_1 - I_{j+1} \quad \text{by (13.18)}.$$

Thus,

$$\sum_{\ell=1}^{L} \frac{1}{z_\ell} \int_{e_\ell} \left(\sum_{k=N+1}^{N+(L-1)} \beta_{(k-N)} \vec{n}_{(k-N)} \right)_\ell (\bar{\vec{n}}_j)_\ell \, ds = I_1 - I_{j+1}, \quad 1 \le j \le N, \qquad (13.36)$$

and

$$D_{kj} = \sum_{\ell=1}^{L} \frac{1}{z_\ell} \int_{e_\ell} \left(\vec{n}_{(k-N)} \right)_\ell (\bar{\vec{n}}_j)_\ell \, ds$$

$$= \sum_{\ell=1}^{L} \frac{1}{z_\ell} \int_{e_\ell} \left(\vec{n}_{(k-N)} \right)_\ell (\vec{n}_j)_\ell \, ds. \qquad (13.37)$$

If $j = k - N$, we have

$$D_{jj} = \sum_{\ell=1}^{L} \frac{1}{z_\ell} \int_{e_\ell} (\vec{n}_j)_\ell (\vec{n}_j)_\ell \, ds$$

$$= \frac{1}{z_1} \int_{e_1} dS + \frac{1}{z_{j+1}} \int_{e_{j+1}} dS$$

$$= \frac{|e_1|}{z_1} + \frac{|e_{j+1}|}{z_{j+1}} dS, \qquad (13.38)$$

where $|e_1|$ and $|e_{j+1}|$ denote the area of the 1st and $(j+1)$st electrodes, respectively. If $j \ne k - N$,

$$D_{kj} = \frac{1}{z_1} \int_{e_1} dS - 0 = \frac{|e_1|}{z_1}.$$

Thus, for $N \le k, j \le N + (L-1)$,

$$D_{kj} = \begin{cases} \frac{|e_1|}{z_1} + \frac{|e_{j+1}|}{z_{j+1}} dS, & j = k - N, \\ \frac{|e_1|}{z_1}, & j \ne k - N. \end{cases} \qquad (13.39)$$

Solving (13.23) gives us the coefficients $\beta_{(k-N)}$ required for the voltages U^h on the electrodes.

13.3.3 Computing the DN matrix

As in Section 13.2.3, we first consider $\mathbf{R}_\sigma$, the ND map, since that is the usual data-collection situation. To compute $\mathbf{R}_\sigma$, we require an orthonormal basis of current patterns, which we will denote here by $\{\phi^n\}_{n=1}^N$, $N < L$, and knowledge of the voltage on the electrodes U_ℓ^n arising from each of these current patterns. Examples of current patterns include those defined in Chapter 12 by (12.40)–(12.43), but note that not all of these patterns are orthonormal (see Exercise 13.3.2). In such a case, it is necessary to normalize both the currents and voltages accordingly.

Now, the matrix approximation $\mathbf{R}_\sigma$ to $\mathcal{R}_\sigma$ has entries

$$\mathbf{R}_\sigma(m,n) = (s_\ell \vec{\phi}^m, \vec{U}^n)_L,$$

where s_ℓ is the arc length of the boundary segment connecting the centers of electrodes ℓ and $\ell+1$. Note that it is essential that the measurements sum to zero in order for Λ and $\mathcal{R}$ to be both right and left inverses. That satisfied, we can define

$$\mathbf{L}_\sigma := [\mathbf{R}_\sigma]^{-1}. \qquad (13.40)$$

Note that $\mathbf{L}_\sigma$ is a matrix of size $N \times N$.

Exercise 13.3.1. *Determine the number of linearly independent current patterns for each of the current patterns defined in Chapter 12 by (12.40), (12.41), (12.42), and (12.43). What is the rank of the DN map for each of these choices of current pattern?*

Exercise 13.3.2. *Of the current patterns studied in Exercise 13.3.1, which are orthogonal? Give the orthonormal formulation of each orthogonal pattern. How must the voltages be normalized correspondingly?*

Exercise 13.3.3. *For the current patterns studied in Exercise 13.3.1 that are not orthogonal, use the Gram–Schmidt algorithm to find an orthonormal formulation of each pattern. How must the voltages be modified to form the ND map?*

Exercise 13.3.4. *The number of degrees of freedom in the inverse problem is equal to the number of linearly independent measurements. Keeping in mind the DN map is self-adjoint, compute the number degrees of freedom for a 32 electrode system when trigonometric current patterns are applied and when "skip 3" current patterns are applied.*

13.4 Adding noise to EIT data matrices

In practice, contributions to the noise come from measurement errors in the input current and measured voltages, from hardware problems, and from modeling errors such as the approximation of an infinite-dimensional operator by a discrete matrix. Bounds for the contributions to the noise from the hardware can often be estimated, but contributions from the modeling are more difficult to quantify, and we do not address that problem here.

We explain one quite useful and flexible way of adding noise to the DN matrix of size $(2N+1) \times (2N+1)$ constructed as in Section 13.2.3. We assume that the ND matrix $\mathbf{R}_\sigma$ has been simulated with high accuracy.

Denote by $\mathcal{E} = [\mathcal{E}_{m,n}]$ an $2N \times 2N$ random matrix whose elements are independent Gaussian random variables with zero mean and unit standard deviation: $\mathcal{E}_{m,n} \sim \mathcal{N}(0,1)$. Set

$$\widetilde{\mathbf{L}}'_\sigma := \mathbf{R}_\sigma^{-1} + c\mathcal{E}, \qquad (13.41)$$

where $c > 0$ is a constant that is adjusted below to obtain the desired relative noise level. Since the DN map is theoretically known to be self-adjoint, we use the additional step

$$\mathbf{L}'_\sigma = \frac{1}{2}(\widetilde{\mathbf{L}}'_\sigma + (\widetilde{\mathbf{L}}'_\sigma)^T). \qquad (13.42)$$

Now construct the self-adjoint simulated noisy data matrix $\mathbf{L}_\sigma^\delta$ by substituting $\mathbf{L}_\sigma'$ into (13.11) and (13.12). Note also that the zero blocks in (13.12) enforce in the noisy data $\mathbf{L}_\sigma^\delta$ the following properties of the DN map: $\Lambda_\sigma 1 = 0$ and $\int_{\partial\Omega} \Lambda_\sigma f \, ds = 0$ for every $f \in H^{1/2}(\partial\Omega)$.

In most cases it is necessary to quantify the noise amplitude. Sometimes this is needed for accurate modeling of the noise level of some actual measurement system, and sometimes it is necessary to study the behavior of a regularized EIT method under controlled decrease of the noise level. Denote by $\mathbf{L}_\sigma$ the noise-free matrix approximation of Λ_σ and define

$$\delta = \|\mathbf{L}_\sigma^\delta - \mathbf{L}_\sigma\|_Y, \qquad (13.43)$$

where $\|\cdot\|_Y$ denotes the norm of the data space Y. Now the noise level δ can be tuned by adjusting the constant c in (13.41) in such a way that δ has the desired value.

Typically in EIT studies we take $Y = \mathcal{L}(H^{1/2}(\partial\Omega), H^{-1/2}(\partial\Omega))$. Let us record the expression for computing those Y-norms for matrices. Denote $\mathbf{E} := \mathbf{L}_\sigma^\delta - \mathbf{L}_\sigma$. Then by formula (C.20) in Appendix C we get

$$\|\mathbf{E}\|_Y = \|[(1+|m|^2)^{-\frac{1}{4}}(1+|n|^2)^{-\frac{1}{4}} \mathbf{E}_{m,n}]\|_{\mathcal{L}(\mathbb{C}^{2N+1})}. \qquad (13.44)$$

The above method of adding noise can be easily modified to cover nonuniform noise sources by allowing different standard deviations for the elements in the noise matrix $\mathcal{E}$.

Chapter 14

Complex geometrical optics solutions

The inverse conductivity problem is very ill-posed, as explained in Chapter 12, and the reader may even have doubts as to whether EIT data uniquely specifies the conductivity or not. We can quickly put these fears to rest—it does, at least for infinite-precision data, but certain stipulations must be made on the conductivity σ and the domain Ω. The history of the proofs of the uniqueness results is, to a large extent, proofs for conductivities with less and less regularity. The crucial technical tools in these works are *complex geometrical optics (CGO) solutions*, sometimes also called exponentially growing solutions.

Having their origins in optics, the complex-valued CGO solutions have exponential growth in certain directions and exponential decay in others. They were first introduced by Faddeev in 1966 [138] and later rediscovered in the context of inverse problems. CGO solutions are a valuable tool both theoretically and computationally since many of the proofs involving them are constructive and lend themselves well to computational algorithms. We give a list of references to important developments in the use of CGO solutions for the inverse conductivity problem in Table 14.1. The list is not meant to be exhaustive as it is focused on the themes in this book. For a thorough survey, see [446].

The rest of this chapter is organized as follows. We discuss Calderón's original analysis of the inverse conductivity problem in Section 14.1. Section 14.2 is devoted to discussion of the D-bar operator, its inverse, and related operators needed in the construction of CGO solutions. In Sections 14.3 and 14.4 we describe numerical computation of the CGO solutions for the Schrödinger equation [350] and for the Beltrami equation [19], respectively. Such computations can be utilized in EIT imaging, as shown below in Chapter 15 and Section 16.3, and will probably find applications in many areas of applied mathematics as they enable the construction of tailored nonlinear Fourier transforms. There is also a third construction available, based on a 2×2 system [59], but we do not discuss that approach in this book. Instead, we refer the reader to [274].

14.1 Calderón's pioneering work

In [64], Calderón makes the important assumption that σ is a constant plus a perturbation $\delta(x)$ and for convenience sets the constant to 1 so that $\sigma(x) = 1 + \delta(x)$. He can then prove that the boundary data uniquely determine the small perturbation δ.

Table 14.1. History of the use of CGO solutions for the inverse conductivity problem (mainly) in dimension two. This collection of references is not intended to be exhaustive.

Analysis, infinite-precision data	Computation, practical data
1980 Calderón [64]	2008 Bikowski & Mueller [48]
Dimension three and higher	
1987 Sylvester & Uhlmann [431]	2008 Boverman, Isaacson, Kao, Saulnier & Newell [55]
1988 Nachman [348]	
1988 R G Novikov [366]	2010 Bikowski, Knudsen & Mueller [47]
	2011 Delbary, Hansen & Knudsen [116]
Regularity $\sigma \in C^2(\Omega)$, $\Omega \subset \mathbb{R}^2$	
1996 Nachman [350]	2000 Siltanen, Mueller & Isaacson [415]
1997 Liu [318]	2003 Mueller & Siltanen [344]
	2004 Knudsen, Mueller & Siltanen [279]
	2004 Isaacson, Mueller, Newell & Siltanen [234]
	2006 Isaacson, Mueller, Newell & Siltanen [233]
	2007 Murphy, Mueller & Newell [347]
	2008 Knudsen, Lassas, Mueller & Siltanen [276]
	2009 Knudsen, Lassas, Mueller & Siltanen [277]
	2009 Murphy & Mueller [346]
	2009 Siltanen & Tamminen [416]
Regularity $\sigma \in C^1(\Omega)$, $\Omega \subset \mathbb{R}^2$	
1997 Brown & Uhlmann [59]	2001 Knudsen & Tamasan [280]
2001 Barceló, Barceló & Ruiz [28]	2003 Knudsen [274]
Complex σ	
2000 Francini [148]	2012 Hamilton, Herrera, Mueller & Von Herrmann [189]
2010 Beretta & Francini [44]	
Regularity $\sigma \in L^\infty(\Omega)$, $\Omega \subset \mathbb{R}^2$	
2003 Astala & Päivärinta [19, 18]	2010 Astala, Mueller, Päivärinta & Siltanen [17]
2005 Astala, Lassas & Päivärinta [15]	2011 Astala, Mueller, Päivärinta, Perämäki & Siltanen [16]
2007 Barceló, Faraco & Ruiz [29]	
2008 Clop, Faraco & Ruiz [87]	

Let us introduce some notation that is used throughout Calderón's paper [64]. Let $L_\sigma = \nabla \cdot (\sigma \nabla \cdot)$, and assume w satisfies the conductivity problem $L_\sigma w = 0$. Then

$$\nabla \cdot (\sigma \nabla w) = 0 \quad \text{in } \Omega \in \mathbb{R}^n, \quad n \geq 2,$$
$$w|_{\partial\Omega} = \Phi.$$

As in (12.23), the quadratic form of the DN map is defined by

$$Q_\sigma(\Phi) = \int_\Omega \sigma |\nabla w|^2 dx.$$

Calderón writes w as $w = u + v$, where

$$\Delta u = 0, \quad \text{in } \Omega,$$
$$u|_{\partial\Omega} = \Phi.$$

14.1. Calderón's pioneering work

Figure 14.1. Alberto Calderón. Left: Calderón as a young man. Right: Calderón at work in his workshop in Buenos Aires later in life. Images courtesy of Alexandra Bellow.

Since $w|_{\partial\Omega} = \Phi$, it is clear that $v|_{\partial\Omega} = 0$ and hence $v \in H_0^1(D)$. Moreover, since $L_\sigma w = 0$, it is straightforward to see that

$$0 = L_{1+\delta} w = L_\delta u + L_1 v + L_\delta v. \tag{14.1}$$

However, one cannot specify here a PDE satisfied by v.

Calderón (see Figure 14.1) proves that the linearized problem is injective by proving that the Fréchet differential of Q_σ about $\sigma = 1$ is injective.

Definition 14.1.1. *A Fréchet differential of a functional, F, is the term dF that satisfies*

$$F(x + y) = F(x) + dF(x; y) + o(\|y\|). \tag{14.2}$$

Denoting the Fréchet differential of Q about $\sigma = 1$ by $dQ_\sigma(\Phi)|_{\sigma=1}$, the linearization of the quadratic form is

$$Q_\sigma(\Phi) = Q_1(\Phi) + dQ_\sigma(\Phi)|_{\sigma=1}(\sigma - 1).$$

Using the linearization of $Q_\sigma(\Phi)$, the difference in quadratic forms is then

$$Q_{\sigma_1}(\Phi) - Q_{\sigma_2}(\Phi) \approx \delta_1 dQ_{\sigma_1}(\Phi)\big|_{\sigma=1} - \delta_2 dQ_{\sigma_2}(\Phi)\big|_{\sigma=1},$$

and so it is sufficient to show that dQ_σ is injective. Calderón shows that if $dQ_{\sigma_1}(\Phi)\big|_{\sigma=1} = dQ_{\sigma_2}(\Phi)\big|_{\sigma=1}$, then $\delta_1 = \delta_2$.

Since we regard Q_σ as a function of $\sigma = 1 + \delta(x)$, in terms of Q_σ, (14.2) becomes

$$Q_{1+\delta}(\Phi) - Q_1(\Phi) - dQ_\sigma(\Phi)|_{\sigma=1} = o(\|\delta\|). \tag{14.3}$$

We must compute $dQ_{\sigma_1}(\Phi)$. In Exercises 14.1.1 and 14.1.2 the reader is asked to verify

$$Q_1(\Phi) = \int_\Omega |\nabla w|^2 dx = \int_\Omega |\nabla u|^2 dx \qquad (14.4)$$

and

$$Q_{1+\delta}(\Phi) = \int_\Omega (1+\delta)|\nabla u|^2 + \delta \nabla u \cdot \nabla v \, dx. \qquad (14.5)$$

Our candidate for the Fréchet differential will be

$$dQ_\sigma(\Phi)|_{\sigma=1} = \int_\Omega \delta |\nabla u|^2 dx. \qquad (14.6)$$

Using expressions (14.4), (14.5), and (14.6) in (14.3) results in

$$\int_\Omega (1+\delta)|\nabla u|^2 + \delta \nabla u \cdot \nabla v \, dx - \int_\Omega |\nabla u|^2 dx - \int_\Omega \delta |\nabla u|^2 dx = \int_\Omega \delta \nabla u \cdot \nabla v \, dx.$$

To show that (14.6) is the Fréchet differential we must verify $\delta \nabla u \cdot \nabla v$ is $o(\|\delta\|)$. For $v \in H_0^1(\Omega)$, equation (14.1) implies

$$0 = \Delta v + \nabla \cdot \delta \nabla v + \nabla \cdot \delta \nabla u.$$

Multiplying by v, integrating over the domain, and integrating by parts yields

$$0 = \int_\Omega (\nabla \cdot \nabla v + \nabla \cdot \delta \nabla v) v \, dx + \int_\Omega (\nabla \cdot \delta \nabla u) v \, dx$$

$$= -\int_\Omega \nabla v \cdot \nabla v + \delta \nabla v \cdot \nabla v \, dx - \int_\Omega \delta \nabla u \cdot \nabla v \, dx$$

$$= -\int_\Omega (1+\delta)|\nabla v|^2 dx - \int_\Omega \delta \nabla u \cdot \nabla v \, dx. \qquad (14.7)$$

For small perturbations such that $\|\delta\|_\infty < 1$ the first term in (14.7) can be bounded below as follows:

$$\left| -\int_\Omega (1+\delta)|\nabla v|^2 dx \right| \geq (1 - \|\delta\|_\infty) \|\nabla v\|_{L^2}^2,$$

and the second term in (14.7) can be bounded above with Hölder's inequality:

$$\left| \int_\Omega \delta \nabla u \cdot \nabla v \, dx \right| \leq \|\delta\|_\infty \|\nabla u\|_{L^2} \|\nabla v\|_{L^2}.$$

From (14.7), these bounds then yield the inequality

$$(1 - \|\delta\|_\infty) \|\nabla v\|_{L^2} \leq \|\delta\|_\infty \|\nabla u\|_{L^2}.$$

Now write δ as $\delta = \epsilon \delta_e$, where ϵ is a scalar and $\|\delta_e\| = 1$ and let $\epsilon \to 0$ in (14.1). Then

$$0 \leq \lim_{\|\delta\| \to 0} \frac{1}{\|\delta\|} \left| \int_\Omega \delta \nabla u \cdot \nabla v \, dx \right| = \lim_{\epsilon \to 0} \left| \int_\Omega \delta_e \nabla u \cdot \nabla v \, dx \right|$$

$$\leq \lim_{\epsilon \to 0} \|\nabla u\|_{L^2} \|\nabla v\|_{L^2}$$

$$\leq \lim_{\epsilon \to 0} \frac{\epsilon \|\nabla u\|_{L^2}^2}{1 - \epsilon}$$

$$= 0.$$

Therefore, $\delta \nabla u \cdot \nabla v$ is $o(\|\delta\|)$, and the Fréchet differential is given by (14.6).

14.2. The $\bar{\partial}$ operator and its kin

Calderón shows the injectivity of $dQ_\sigma|_{\sigma=1}$ by showing an even stronger result: If $\int_\Omega \delta |\nabla u|^2 dx = 0$ for all harmonic functions u, then δ must be zero. To do this, he uses a family of CGO solutions to $\Delta u = 0$ of the form

$$u_1(x) = e^{\pi i(z \cdot x) + \pi(a \cdot x)} \quad \text{and} \quad u_2(x) = e^{\pi i(z \cdot x) - \pi(a \cdot x)}, \tag{14.8}$$

where $a, z \in \mathbb{R}^n$ with $z \cdot a = 0$ and $|z| = |a|$.

Note that if (14.6) vanishes for harmonic functions u_1 and u_2, then

$$0 = \int_D \delta |\nabla (u_1 + u_2)|^2 dx = \int_D \delta (\nabla u_1 + \nabla u_2) \cdot (\nabla u_1 + \nabla u_2) dx$$

$$= \int_D \delta (|\nabla u_1|^2 + 2\nabla u_1 \cdot \nabla u_2 + |\nabla u_2|^2) dx$$

$$= 2 \int_D \delta \nabla u_1 \cdot \nabla u_2 \, dx. \tag{14.9}$$

For the two particular harmonic functions u_1 and u_2 defined in (14.8), (14.9) becomes

$$\int_D \delta \nabla u_1 \cdot \nabla u_2 \, dx = \int_D \delta (\pi i z + \pi a) \cdot (\pi i z - \pi a) e^{2\pi i(z \cdot x)} dx$$

$$= -2\pi^2 |z|^2 \int_D \delta e^{2\pi i(z \cdot x)} dx. \tag{14.10}$$

Equation (14.10) can be zero only if $\delta = 0$ (excluding the trivial case when $z = 0$). Hence if $dQ_\gamma(\Phi)|_{\gamma=1}$ vanishes for all u with $\Delta u = 0$, then δ must be zero and therefore $dQ_\gamma(\Phi)|_{\gamma=1}$ is injective.

Exercise 14.1.1. *Verify that $Q_1(\Phi) = \int_\Omega |\nabla w|^2 dx = \int_\Omega |\nabla u|^2 dx$ as in (14.4).*

Exercise 14.1.2. *Verify that $Q_{1+\delta}(\Phi) = \int_\Omega (1+\delta)|\nabla u|^2 + \delta \nabla u \cdot \nabla v \, dx$ as in (14.5).*

Exercise 14.1.3. *Verify that if $z = |z|e^{i\phi}$, then $a = |z|e^{i(\phi \pm \pi/2)}$. Suppose Ω is a disc of radius R. Show that if $x|_{\partial \Omega} = Re^{i\theta}$, then $\pi(a \cdot x + i(z \cdot x))|_{\partial \Omega} = |z|\pi Re^{\mp i(\theta - \phi)}$. One can then derive a power series expansion for u_1 and u_2:*

$$\sum_{j=0}^\infty a_j(z) e^{\mp ij\theta}. \tag{14.11}$$

Compute the exponentially growing solutions u_1 and u_2 on the boundary of Ω for a vector of z values formed by taking a discrete sequence of $0 \leq \phi < 2\pi$. Plot your results.

14.2 The $\bar{\partial}$ operator and its kin

Definition 14.2.1. *For a complex variable $z = x + iy$ define*

$$\bar{\partial}_z = \frac{1}{2}\left(\frac{\partial}{\partial x} + i\frac{\partial}{\partial y}\right), \quad \partial_z = \frac{1}{2}\left(\frac{\partial}{\partial x} - i\frac{\partial}{\partial y}\right).$$

It can be seen from the Cauchy–Euler equations that a function $f(z,\bar{z})$ is analytic if and only if $\bar{\partial}_z f = 0$ and so the $\bar{\partial}$ derivative of a function is a measure of its departure from analyticity.

The calculus of the $\bar{\partial}$ and ∂ operators is straightforward. The product and quotient rules hold as usual. The complex chain rule takes the form

$$\bar{\partial}_z(f \circ F) = ((\partial_z f) \circ F) \cdot \bar{\partial}_z F + ((\bar{\partial}_z f) \circ F) \cdot \overline{\partial_z F}.$$

Denote the outward normal vector to $\partial\Omega$ at $z = (x,y)$ by $\nu = (\nu_1(z), \nu_2(z))$, or as a complex number, $\nu(z) = \nu_1(z) + i\nu_2(z)$. Then the divergence theorem in $\mathbb{R}^2$ states that

$$\int_\Omega \frac{\partial}{\partial x_j} u\, dx_1 dx_2 = \int_{\partial\Omega} u\nu_j\, ds, \tag{14.12}$$

where ds denotes the arc length measure on the boundary curve $\partial\Omega$.

The reader can use (14.12) to verify that integration by parts holds as follows. Let Ω be a bounded domain in $\mathbb{R}^2$ with boundary $\partial\Omega$ that forms a simple closed contour C oriented counterclockwise. The tangent vector τ to $\partial\Omega$ at z is given by $\tau(z) = (-\nu_2(z), \nu_1(z))$ so that $\nu \cdot \tau = 0$, or $\tau(z) = -\nu_2(z) + i\nu_1(z)$. Then for functions $u(x,y)$ and $w(x,y)$ defined in the closure $\bar{\Omega}$ of Ω,

$$\int_\Omega u\bar{\partial}w\, dxdy = \frac{1}{2}\int_\Omega u(\partial_x w + i\partial_y w)\, dxdy$$
$$= \frac{1}{2}\int_{\partial\Omega} wu(\nu_1 + i\nu_2)\, ds - \int_\Omega w\bar{\partial}u\, dxdy \tag{14.13}$$

and

$$\int_\Omega u\partial w\, dxdy = \frac{1}{2}\int_\Omega u(\partial_x w - i\partial_y w)\, dx dy$$
$$= \frac{1}{2}\int_{\partial\Omega} wu(\nu_1 - i\nu_2)\, ds - \int_\Omega w\partial u\, dxdy$$
$$= \frac{i}{2}\int_{\partial\Omega} wu\bar{\tau}\, ds - \int_\Omega w\partial u\, dxdy. \tag{14.14}$$

Integrals in the complex plane are sometimes expressed with the differential as an iterated integral $dxdy$ and sometimes as a wedge product $dz \wedge d\bar{z}$. We will use the former notation, but the connection is given by

$$dz \wedge d\bar{z} = (dx + idy) \wedge (dx - idy) = -2i\, dxdy,$$

where we used the identities $dx \wedge dx = 0 = dy \wedge dy$ and $dx \wedge dy = -dy \wedge dx$.

For a continuous function f of a complex variable $z = x + iy$, the solution to the $\bar{\partial}$ equation $\bar{\partial}_z u(z,\bar{z}) = f(z,\bar{z})$ on Ω is given by

$$u(z) = \frac{1}{2\pi i}\int_{\partial\Omega} \frac{u(\zeta)}{\zeta - z}d\zeta - \frac{1}{\pi}\int_\Omega \frac{f(\zeta)}{\zeta - z}d\zeta_1 d\zeta_2. \tag{14.15}$$

This is called the generalized Cauchy integral formula, the Cauchy–Pompeiu formula, or Bochner–Martinelli formula. See, for example, [1] for a proof. The complex delta function

14.3. CGO solutions for the Schrödinger equation

$\delta(z)$ is defined as a distribution over $\mathbb{C}$ and possesses the same property as the real Dirac delta function:

$$\int_{\mathbb{C}} \delta(z-z_0) f(z) \, dx \, dy = f(z_0).$$

Then one can show

$$F(z_0) = \frac{1}{\pi} \int_{\Omega} F(z) \overline{\partial}_z \left(\frac{1}{z-z_0} \right) dx \, dy, \qquad (14.16)$$

and so $\frac{1}{\pi z}$ is a fundamental solution for the $\overline{\partial}_z$ operator.

Note that the Laplacian $\Delta = \partial_{xx} + \partial_{yy}$ can be expressed in terms of $\overline{\partial}_z$ and ∂_z by $\Delta = 4\overline{\partial}_z \partial_z = 4\partial_z \overline{\partial}_z$, where $z = x + iy$.

Exercise 14.2.1. *Show that the unique solution of the equation $\overline{\partial}_z u(z) = f(z)$ with the asymptotic condition $\lim_{|z| \to \infty} u(z) = 1$ is given by*

$$u(z) = 1 - \frac{1}{\pi} \int_{\Omega} \frac{f(\zeta)}{\zeta - z} d\zeta_1 d\zeta_2.$$

14.3 CGO solutions for the Schrödinger equation

Nachman's uniqueness proof [350] for the two-dimensional inverse conductivity problem with twice differentiable conductivities is constructive. That is, the proof provides equations for computing $\sigma(z)$ directly from the data Λ_σ via a nonlinear Fourier transform known as the scattering transform. This nonlinear Fourier analysis is based on the use of certain CGO solutions for a Schrödinger equation, whose definition and numerical computation we explain here. These are actually the first CGO solutions ever to be computed numerically; see [413] for the original implementation. We explain below a more efficient algorithm.

Apart from the D-bar method for EIT, the computational CGO solutions have been used in inversion methods for electrical inclusion detection [227, 226, 220, 447, 219] and the Cauchy problem [229]. The solutions and the corresponding nonlinear Fourier transform are useful also outside the context of inverse boundary-value problems. They are related for example to the Novikov–Veselov equation, which is a (2+1)-dimensional analogue of the celebrated Korteweg–de Vries equation; see [459, 367, 51, 442, 443, 179, 300, 301, 302].

14.3.1 The generalized Lippmann–Schwinger equation

Let $\sigma \in C^2(\Omega)$. Assume that $0 < c \leq \sigma(z)$ for all $z \in \Omega$ and that $\sigma(z) \equiv 1$ for all z in a neighborhood of the boundary $\partial \Omega$. Define

$$q(z) = \frac{\Delta \sqrt{\sigma(z)}}{\sqrt{\sigma(z)}}. \qquad (14.17)$$

The differentiability and positivity assumptions on σ ensure that (14.17) makes sense. The assumption that σ is constant in a neighborhood of $\partial \Omega$ also implies $q = 0$ in a neighborhood of $\partial \Omega$, and so we can smoothly extend $\sigma \equiv 1$ and $q \equiv 0$ outside Ω.

Now the change of variables $\widetilde{u} = \sigma^{1/2} u$ transforms the conductivity equation $\nabla \cdot \sigma \nabla u = 0$ into the Schrödinger equation

$$(-\Delta + q(z))\widetilde{u}(z) = 0, \quad z \in \mathbb{R}^2,$$

and due to the positivity of σ we can also write $u = \sigma^{-1/2}\widetilde{u}$.

To define the CGO solutions we introduce a complex parameter k and look for solutions $\psi(z,k)$ of the Schrödinger equation

$$(-\Delta + q)\psi(\cdot,k) = 0 \tag{14.18}$$

satisfying the asymptotic condition

$$e^{-ikz}\psi(z,k) - 1 \in W^{1,\tilde{p}}(\mathbb{R}^2) \tag{14.19}$$

for any $2 < \tilde{p} < \infty$.

Note that we will denote a point in the plane by z, writing $z = x + iy$ or $z = (x,y)$ depending upon the context; for example, we have $e^{ikz} = \exp(i(k_1 + ik_2)(x + iy))$ in (14.19).

The space $W^{1,\tilde{p}}(\mathbb{R}^2)$ above is the Sobolev space consisting of $L^{\tilde{p}}(\mathbb{R}^2)$ functions whose weak derivatives belong to $L^{\tilde{p}}(\mathbb{R}^2)$ as well. Note that by the Sobolev imbedding theorem $W^{1,\tilde{p}}(\mathbb{R}^2)$ functions are bounded and continuous [4]. By [350, Theorem 1.1] for any $k \in \mathbb{C} \setminus 0$ there is a unique solution ψ of (14.18) satisfying (14.19).

Let us discuss the construction of the solutions ψ. Note that $\psi(z,k)$ is asymptotic to the exponentially growing function e^{ikz} in the sense of (14.19). Therefore it is convenient to define a bounded function $m(z,k)$ by

$$m(z,k) := e^{-ikz}\psi(z,k), \tag{14.20}$$

and construct m. Then we get ψ from the formula $\psi(z,k) = e^{ikz}m(z,k)$.

We derive a PDE for m. Calculate using (14.18) and (14.20)

$$\begin{aligned} q(z)e^{ikz}m(z,k) &= \Delta\psi(z,k) \\ &= 4\partial\overline{\partial}(e^{ikz}m(z,k)) \\ &= 4\partial(e^{ikz}\overline{\partial}m(z,k)) \\ &= 4(ike^{ikz}\overline{\partial}m(z,k) + e^{ikz}\partial\overline{\partial}m(z,k)) \\ &= e^{ikz}(4ik\overline{\partial} + \Delta)m(z,k). \end{aligned}$$

So, $m(z,k)$ satisfies

$$(-\Delta - 4ik\overline{\partial} + q(z))m(z,k) = 0. \tag{14.21}$$

We need to construct a solution of (14.21) satisfying the asymptotic condition $m(z,k) - 1 \in W^{1,\tilde{p}}(\mathbb{R}^2)$.

Now assume we are given a fundamental solution g_k satisfying

$$(-\Delta - 4ik\overline{\partial})g_k(z) = \delta(z). \tag{14.22}$$

Then it is easy to see that a solution of the Lippmann–Schwinger-type

$$m = 1 - g_k * (qm) \tag{14.23}$$

14.3. CGO solutions for the Schrödinger equation

yields a (formal) solution to (14.21) with the desired asymptotic condition; just apply $-\Delta - 4ik\bar{\partial}$ to both sides of (14.23). As explained below, equation (14.23) is Fredholm equation of the second kind in the space $W^{1,\tilde{p}}(\mathbb{R}^2)$. The convolution in (14.23) is defined by

$$(h * f)(z) := \int_{\mathbb{R}^2} h(z-w) f(w) \, dw_1 dw_2. \tag{14.24}$$

Rigorous solvability analysis for (14.23) involves several steps; details can be found in [349, 350]. First, multiplication by $q \in L^p(\mathbb{R}^2)$ is a compact linear operator from $W^{1,\tilde{p}}(\mathbb{R}^2)$ to $L^p(\mathbb{R}^2)$; the proof is based on the boundedness of $W^{1,\tilde{p}}(\mathbb{R}^2)$ functions. Here p is related to $\tilde{p}$ via $1/\tilde{p} = 1/p - 1/2$. Second, the convolution operator $g_k *$ is bounded from $L^p(\mathbb{R}^2)$ to $W^{1,\tilde{p}}(\mathbb{R}^2)$. Third, a solution to (14.23) is given by the formula

$$m - 1 = [I + g_k * (q \cdot)]^{-1} (g_k * q), \tag{14.25}$$

provided that the Fredholm operator $[I + g_k * (q \cdot)]$ is invertible in the space $W^{1,\tilde{p}}(\mathbb{R}^2)$. Nonzero complex numbers k for which invertibility fails are called exceptional points; it is one of the breakthroughs of Nachman's article [350] to show that there are no exceptional points for potentials of the form (14.17).

We discuss the definition and properties of the fundamental solution $g_k(z)$ in Section 14.3.2.

Exercise 14.3.1. Fix $k \in \mathbb{C} \setminus 0$. In what part of the z-plane does the inequality $|e^{ikz}| > 1$ hold?

Exercise 14.3.2. Assume that $\nabla \cdot \sigma \nabla u = 0$. Show that the function $\tilde{u} = \sigma^{1/2} u$ satisfies the Schrödinger equation $(-\Delta + q(z))\tilde{u}(z) = 0$.

Exercise 14.3.3. Show that the function m defined by (14.25) is a solution of (14.23). Furthermore, show that m is also a solution of the PDE (14.21) satisfying $m(\cdot, k) - 1 \in W^{1,\tilde{p}}(\mathbb{R}^2)$. You can make use of the mapping properties mentioned in the text above.

14.3.2 Faddeev Green's function in dimension two

A special family of Green's functions for the Laplacian plays an important role in the D-bar method and in the definition of the nonlinear Fourier transform. The family is constructed using the fundamental solution g_k appearing in (14.22).

A fundamental solution for the operator $-\Delta - 4ik\bar{\partial}$ can be derived using Fourier transforms. We use the following definition of two-dimensional Fourier transform and its inverse:

$$\widehat{f}(\xi) = (\mathcal{F} f)(\xi) = \int_{\mathbb{R}^2} e^{-iz \cdot \xi} f(z) \, dx dy,$$

$$f(z) = (\mathcal{F}^{-1} \widehat{f})(z) = \frac{1}{(2\pi)^2} \int_{\mathbb{R}^2} e^{iz \cdot \xi} \widehat{f}(\xi) \, d\xi_1 d\xi_2.$$

Here $\xi = (\xi_1, \xi_2) \in \mathbb{R}^2$ and $z \cdot \xi = x\xi_1 + y\xi_2$.

Let us compute formally the Fourier symbol of the partial differential operator $-\Delta - 4ik\bar{\partial}$:

$$\mathcal{F}(-\Delta f(z,k) - 4ik\bar{\partial}f(z,k))(\xi)$$
$$= \left(|\xi|^2 - 4ik\frac{1}{2}(i\xi_1 + i^2\xi_2)\right)\widehat{f}(\xi,k)$$
$$= \left(|\xi|^2 + 2k(\xi_1 + i\xi_2)\right)\widehat{f}(\xi,k)$$
$$= \left(|\xi|^2 + 2k\xi\right)\widehat{f}(\xi,k)$$
$$\equiv P(\xi)\widehat{f}(\xi,k).$$

Now we can define g_k as a formal integral,

$$g_k(z) \equiv \mathcal{F}^{-1}\left(\frac{1}{P(\xi)}\right)(z) = \frac{1}{4\pi^2}\int_{\mathbb{R}^2}\frac{e^{iz\cdot\xi}}{\xi(\bar{\xi}+2k)}d\xi_1 d\xi_2, \qquad (14.26)$$

where we abuse notation by writing $\xi(\bar{\xi}+2k) = (\xi_1 + i\xi_2)(\xi_1 - i\xi_2 + 2k)$. The integral in (14.26) is formal because $|\xi(\bar{\xi}+2k)|^{-1}$ is not integrable over $\mathbb{R}^2$ due to large $|\xi|$ asymptotics of the form $|\xi|^{-2}$. On the other hand, interpreting (14.26) using $L^2(\mathbb{R}^2)$ properties of the Fourier transform is not possible either since the two poles of the integrand at $\xi = 0$ and $\xi = -2\bar{k}$ are not square integrable.

However, formula (14.26) can be interpreted in the sense of tempered distributions. Alternatively, noting that the integrand belongs to $L^p(\mathbb{R}^2)$ for $1 < p < 2$, we can use the interpolation theorem of Riesz and Thorin to see that $g_k \in L^{p'}(\mathbb{R}^2)$ with $1/p' + 1/p = 1$. See [413] for more details.

Coordinate changes in (14.26) give the following symmetries:

$$g_k(z) = g_1(kz), \qquad (14.27)$$
$$g_k(z) = \overline{g_{\bar{k}}(-\bar{z})}, \qquad (14.28)$$
$$g_k(z) = e_{-k}(z)\overline{g_k(z)}. \qquad (14.29)$$

The proof of equalities (14.27)–(14.29) is left as an exercise. In particular, (14.27) shows that for numerical evaluation of $g_k(z)$ for arbitrary $k \neq 0$ and $z \neq 0$ it is enough to design an algorithm for $g_1(z)$ for $z \neq 0$.

Several quite complicated algorithms for evaluating $g_1(z)$ were introduced in [415, 413, 229]. However, modern mathematical software, including MATLAB, provides efficient implementations of the exponential-integral function Ei(z). Then one can make use of [51, formula (3.10)] and compute $g_1(z)$ as follows:[2]

$$g_1(z) = \frac{1}{4\pi}e^{-iz}\text{Re}(\text{Ei}(iz)). \qquad (14.30)$$

See Figure 14.2 for a plot of the fundamental solution $g_1(x)$.

Define the Faddeev Green's function by

$$G_k(z) := e^{ikz}g_k(z). \qquad (14.31)$$

---

[2]Thanks to Allan Perämäki for pointing this out!

14.3. CGO solutions for the Schrödinger equation

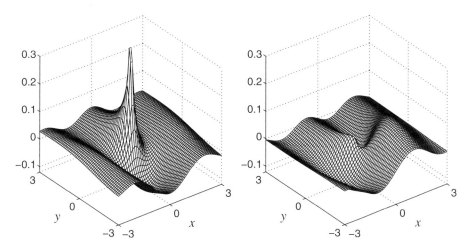

Figure 14.2. Plot of the fundamental solution $g_1(z)$ in the square defined by $-3 \leq x \leq 3$ and $-3 \leq y \leq 3$, where $z = x + iy$. Left: real part of $g_1(z)$. Note the logarithmic singularity of $\text{Re}(g_1(z))$ at $z = 0$. Right: imaginary part of $g_1(z)$. The symmetry $g_1(z) = \overline{g_1(-\bar{z})}$ holds by formula (14.28), so $\text{Im}(g_1(iy)) = 0$ for all $y \in \mathbb{R}$. Hence $\text{Im}(g_1(z))$ is continuous in the whole z-plane.

Note that

$$\begin{aligned}
-\Delta G_k(z) &= -4\partial\bar{\partial}(e^{ikz}g_k(z)) \\
&= -4\bar{\partial}(ike^{ikz}g_k(z) + e^{ikz}\partial g_k(z)) \\
&= e^{ikz}(-4ik\bar{\partial}g_k(z) - 4\bar{\partial}\partial g_k(z)) \\
&= e^{ikz}(-\Delta - 4ik\bar{\partial})g_k(z) \\
&= \delta(z).
\end{aligned}$$

Recall that the standard Green's function for $-\Delta$ is

$$G_0(z) = -\frac{1}{2\pi}\log|z|.$$

We can write $G_k(z) = G_0(z) + H_k(z)$, where $H_k(z)$ is harmonic since $\Delta H_k = -\delta + \delta = 0$, and by Weyl's lemma H_k is a smooth harmonic function. Plots of $G_0(z)$ and $G_1(z)$ are found in Figure 14.3. The functions $G_k(z)$ and $H_k(z)$ are real-valued for all $k \in \mathbb{C} \setminus 0$. This is seen by computing

$$\overline{G_k(z)} = e^{-i\bar{k}\bar{z}}\overline{g_k(z)} = e^{-i\bar{k}\bar{z}}e_k(z)g_k(z) = e^{ikz}g_k(z) = G_k(z).$$

Let us study the behavior of $g_k(z)$ and $G_k(z)$ for z and k near zero. The computation

$$H_k(z) = e^{ikz}g_1(kz) + \frac{1}{2\pi}\log|k||z| - \frac{1}{2\pi}\log|k| = H_1(kz) - \frac{1}{2\pi}\log|k| \quad (14.32)$$

shows that

$$G_k(z) = -\frac{1}{2\pi}\log|z| + H_k(z) = -\frac{1}{2\pi}\log|z| + H_1(kz) - \frac{1}{2\pi}\log|k|.$$

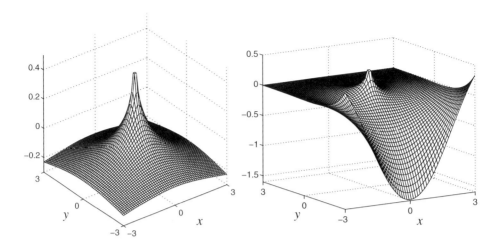

Figure 14.3. Left: plot of the standard Green's function $G_0(z)$ in the square defined by $-3 \leq x \leq 3$ and $-3 \leq y \leq 3$, where $z = x + iy$. Right: Faddeev Green's function $G_1(z)$. Both functions are real-valued.

Therefore, both $G_k(z)$ and $g_k(z) = e^{-ikz} G_k(z)$ have logarithmic singularities with respect to z at $z = 0$ and with respect to k at $k = 0$.

Exercise 14.3.4. *Prove the symmetry properties* (14.27)–(14.29) *of Faddeev's fundamental solution* $g_k(z)$.

14.3.3 Reduction to a periodic integral equation

We construct a numerical solution method for the generalized Lippmann–Schwinger equation (14.23). That equation is defined in the whole plane $\mathbb{R}^2$, however, so some sort of truncation is needed for producing a computer program manipulating finite quantities. We use a periodization technique introduced by Gennadi Vainikko in [449]; see also [404, Section 10.5]. The adaptation of Vainikko's method to (14.23) was first introduced in [344].

We assume here that Ω is the unit disc and $\sigma \equiv 1$ in a neighborhood of $\partial \Omega$. Then the continuous potential $q : \mathbb{R}^2 \to \mathbb{R}$ defined by (14.17) is supported in $\Omega = B(0,1)$. Take $s > 2$ and define a square $Q \subset \mathbb{R}^2$ by

$$Q := \{(x, y) \in \mathbb{R}^2 : -s \leq x < s, \, -s \leq y < s\}. \tag{14.33}$$

We consider tiling the plane by translated copies of Q and work with $2s$-periodic functions $f : \mathbb{R}^2 \to \mathbb{C}$ satisfying

$$\widetilde{f}(x + 2j_1 s, y + 2j_2 s) = \widetilde{f}(x, y) \qquad \text{for } j_1, j_2 \in \mathbb{Z}. \tag{14.34}$$

We indicate $2s$-periodic functions adding $\widetilde{}$ on top of symbols. One way to think about the periodic setting is to consider functions defined on the torus $\mathbb{T}^2$.

Choose a smooth cutoff function η satisfying

$$\eta(z) = \begin{cases} 1 & \text{for } |z| \leq 2, \\ 0 & \text{for } |z| \geq 2 + (s-2)/2, \end{cases} \tag{14.35}$$

14.3. CGO solutions for the Schrödinger equation

and $0 \leq \eta(z) \leq 1$ for all $z \in \mathbb{C}$. Define a $2s$-periodic approximate Green's function $\widetilde{g}_k$ by setting it to $\eta(z)g_k(z)$ inside Q and extending periodically:

$$\widetilde{g}_k(x + 2j_1 s, y + 2j_2 s) = \eta(x,y) g_k(x,y) \tag{14.36}$$

for $(x,y) \in Q \setminus 0$ and $j_1, j_2 \in \mathbb{Z}$. The periodic counterpart of the Lippmann–Schwinger-type equation (14.23) is

$$\widetilde{m} = 1 - \widetilde{g}_k \widetilde{*} (q\widetilde{m}), \tag{14.37}$$

where $\widetilde{*}$ denotes convolution on the torus:

$$(\widetilde{f} \widetilde{*} \widetilde{h})(z) = \int_Q \widetilde{f}(z-w) \widetilde{h}(w) \, dw_1 dw_2. \tag{14.38}$$

The motivation behind introducing (14.37) is that it can be solved numerically in an effective way and that

$$\widetilde{m}(z,k) = m(z,k) \quad \text{for } |z| < 1. \tag{14.39}$$

The proof of the crucial identity (14.39) is left as an exercise. Note that if we know $m(z)$ for $|z| < 1$, we can substitute it to the right-hand side of (14.23) and obtain $m(z,k)$ for all $z \in \mathbb{C}$ from the left-hand side of (14.23).

We next discuss the solvability of the periodic equation (14.37). Fix a complex parameter $k \neq 0$. The linear operator $I + \widetilde{g}_k \widetilde{*} (q \cdot) : W^{1,\widetilde{p}}(\mathbb{T}^2) \to W^{1,\widetilde{p}}(\mathbb{T}^2)$ can be seen to be of the form "identity + compact" using an argument similar to what was used in the nonperiodic case. Therefore, by the Fredholm alternative it is enough to show that $I + \widetilde{g}_k \widetilde{*} (q \cdot)$ is one-to-one. The proof of this fact is left as an exercise.

Exercise 14.3.5. *Show the validity of (14.39). Hint: the Green's functions $g_k(z)$ and $\widetilde{g}_k(z)$ coincide for $|z| \leq 2$, and $|z - w| \leq 2$ if $|z| \leq 1$ and $|w| \leq 1$.*

Exercise 14.3.6. *Assume that $\widetilde{h} \in W^{1,\widetilde{p}}(\mathbb{T}^2)$ satisfies $\widetilde{h} = -\widetilde{g}_k \widetilde{*} (q\widetilde{h})$. Define a nonperiodic function by formula $h = -g_k * (q\widetilde{h}|_\Omega)$ and show that $h \in W^{1,\widetilde{p}}(\mathbb{R}^2)$. Furthermore, prove that $[I + g_k * (q \cdot)]h = 0$. Then the invertibility of $[I + g_k * (q \cdot)]$ implies $h = 0$. Finally, show that $\widetilde{h}|_\Omega = h|_\Omega = 0$.*

14.3.4 Fast solver for the periodic equation

Here we explain how to solve (14.37) numerically following the ideas in [449]. Write (14.37) in the form

$$[I + \widetilde{g}_k \widetilde{*} (q \cdot)] \widetilde{m} = 1. \tag{14.40}$$

Let $s > 2$ and define the square $Q \subset \mathbb{R}^2$ by (14.33). Choose a positive integer $m > 0$, denote $M = 2^m$, and set $h = 2s/M$. Define an integer grid $\mathbb{Z}_m^2 \subset \mathbb{Z}^2$ by

$$\mathbb{Z}_m^2 = \{(j_1, j_2) \in \mathbb{Z}^2 \mid -2^{m-1} \leq j_1 < 2^{m-1}, \; -2^{m-1} \leq j_2 < 2^{m-1}\} \tag{14.41}$$

and a computational grid $\mathcal{G}_m \subset Q$ by

$$\mathcal{G}_m = \{jh \mid j \in \mathbb{Z}_m^2\}. \tag{14.42}$$

Note that the number of points in $\mathcal{G}_m$ is M^2. See Figure 14.4 for an illustration of the 4×4 grid $\mathcal{G}_2$.

Let $\widetilde{\varphi}$ be a $2s$-periodic complex-valued function in the sense of (14.34), and fix a grid $\mathcal{G}_m$ with some $m > 0$. We define a grid approximation of $\widetilde{\varphi}$ based on representing the values

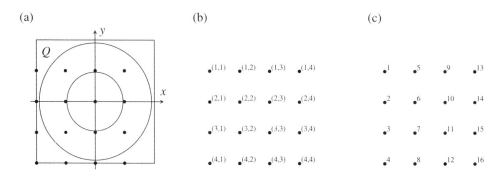

Figure 14.4. (a) Computational 4×4 grid $\mathcal{G}_2$, drawn as black dots, in the square Q given by (14.33). The two circles shown have radii 1 and 2. (b) Matrix-style numbering of the grid points. (c) Vector-style numbering of the grid points.

of $\widetilde{\varphi}$ at the grid points as a complex $M \times M$ matrix. The matrix-style grid approximation $\widetilde{\varphi}_h^M \in \mathbb{C}^{M \times M}$ is defined by the formula

$$\widetilde{\varphi}_h^M(\ell_1, \ell_2) = \widetilde{\varphi}\Big(h(-2^{m-1} - 1 + \ell_1), h(2^{m-1} - \ell_2)\Big), \qquad (14.43)$$

where $1 \leq \ell_1 \leq M$ is row index and $1 \leq \ell_2 \leq M$ is column index. This indexing is illustrated in Figure 14.4(b).

We use the fast Fourier transform (FFT) to implement the convolution operation appearing in (14.37). The idea is to compute the periodic convolution on the torus as multiplication in the frequency domain. We represent the periodic approximate fundamental solution $\widetilde{g}_k$ defined in (14.36) using the matrix-style grid approximation $(\widetilde{g}_k)_h^M$ given by (14.43). There is one difference in the grid approximation of $\widetilde{g}_k$ compared to using formula (14.43) for smooth functions: we substitute zero as the element of the matrix $(\widetilde{g}_k)_h^M$ corresponding to $z = 0$, where $\widetilde{g}_k(z)$ has logarithmic singularity.

We are interested in the approximate numerical solution of the linear equation (14.40) using a matrix-free iterative method as discussed in Appendix D. To this end, we introduce a variant of the grid approximation based on vector representation of the unknown and define two functions, "mat" and "vec," for moving between the two discrete representations. The only difference between the two representations is the numbering of the very same grid points: one uses two indices (see Figure 14.4(b)) and the other uses only one index (see Figure 14.4(c)).

Let $\mathbf{f} = [\mathbf{f}(1), \mathbf{f}(2), \mathbf{f}(3), \ldots, \mathbf{f}(M^2)]^T \in \mathbb{C}^{M^2}$ be an arbitrary vertical vector with complex-valued elements. Define the function mat : $\mathbb{C}^{M^2} \to \mathbb{C}^{M \times M}$ by the formula

$$(\mathrm{mat}(\mathbf{f}))(\ell_1, \ell_2) = \mathbf{f}((\ell_2 - 1)M + \ell_1).$$

Also, define the function vec : $\mathbb{C}^{M \times M} \to \mathbb{C}^{M^2}$ uniquely by the requirement $\mathrm{vec}(\mathrm{mat}(\mathbf{f})) = \mathbf{f}$ for all vectors $\mathbf{f} \in \mathbb{C}^{M^2}$.

Now we can compute $\widetilde{m}(\cdot, k)_h^M$ as the solution of (14.40) using the following algorithm:

1. Construct initial guess $\mathbf{m}_0 \in \mathbb{C}^{M^2}$. A suitable choice is $\mathbf{m}_0 = \mathbf{1}$, the vertical vector with all M^2 elements equal to 1.

14.3. CGO solutions for the Schrödinger equation

2. Call GMRES with initial guess $\mathbf{m}_0$ and the linear operator in the left-hand side of (14.40) described by the routine

$$\mathbf{f} \mapsto \mathbf{f} + \text{vec}\left(h^2 \, \text{IFFT}\left(\text{FFT}((\widetilde{g}_k)_h^M) \cdot \text{FFT}(q_h^M \cdot \text{mat}(\mathbf{f}))\right)\right),$$

where $\cdot$ denotes elementwise multiplication of matrices. See Figure 14.5. Denote the output vector of GMRES by $\mathbf{m} \in \mathbb{C}^{M^2}$.

3. Obtain solution of (14.40) by $\widetilde{m}(\cdot, k)_h^M = \text{mat}(\mathbf{m})$.

Note that the linear operator in the left-hand side of (14.40) is not constructed explicitly as a matrix. This trick results in the above method having computational complexity $M^2 \log M$ when the solutions are computed on an $M \times M$ grid. In contrast, the first numerical CGO solver [413] has complexity M^4.

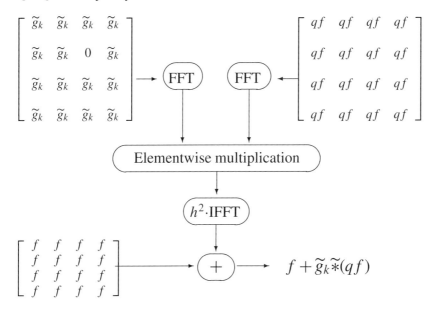

Figure 14.5. Schematic illustration of the computational routine passed on to the GMRES solver. The grid in this simple example is the 4×4 grid shown in Figure 14.4. It is crucial that q is supported in the unit disc. This is only the matrix-based part of the routine; the necessary switching between vector and matrix representations is not shown.

Finally we have

$$m(z_{(\ell_1, \ell_2)}, k) \approx \widetilde{m}_h^M(\ell_1, \ell_2) \tag{14.44}$$

for all grid points $z_{(\ell_1, \ell_2)} \in \mathcal{G}_m$ satisfying $|z_{(\ell_1, \ell_2)}| < 1$.

14.3.5 Numerical examples

Let us demonstrate the above methods with a simple example. We take a strictly positive conductivity $\sigma : \Omega \to \mathbb{R}$ modeling roughly a cross-section of the human chest. The heart, filled with blood, is more conductive than background, whereas the lungs contain air and

214 Chapter 14. Complex geometrical optics solutions

are less conductive. The background conductivity is taken to be one, and in particular $\sigma(z) \equiv 1$ for z near the boundary $\partial \Omega$. The various features in σ are constructed using polynomials and ellipses in such a way that $\sigma \in C^2(\overline{\Omega})$. We continue σ smoothly as one outside Ω. See Figure 14.6 for a plot of the conductivity.

We construct the potential q corresponding to σ using formula (14.17) and implementing the Laplace operator numerically by finite differences. See Figure 14.7 for a plot of the potential.

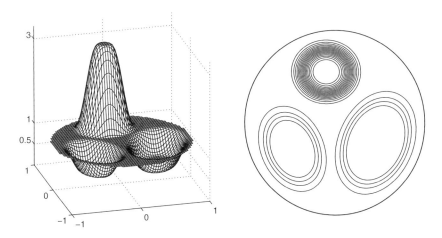

Figure 14.6. Smooth heart-and-lungs conductivity (simulated phantom). Left: three-dimensional mesh plot of $\sigma(x, y)$. Right: contour plot of $\sigma(x, y)$ inside the unit disc. See Figure 14.7 for a plot of the corresponding Schrödinger potential.

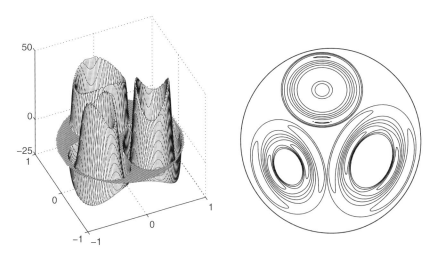

Figure 14.7. Potential corresponding to the smooth heart-and-lungs conductivity shown in Figure 14.6. Left: three-dimensional mesh plot of $q(x, y)$. Right: contour plot of $q(x, y)$ inside the unit disc.

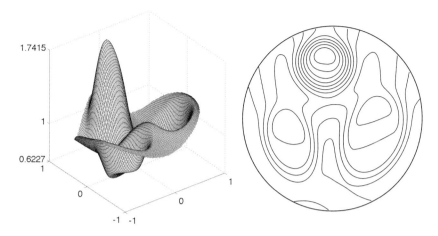

Figure 14.8. CGO solution corresponding to the smooth heart-and-lungs conductivity shown in Figure 14.6. Here $k = 2$. Left: three-dimensional mesh plot of the real part of $m(z, 2)$. Right: contour plot of the real part of $m(z, 2)$ inside the unit disc.

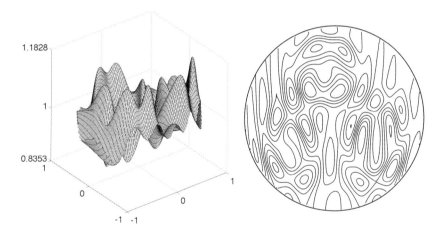

Figure 14.9. CGO solution corresponding to the smooth heart-and-lungs conductivity shown in Figure 14.6. Here $k = 10$. Left: three-dimensional mesh plot of the real part of $m(z, 10)$. Right: contour plot of the real part of $m(z, 10)$ inside the unit disc.

We compute CGO solutions corresponding to q using the periodic framework described above. We take $m = 3$, $M = 8$, and $s = 2.3$. See Figures 14.8 and 14.9 for plots of $m(z, k)$ for z ranging in the unit disc and $k = 2$ and $k = 10$, respectively.

14.4 CGO solutions for the Beltrami equation

The construction of CGO solutions discussed in Section 14.3 relies on the two derivatives available in σ. Astala and Päivärinta introduced another construction in [19] requiring no smoothness in the conductivity; it is enough to assume that $\sigma \in L^\infty(\Omega)$ is strictly positive. The approach is based on a Beltrami equation.

14.4.1 The real-linear Beltrami equation

Let $\Omega \subset \mathbb{R}^2$ be the unit disc and let $\sigma : \Omega \to (0, \infty)$ be an essentially bounded measurable function satisfying $\sigma(z) \geq c > 0$ for almost every $z \in \Omega$.

Following [19], we discuss unique CGO solutions u_1 and u_2 of the conductivity equations
$$\nabla \cdot \sigma \nabla u_1(\cdot, k) = 0, \qquad \nabla \cdot \sigma^{-1} \nabla u_2(\cdot, k) = 0, \tag{14.45}$$
where k is a complex parameter and the solutions have asymptotic behavior $u_1 \sim e^{ikz}$ and $u_2 \sim ie^{ikz}$ when $|z| \to \infty$. To define these global solutions we have set $\sigma(z) \equiv 1$ outside Ω. In particular, this makes the CGO solutions harmonic in $\mathbb{R}^2 \setminus \Omega$.

To construct the CGO solutions (14.45), we first define a real-valued function $\mu : \mathbb{R}^2 \to (-1, 1)$ by
$$\mu := \frac{1 - \sigma}{1 + \sigma}. \tag{14.46}$$

Note that μ is supported in Ω. We then consider the CGO solutions $f_\mu = f_\mu(z, k)$ of the Beltrami equation
$$\overline{\partial}_z f_\mu = \mu \overline{\partial_z f_\mu}, \tag{14.47}$$
where the solutions can be written in the form
$$f_\mu(z, k) = e^{ikz}(1 + \omega(z, k)) \tag{14.48}$$
and the functions ω_μ have the asymptotics
$$\omega(z, k) = \mathcal{O}(z^{-1}) \quad \text{as } |z| \to \infty. \tag{14.49}$$

Equation (14.47) is only real-linear and not complex-linear because the solution f_μ is complex-conjugated on the right-hand side.

We remark that the following key relations for a complex function $f = u + iv$ connect the PDEs (16.27) and (14.47):
$$\overline{\partial}_z f_\mu = \mu \overline{\partial_z f_\mu} \quad \Leftrightarrow \quad \nabla \cdot \sigma \nabla u = 0 \text{ and } \nabla \cdot \sigma^{-1} \nabla v = 0.$$

Set
$$h_+ = \frac{1}{2}(f_\mu + f_{-\mu}), \qquad h_- = \frac{i}{2}(\overline{f_\mu} - \overline{f_{-\mu}}). \tag{14.50}$$

Then the functions $u_1(z, k)$ and $u_2(z, k)$ in formula (14.45) are given by
$$u_1 = h_+ - ih_-, \qquad u_2 = i(h_+ + ih_-). \tag{14.51}$$

In the case of C^2 smooth conductivities we have the identity $u_1(z, k) = \sigma^{-1/2} \psi(z, k)$ with ψ being as in Section 14.3.

The existence and uniqueness of the CGO solutions f_μ was proved in [19]. Numerical computation of ω was first introduced in [17], directly based on the original construction in [19]. The lack of complex-linearity in the equations was compensated for by keeping the real and imaginary parts of the solution separate in a real-linear solution process. However, a more efficient algorithm was developed in [217], which will be explained in Section 14.4.2.

14.4.2 Reformulation with a complex-linear equation

Huhtanen and Perämäki introduced in [217] an efficient method for the computation of the CGO solutions f_μ. The rest of this section is devoted to a discussion of that method.

By [19, formula (4.8)] we know that the function ω defined in (14.49) satisfies the equation
$$\overline{\partial}\omega - \nu\overline{\partial\omega} - \alpha\overline{\omega} - \alpha = 0, \tag{14.52}$$
where the derivatives are taken with respect to z and
$$\alpha(z,k) = -i\overline{k}e_{-k}(z)\mu(z), \tag{14.53}$$
$$\nu(z,k) = e_{-k}(z)\mu(z),$$
with $e_k(z) := \exp(i(kz + \overline{k}\overline{z}))$. Rewriting (14.52) as $\overline{\partial}\omega = \nu\overline{\partial\omega} + \alpha\overline{\omega} + \alpha$ reveals that $\overline{\partial}\omega$ is supported in Ω.

Define $u \in L^p(\Omega)$ by $\overline{u} = -\overline{\partial}\omega$. Then $\omega = -P\overline{u}$ and $\partial\omega = -S\overline{u}$, where the solid Cauchy transform $P = \overline{\partial}^{-1}$ and the Beurling transform $S = \partial\overline{\partial}^{-1}$ are defined by
$$Pf(z) = -\frac{1}{\pi}\int_{\mathbb{C}} \frac{f(\lambda)}{\lambda - z} d\lambda_1 d\lambda_2, \tag{14.54}$$
$$Sg(z) = -\frac{1}{\pi}\int_{\mathbb{C}} \frac{g(\lambda)}{(\lambda - z)^2} d\lambda_1 d\lambda_2. \tag{14.55}$$

We have identified the points $(\lambda_1, \lambda_2) \in \mathbb{R}^2$ and $\lambda_1 + i\lambda_2 \in \mathbb{C}$ in (14.54) and (14.55). The integral in (14.55) should be understood as a principal value integral.

Substituting u to (14.52) leads to the real-linear integral equation
$$-\overline{u} - \nu\overline{(-S\overline{u})} - \alpha\overline{(-P\overline{u})} = \alpha.$$

We can further simplify using (14.53):
$$u + (-\overline{\nu}S - \overline{\alpha}P)\overline{u} = -\overline{\alpha}. \tag{14.56}$$

Denote complex conjugation in operator form as $\overline{f} = \rho(f)$. Then (14.56) takes the form
$$(I + A\rho)u = -\overline{\alpha}, \tag{14.57}$$
where $A := (-\overline{\nu}S - \overline{\alpha}P)$. It is shown in [217] that $I + A$ is invertible in the space $L^p(\Omega)$.

A special preconditioning step is introduced in [217] for the first time, transforming the real-linear equation (14.57) into a complex-linear equation allowing standard iterative solution by GMRES. Consider the following equation in the space $L^p(\Omega)$:
$$(I - A\overline{A})v = \overline{\alpha}. \tag{14.58}$$

Now (14.58) is complex-linear, and the solution u of (14.57) can be written as $u = (I - A\rho)v$.

Summarizing, we can construct ω as follows:

1. Solve for v from (14.58). Note that v is supported in Ω.

2. Calculate $u = (I - A\rho)v$. Note that u is supported in Ω.

3. Compute $\omega = -P\overline{u}$.

Exercise 14.4.1. *Derive* (14.52) *from the definitions of Section* 14.4.

14.4.3 Reduction to a periodic integral equation

As shown in [217] and discussed in Section 14.4.2, the computation of CGO solutions to the real-linear Beltrami equation can reduced to the solution of the complex-linear equation (14.58). Furthermore, one can use the iterative GMRES method for the solution of periodized and discretized version of (14.58). To that end, we need to introduce a periodic version of the operator $A := (-\overline{\nu}S - \overline{\alpha}P)$, where the Cauchy and Beurling transforms are defined in (14.54) and (14.55).

Let $s > 2$ and define the square Q and cutoff function η as in Section 14.3.3. Define a $2s$-periodic approximate Green's function $\widetilde{g}$ for the D-bar operator by setting it to $\eta(z)/(\pi z)$ inside Q and extending periodically:

$$\widetilde{g}(z + 2j_1 s + i 2 j_2 s) = \frac{\eta(z)}{\pi z} \tag{14.59}$$

for $z \in Q \setminus 0$ and $j_1, j_2 \in \mathbb{Z}$. Define a periodic approximate Cauchy transform by

$$\widetilde{P} f(z) = (\widetilde{g} \widetilde{*} f)(z) = \int_Q \widetilde{g}(z - w) f(w) \, dw_1 dw_2, \tag{14.60}$$

where $\widetilde{*}$ denotes convolution on the torus.

The Beurling transform (14.55) is approximated in the periodic context by writing

$$\widetilde{\beta}(z + 2 j_1 s + i 2 j_2 s) = \frac{\eta(z)}{\pi z^2}$$

for $z \in Q \setminus 0$ and $j_1, j_2 \in \mathbb{Z}$, and defining

$$\widetilde{S} g(z) := (\widetilde{\beta} \widetilde{*} g)(z) = \int_Q \widetilde{\beta}(z - w) g(w) \, dw_1 dw_2. \tag{14.61}$$

Set $\widetilde{A} := (-\overline{\widetilde{\nu}}\widetilde{S} - \overline{\widetilde{\alpha}}\widetilde{P})$ with the functions $\widetilde{\alpha}$ and $\widetilde{\nu}$ being trivial periodic extensions of α and ν, which are both supported in the unit disc. The periodic version of (14.58) takes the form

$$(I - \widetilde{A}\overline{\widetilde{A}})\widetilde{v} = \overline{\widetilde{\alpha}}. \tag{14.62}$$

Exercise 14.4.2. *Show that* $v|_\Omega = \widetilde{v}|_\Omega$, *where* v *is the solution of* (14.58) *and* $\widetilde{v}$ *is the solution of* (14.62).

14.4.4 Fast solver for the periodic equation

Our strategy is to use the iterative GMRES method for the solution of the discretized version of the periodic equation (14.62) analogously to Section 14.3.4. To that end, we need

14.4. CGO solutions for the Beltrami equation

to discretize the periodic Cauchy and Beurling transforms defined in (14.60) and (14.60), respectively.

Set

$$\widetilde{g}_h(j) = \begin{cases} \widetilde{g}(jh) & \text{for } j \in \mathbb{Z}_m^2 \setminus 0, \\ 0 & \text{for } j = 0, \end{cases} \quad (14.63)$$

and

$$\widetilde{\beta}_h(j) = \begin{cases} \widetilde{\beta}(jh) & \text{for } j \in \mathbb{Z}_m^2 \setminus 0, \\ 0 & \text{for } j = 0; \end{cases} \quad (14.64)$$

note that here the point $jh \in \mathbb{R}^2$ is interpreted as the complex number $hj_1 + ihj_2$. Now $\widetilde{g}_h$ and $\widetilde{\beta}_h$ are $M \times M$ matrices with complex entries. Given a periodic function φ, the discrete transforms $\widetilde{P}\varphi$ are defined by

$$(\widetilde{P}\varphi_h)_h = h^2 \, \text{IFFT}\big(\text{FFT}(\widetilde{g}_h) \cdot \text{FFT}(\varphi_h)\big), \quad (14.65)$$

$$(\widetilde{S}\varphi_h)_h = h^2 \, \text{IFFT}\big(\text{FFT}(\widetilde{\beta}_h) \cdot \text{FFT}(\varphi_h)\big), \quad (14.66)$$

and all the ingredients for numerical solution are in place.

14.4.5 Numerical examples

Let us demonstrate the above methods with a simple example. Analogously to Section 14.3.5, we take a strictly positive conductivity $\sigma : \Omega \to \mathbb{R}$ modeling an idealized cross-section of the human chest. The background conductivity is again taken to be one, but the conductive heart and resistive lungs are separated from the background by a jump discontinuity, chosen since the Astala–Päivärinta theory is developed for nonsmooth conductivities $\sigma \in L^\infty(\Omega)$. See Figure 14.10 for plots of the conductivity.

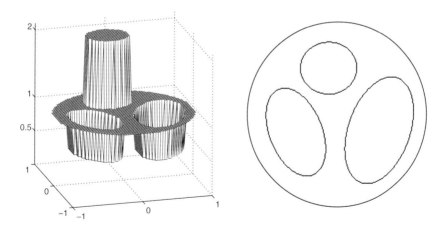

Figure 14.10. Discontinuous heart-and-lungs conductivity (simulated phantom). Left: three-dimensional mesh plot of $\sigma(x, y)$. Right: contour plot of $\sigma(x, y)$ inside the unit disc. See Figure 14.11 for a plot of the corresponding function μ.

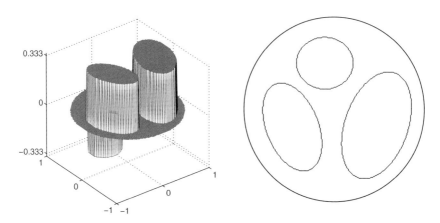

Figure 14.11. Function μ defined by (14.46) corresponding to the discontinuous heart-and-lungs conductivity shown in Figure 14.10. Left: three-dimensional mesh plot of $\mu(x, y)$. Right: contour plot of $\mu(x, y)$ inside the unit disc.

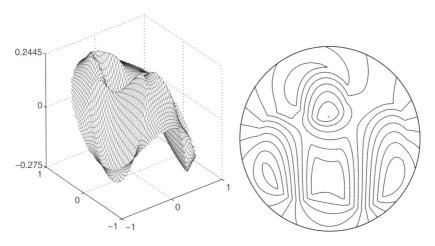

Figure 14.12. CGO solution corresponding to the smooth heart-and-lungs conductivity shown in Figure 14.10. Here $k = 2$. Left: three-dimensional mesh plot of the real part of $\omega(z, 2)$. Right: contour plot of the real part of $\omega(z, 2)$ inside the unit disc.

We construct the function $\mu(z) = \frac{1-\sigma(z)}{1+\sigma(z)}$ and show plots of it in Figure 14.11.

Finally, we use the numerical algorithm described in Section 14.4.4 to compute $\omega(z, 2)$ (see Figure 14.12) and $\omega(z, 10)$ (see Figure 14.13) for z ranging in the unit disc. Computational parameters were $s = 2.1$ and $M = 9$.

14.4. CGO solutions for the Beltrami equation

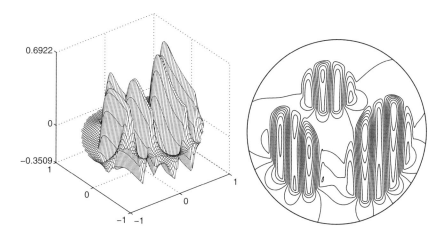

Figure 14.13. CGO solution corresponding to the smooth heart-and-lungs conductivity shown in Figure 14.10. Here $k = 10$. Left: three-dimensional mesh plot of the real part of $\omega(z, 10)$. Right: contour plot of the real part of $\omega(z, 10)$ inside the unit disc.

Chapter 15
A regularized D-bar method for direct EIT

The image formation task in EIT is a nonlinear ill-posed inverse problem, as explained in Chapter 12. Thus the inversion methods discussed in Part I of this book are not directly applicable to EIT. It is possible to apply the methods of Part I to a linearized approximation to EIT, but this approach is not advisable if the conductivity has high contrast. Also, linearized methods need reliable a priori information about the coarse structure of the target, as they aim to recover the deviation of the conductivity from a known background.

Another approach for interpreting EIT measurements is to solve a nonlinear minimization problem of the form

$$\arg\min_{\widetilde{\sigma}\in X}\{\|\Lambda_\sigma - \Lambda_{\widetilde{\sigma}}\|_Y + \alpha\|\widetilde{\sigma}\|_X\}, \tag{15.1}$$

where $\alpha > 0$ is a regularization parameter, Λ_σ is the measured data, and $\widetilde{\sigma}$ ranges over a finite-dimensional space X of suitably discretized conductivity distributions. The approach (15.1) is a nonlinear version of the regularization techniques discussed in Part I for linear inverse problems. However, practical iterative minimization of (15.1) is prone to get stuck to local minima unless the initial guess is close to the actual (unknown) conductivity.

It is the spirit of Part II of this book to discuss noniterative, noise-robust solution methods for the full nonlinear EIT problem instead of the two approaches mentioned above.

There is currently only one direct EIT method that allows a proper regularization analysis [277], and that is the topic of this chapter. The regularization is provided by applying a low-pass filter in the nonlinear Fourier transform domain, and the cutoff frequency depends explicitly on the noise level. This approach combines the PDE-based research tradition and the regularization-based school of inverse problems research. Moreover, it provides a practical noise-robust reconstruction algorithm.

Throughout this chapter we consider a strictly positive and twice continuously differentiable conductivity σ defined in the unit disc $\Omega \subset \mathbb{R}^2$. Furthermore, we assume that $\sigma(z) \equiv 1$ for z near the boundary $\partial\Omega$. These assumptions are needed in the rigorous proof of the regularization properties of the D-bar method.

The above assumptions are somewhat unrealistic from the point of view of medical applications. In particular, tissues and organs in the body usually have different conductivities and are separated by crisp boundaries, leading to discontinuous σ. However, these shortcomings can be overcome in practice by using algorithms whose regularization properties do not yet have a full theoretical framework. In Chapter 16 we discuss some of these extensions in detail from both a theoretical and a numerical perspective.

We remark that there are some regularized EIT algorithms for detecting inclusions in a known background; see [222, 228, 308]. In contrast, the regularized D-bar method discussed here gives a global reconstruction of the unknown conductivity coefficient.

The D-bar method gets its name from the differential operator known as the D-bar operator $\bar{\partial}$ that arises in the PDEs associated with the method. When a subscript is given on the operator, it indicates with respect to which complex variable the operation is performed. See Section 14.2 for definitions and basic results regarding the D-bar operator. The reader is referred to the sources [1, 38, 14] as well, which provide excellent introductions to $\bar{\partial}$ problems.

15.1 Reconstruction with infinite-precision data

The regularized D-bar method is based on Nachman's constructive uniqueness proof [350] for the two-dimensional inverse conductivity problem, which we recall here. Throughout this section we assume that we know the DN map $\Lambda_\sigma : H^{1/2}(\partial\Omega) \to H^{-1/2}(\partial\Omega)$ representing all possible voltage-to-current measurements with infinite precision.

The method uses three intermediate functions: the Schrödinger potential $q(x) = \sigma^{-1/2}\Delta\sigma^{1/2}$, the CGO solutions $\psi(z,k) = e^{ikz}m(z,k)$ discussed in Section 14.3, and the *scattering transform* $\mathbf{t}: \mathbb{C} \to \mathbb{C}$. Note that the strict positivity of σ, the continuity of partial derivatives of σ up to order two, and the assumption $\sigma(z) \equiv 1$ near $\partial\Omega$ imply that q is compactly supported, continuous, and can be extended continuously as zero into $\mathbb{R}^2 \setminus \Omega$.

The scattering transform was not introduced in Section 14.3 since the purpose there was to introduce the CGO solutions and compute them from the forward problem. From this section we saw that they are very oscillatory functions and require a somewhat fine grid for their representation. Recall that the CGO solutions $\psi(z,k)$ satisfy

$$(-\Delta + q)\psi(\cdot,k) = 0$$

with the asymptotic condition

$$e^{-ikz}\psi(z,k) - 1 \in W^{1,\tilde{p}}(\mathbb{R}^2)$$

for any $2 < \tilde{p} < \infty$, and the functions $m(z,k)$ satisfy

$$(-\Delta - 4ik\bar{\partial} + q)m(\cdot,k) = 0$$

with the asymptotic condition

$$m(z,k) - 1 \in W^{1,\tilde{p}}(\mathbb{R}^2).$$

The scattering transform for the inverse conductivity problem is not directly measurable from the physical experiments. For that reason, in order to use the D-bar method in this context, an explicit connection is needed between the scattering transform $\mathbf{t}(k)$ and the DN map Λ_σ. We introduce the scattering transform in Section 15.1.1 and the connection to the DN map in Section 15.1.3. That connection is used in Steps 1 and 2 in the reconstruction process outlined here:

$$\Lambda_\sigma \xrightarrow{\text{Step 1}} \psi(\cdot,k)|_{\partial\Omega} \xrightarrow{\text{Step 2}} \mathbf{t}(k) \xrightarrow{\text{Step 3}} m(z,k) \xrightarrow{\text{Step 4}} \sigma. \quad (15.2)$$

15.1. Reconstruction with infinite-precision data

Step 3 involves solving a D-bar equation with respect to k, and Step 4 is a trivial function evaluation. Figure 15.1 shows the process (15.2) in the form of a flowchart, and we discuss Steps 1–4 in detail in Sections 15.1.2–15.1.5, respectively.

We remark that in [350, 415, 344, 277] and other related works the function m was denoted by μ, but we have changed the notation here to avoid confusion with the function μ used in the Astala–Päivärinta theory discussed in Section 16.3.

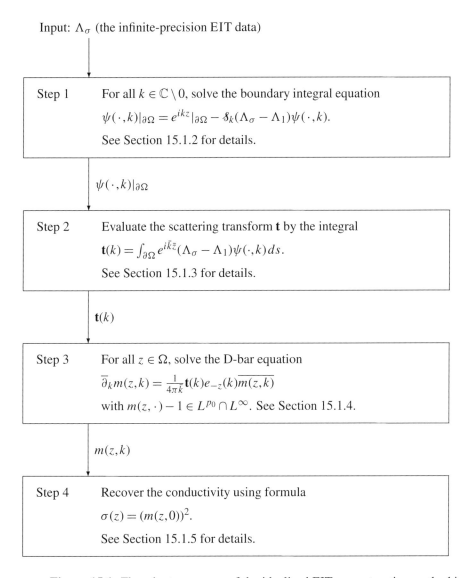

Figure 15.1. Flowchart summary of the idealized EIT reconstruction method introduced in [350]. The starting point of the process is the infinite-precision measurement data model, the DN map Λ_σ.

15.1.1 The scattering transform

The scattering transform is defined by

$$\mathbf{t}(k) := \int_{\mathbb{R}^2} e^{i\bar{k}\bar{z}} q(z)\psi(z,k)\,dx\,dy \qquad (15.3)$$

or, equivalently,

$$\mathbf{t}(k) := \int_{\mathbb{R}^2} e_k(z) q(z) m(z,k)\,dx\,dy, \qquad (15.4)$$

where the exponential function e_k satisfying $|e_k(z)| = 1$ is defined by

$$e_k(z) := e^{i(kz+\bar{k}\bar{z})} = e^{-i(-2k_1, 2k_2)\cdot(x,y)}. \qquad (15.5)$$

Note that since m is asymptotically close to 1, substituting (15.5) into (15.4) shows that the scattering transform $\mathbf{t}$ is approximately the Fourier transform of q evaluated at $(-2k_1, 2k_2)$. Indeed, by the proof of [350, Lemma 2.6] we have for some $0 < s < 1$ and large enough $|k|$ the inequality

$$|\mathbf{t}(k_1, k_2) - \hat{q}(-2k_1, 2k_2)| \leq C|k|^{-s}. \qquad (15.6)$$

In addition to (15.6), there are qualitative similarities between $\mathbf{t}$ and $\hat{q}$. It was proved in [415, Theorem 3.3] that, roughly,

(i) $\mathbf{t}(k) = \mathbf{t}(|k|)$ for all $k \in \mathbb{C}$ if and only if $\sigma(z) = \sigma(|z|)$ for all $z \in \mathbb{R}^2$;

(ii) dilating the conductivity by $\sigma(\lambda z)$ for some $\lambda > 0$ corresponds to dilating the scattering transform by $\mathbf{t}(k/\lambda)$;

(iii) reflectional symmetry in σ corresponds to reflectional symmetry in $\mathbf{t}$; and

(iv) translating σ corresponds to multiplication of $\mathbf{t}$ by a certain exponential function of modulus one.

Moreover, it was proved in [415, Theorem 3.2] that extra smoothness in σ allows stronger decay estimates for $|\mathbf{t}(k)|$ as $|k| \to \infty$.

As an example we prove that a rotationally symmetric conductivity leads to a rotationally symmetric and real-valued scattering transform. This is part of [415, Theorem 3.3]. Assume that $\sigma(z) = \sigma(|z|)$ for all $z \in \mathbb{R}^2$; then clearly $q(z) = q(|z|)$ for all $z \in \mathbb{R}^2$. Note that we set $\sigma(z) = 1$ and $q(z) = 0$ for $z \in \mathbb{R}^2 \setminus \Omega$. We wish to prove that

$$\mathbf{t}(k) = \mathbf{t}(|k|), \qquad \overline{\mathbf{t}(k)} = \mathbf{t}(k). \qquad (15.7)$$

Using uniqueness of solutions to the Schrödinger equation (14.18) with the asymptotic condition (14.19) it is easy to see that

$$m(z,k) = m(e^{i\varphi}z, e^{-i\varphi}k), \qquad (15.8)$$

$$m(z,k) = \overline{m(-\bar{z},\bar{k})} \qquad (15.9)$$

for all $z \in \mathbb{R}^2$ and $k \in \mathbb{C} \setminus 0$ and $\varphi \in \mathbb{R}$. Then $\mathbf{t}(k) = \mathbf{t}(e^{i\varphi}k)$ and $\mathbf{t}(k) = \overline{\mathbf{t}(\bar{k})}$ by substituting (15.8) and (15.9), respectively, to formula (15.4). Now (15.7) follows.

15.1. Reconstruction with infinite-precision data

Exercise 15.1.1. *Prove formulas* (15.8) *and* (15.9).

Exercise 15.1.2. *Let $\lambda > 0$ and let $\mathbf{t}_i(k)$ denote the scattering transform defined by* (15.3) *corresponding to $q_i = \sigma_i^{-1/2}\Delta\sigma_i^{1/2}$ for $i = 1, 2$. Prove that $\sigma_2(z) = \sigma_1(\lambda z)$ for all $z \in \mathbb{R}^2$ if and only if $\mathbf{t}_2(k) = \mathbf{t}_1(\lambda^{-1}k)$ for all $k \in \mathbb{C}$.*

Exercise 15.1.3. *Let $e_k(z)$ be defined as in* (15.5), $\zeta \in \mathbb{R}^2$, *and let $\mathbf{t}_i(k)$ denote the scattering transform defined by* (15.3) *corresponding to $q_i = \sigma_i^{-1/2}\Delta\sigma_i^{1/2}$ for $i = 1, 2$. Prove that $\sigma_2(z) = \sigma_1(z - \zeta)$ for all $z \in \mathbb{R}^2$ if and only if $\mathbf{t}_2(k) = e_k(\zeta)\mathbf{t}_1(k)$ for all $k \in \mathbb{C}$.*

15.1.2 From Λ_σ to $\psi|_{\partial\Omega}$ using a boundary integral equation

The relationships between the DN map and $\psi|_{\partial\Omega}$ and $\mathbf{t}(k)$ are derived using Alessandrini's identity [5]. The analysis makes use of the DN map Λ_q of the Schrödinger equation, defined for functions $f \in H^{1/2}(\partial\Omega)$ by

$$\Lambda_q f := \left.\frac{\partial v}{\partial \nu}\right|_{\partial\Omega},$$

where the function $v \in H^1(\Omega)$ is the unique solution of the Dirichlet problem

$$(-\Delta + q)v = 0 \quad \text{in } \Omega, \qquad v|_{\partial\Omega} = f. \tag{15.10}$$

We first include a derivation of Alessandrini's identity.

Theorem 15.1 (see [5]). *For any two solutions $v_\ell \in H^1(\Omega)$, $\ell = 1, 2$, to*

$$(-\Delta + q_\ell)v_\ell = 0 \quad \text{in} \quad \Omega, \tag{15.11}$$

the following identity holds:

$$\int_\Omega (q_1 - q_2)v_1 v_2 \, dx\, dy = \int_{\partial\Omega} v_1(\Lambda_{q_1} - \Lambda_{q_2})v_2 \, ds. \tag{15.12}$$

Proof. From (15.11),

$$0 = \int_\Omega \big(v_2(\Delta v_1 - q_1 v_1) - v_1(\Delta v_2 - q_2 v_2)\big) \, dx\, dy. \tag{15.13}$$

Rearranging terms and applying Green's theorem results in

$$\begin{aligned}\int_\Omega (q_1 - q_2)v_1 v_2 \, dx\, dy &= \int_\Omega \big(v_2 \Delta v_1 - v_1 \Delta v_2\big) \, dx\, dy \\ &= \int_{\partial\Omega} \left(v_2 \frac{\partial v_1}{\partial \nu} - v_1 \frac{\partial v_2}{\partial \nu}\right) ds \\ &= \int_{\partial\Omega} \big(v_2 \Lambda_{q_1} v_1 - v_1 \Lambda_{q_2} v_2\big) \, ds. \end{aligned} \tag{15.14}$$

Consider now the Dirichlet problem

$$(-\Delta + q_1)w = 0 \quad \text{in } \Omega, \qquad w = v_2 \quad \text{on } \partial\Omega. \tag{15.15}$$

Computing similarly to (15.14),

$$\begin{aligned}
0 &= \int_\Omega \big(v_1(\Delta w - q_1 w) - w(\Delta v_1 - q_1 v_1)\big) dx\,dy \\
&= \int_\Omega (v_1 \Delta w - w \Delta v_1) dx\,dy \\
&= \int_{\partial\Omega} \left(v_1 \frac{\partial w}{\partial \nu} - w \frac{\partial v_1}{\partial \nu}\right) ds \\
&= \int_{\partial\Omega} (v_1 \Lambda_{q_1} w - w \Lambda_{q_1} v_1) ds \\
&= \int_{\partial\Omega} (v_1 \Lambda_{q_1} v_2 - v_2 \Lambda_{q_1} v_1) ds,
\end{aligned}$$

we see that

$$\int_{\partial\Omega} v_2 \Lambda_{q_1} v_1 = \int_{\partial\Omega} v_1 \Lambda_{q_1} v_2 \, ds. \tag{15.16}$$

Now by (15.14) and (15.16),

$$\begin{aligned}
\int_\Omega (q_1 - q_2) v_1 v_2 \, dx\,dy &= \int_{\partial\Omega} (v_1 \Lambda_{q_1} v_2 - v_1 \Lambda_{q_2} v_2) ds \\
&= \int_{\partial\Omega} v_1 (\Lambda_{q_1} - \Lambda_{q_2}) v_2 \, ds. \quad \square
\end{aligned}$$

Recall the construction of the solutions ψ, based on the definition $\psi(z,k) = e^{ikz} m(z,k)$. The functions m are defined using a Lippmann–Schwinger-type equation $m = 1 - g_k * (qm)$ derived as (14.23) above. Multiplying through with e^{ikz} yields

$$\psi(z,k) = e^{ikz} - G_k * (q\psi), \tag{15.17}$$

where $G_k(z) = e^{ikz} g_k(z)$ is the Faddeev Green's function for the Laplacian.

Choosing now $q_2 = q = \sigma^{-1/2} \Delta \sigma^{1/2}$, $v_2 = \psi(\cdot, k)$, $q_1 = 0$, and $v_1 = G_k(z - \zeta)$ with $\zeta \in \mathbb{C} \setminus \overline{\Omega}$ in (15.12) gives

$$\begin{aligned}
& \int_{\partial\Omega} G_k(z - \zeta)(\Lambda_0 - \Lambda_q)\psi(\cdot,k)|_{\partial\Omega} \, ds(\zeta) \\
&= -\int_\Omega G_k(z - \zeta) q(\zeta) \psi(\zeta, k) d\zeta \\
&= -e^{ikz} + \psi(z,k),
\end{aligned} \tag{15.18}$$

where the last line follows from (15.17). Taking $z \to \partial\Omega$ in (15.18) results in the boundary integral equation

$$\psi(\cdot,k)|_{\partial\Omega} = e^{ikz}|_{\partial\Omega} - \mathcal{S}_k(\Lambda_q - \Lambda_0)\psi(\cdot,k), \tag{15.19}$$

15.1. Reconstruction with infinite-precision data

where the Faddeev single-layer operator is defined by

$$(\mathcal{S}_k \phi)(z) := \int_{\partial \Omega} G_k(z - \zeta) \phi(\zeta) ds(\zeta). \tag{15.20}$$

See [350] for the details on why z can be taken to $\partial\Omega$; the proof is based on the continuity of the usual single-layer operator.

Notice that (15.19) does not involve the infinite-precision EIT measurement Λ_σ, but rather the seemingly abstract operator Λ_q. The relationship $q = \sigma^{-1/2} \Delta \sigma^{1/2}$ allows one to derive a relationship between the DN maps Λ_q and Λ_σ. Denote by u the solution of the Dirichlet problem

$$\nabla \cdot \sigma \nabla u = 0 \quad \text{in } \Omega, \quad u|_{\partial\Omega} = f,$$

and let v be as in (15.10). Then $v = \sigma^{1/2} u$ and

$$\Lambda_q f = \left. \frac{\partial v}{\partial \nu} \right|_{\partial\Omega} = \sigma^{-1/2} \left(\frac{1}{2} \frac{\partial \sigma}{\partial \nu} + \Lambda_\sigma \right) \sigma^{-1/2} f,$$

and our assumption that $\sigma \equiv 1$ near $\partial\Omega$ implies that

$$\Lambda_q = \Lambda_\sigma. \tag{15.21}$$

Substituting (15.21) into (15.19) gives the desired equation

$$\psi(\cdot, k)|_{\partial\Omega} = e^{ikz}|_{\partial\Omega} - \mathcal{S}_k (\Lambda_\sigma - \Lambda_1) \psi(\cdot, k). \tag{15.22}$$

Exercise 15.1.4. *Justify the choice $v_2 = G_k(z - \zeta)$ that results in the boundary integral formula for $\psi(z, k)$.*

15.1.3 From $\psi|_{\partial\Omega}$ to t via integration over the boundary

Choosing $q_1 = q = \sigma^{-1/2} \Delta \sigma^{1/2}$, $v_1 = \psi(\cdot, k)$, $q_2 = 0$, and $v_2 = e^{i\bar{k}\bar{z}}$ in (15.12) results in an explicit formula for the scattering transform $\mathbf{t}(k)$ from the DN data:

$$\mathbf{t}(k) = \int_\Omega q(z) e^{i\bar{k}\bar{z}} \psi(z, k) \, dx \, dy = \int_{\partial\Omega} e^{i\bar{k}\bar{z}} (\Lambda_\sigma - \Lambda_1) \psi(\cdot, k) \, ds. \tag{15.23}$$

Note that the reconstruction process makes use of the infinite-precision data Λ_σ twice: first in the boundary integral equation (15.22) and then in the integration (15.23).

Exercise 15.1.5. *Justify the choices $v_2 = e^{i\bar{k}\bar{z}}$ and $v_1 = \psi(\cdot, k)$ in (15.12) that result in the boundary integral formula (15.23) for the scattering transform.*

15.1.4 From t to *m*: The D-bar equation

The CGOs solutions $m(z, k) = e^{-ikz} \psi(z, k)$ are parameterized by the complex number k. It is a natural question to ask whether the k-dependence in $m(z, k)$ is complex-analytic. As discussed in Section 14.2, analyticity is equivalent to the D-bar derivative $\bar{\partial}_k m(z, k)$ being zero. It turns out that the result is not zero.

Differentiating the Lippmann–Schwinger equation $m = 1 - g_k * (qm)$ with respect to $\bar{k}$ results in an equation involving the scattering transform. Namely,

$$\frac{\partial}{\partial \bar{k}} m(z,k) = \frac{1}{4\pi \bar{k}} \mathbf{t}(k) e_{-k}(z) \overline{m(z,k)}. \tag{15.24}$$

Thus the mapping $k \mapsto m(z,k)$ is not analytic but only pseudoanalytic in the sense of Vekua [456]. Note that $z \in \mathbb{R}^2$ is a fixed parameter in (15.24), and the equation is over the variable $k \in \mathbb{C}$. The derivation of (15.24) requires computing $\bar{\partial}_k (g_k * f)$ and is found in [350]. Furthermore, [350, Theorem 4.1] shows that (15.24) is uniquely solvable under the following additional assumption on asymptotic behavior of the solution:

$$m(z,k) - 1 \in L^r \cap L^\infty(\mathbb{C}) \tag{15.25}$$

for some Lebesgue exponent $r > 2$.

The method of proof is constructive and uses an analogue to the Lippmann–Schwinger equation. Recall from Section 14.2 that $(\pi k)^{-1}$ is a fundamental solution for $\bar{\partial}_k$. Let $f : \mathbb{C} \to \mathbb{C}$ be an entire function (complex-analytic in the whole of $\mathbb{C}$) and denote

$$T_z(k) := \frac{\mathbf{t}(k)}{4\pi \bar{k}} e_{-k}(z).$$

Applying $\bar{\partial}_k$ to both sides of

$$m = f - \frac{1}{\pi k} * (T_z \overline{m}) \tag{15.26}$$

shows that any solution m of (15.26) is also a solution of (15.24). Now inserting $f \equiv 1$ into (15.26) gives a solution satisfying the asymptotic requirement (15.25), as can be seen by an estimation of the convolution integral (left as an exercise).

Let us write the above in a very explicit form. For any fixed $z \in \mathbb{R}^2$, the solution of (15.24) and (15.25) can be computed as the solution of

$$m(z,k) = 1 + \frac{1}{(2\pi)^2} \int_{\mathbb{R}^2} \frac{\mathbf{t}(k')}{(k - k')\bar{k'}} e_{-z}(k') \overline{m(z,k')} \, dk'_1 dk'_2. \tag{15.27}$$

According to [350], equation (15.27) is a Fredholm equation of the second kind and has a unique solution for any fixed $z \in \mathbb{R}^2$.

The study of D-bar methods such as those above dates back to the work of Beals and Coifman [34, 35, 36, 37, 38] and Ablowitz and Fokas [144, 145] on evolution equations. There, a PDE involving $\bar{\partial}_k$ is typically derived for so-called "Jost solutions" $m(z,k)$, and the present discussion about EIT is a special case of the general framework.

Exercise 15.1.6. *Show that a solution of* (15.26) *with* $f \equiv 1$ *gives a solution of* (15.24) *satisfying the asymptotic requirement* (15.25).

15.1.5 From m to σ

Observe that taking $k \to 0$ in $-\Delta m(z,k) - 4ik\bar{\partial} m(z,k) = q(z) m(z,k)$ and using $q = \frac{\Delta \sqrt{\sigma}}{\sqrt{\sigma}}$ implies

$$-\Delta m(z,0) = \frac{\Delta \sqrt{\sigma}}{\sqrt{\sigma}} m(z,0). \tag{15.28}$$

15.2. Regularization via nonlinear low-pass filtering

Proving formula (15.28) rigorously needs more careful analysis because of the $\log|k|$ singularity of Faddeev's fundamental solution g_k at $k=0$. See [350] for the original proof based on a limit argument as $k \to 0$, and note that a combination of [28, formula (4.6)] and the analysis in [280] shows that one can just substitute $k=0$ without worrying about taking limits.

Now, recalling that σ is continued as $\sigma(z) \equiv 1$ for $z \in \mathbb{R}^2 \setminus \Omega$ we see that

$$m(z,0) = \sigma^{1/2}(z) \tag{15.29}$$

satisfies (15.28) *and* the asymptotic condition $m(\,\cdot\,,0) - 1 \in W^{1,\widetilde{p}}(\mathbb{R}^2)$. Formula (15.29) can be used to recover σ directly from the solution m of (15.27).

15.2 Regularization via nonlinear low-pass filtering

Practical electrode measurements result in a matrix Λ_σ^δ that can be seen as an approximation to the abstract infinite-precision operator Λ_σ. The structure of measurement noise may vary according to the hardware and other particulars; here we just assume that

$$\|\Lambda_\sigma - \Lambda_\sigma^\delta\|_{L(H^{1/2}(\partial\Omega), H^{-1/2}(\partial\Omega))} \leq \delta$$

with a known noise amplitude parameter $\delta > 0$. How to compute an approximation to σ from the finite-dimensional and noisy data Λ_σ^δ?

We will construct a regularized reconstruction method for the nonlinear EIT problem. See Figure 15.2 for a schematic illustration of the concepts and objects involved. The parameter space is taken to be $X = L^\infty(\Omega)$, and the nonlinear forward map $\mathcal{A}(\sigma) = \Lambda_\sigma$ acts between the sets $\mathcal{A} : L^\infty(\Omega) \supset \mathcal{D}(\mathcal{A}) \to Y$. We next define the data space Y and the domain $\mathcal{D}(\mathcal{A})$ of the forward map.

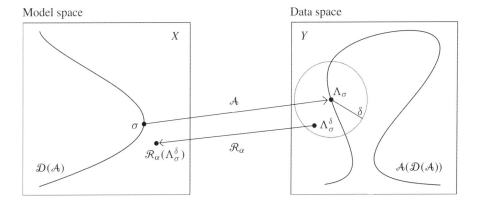

Figure 15.2. Schematic illustration of nonlinear regularization of the EIT problem. Here the forward map is defined as $\mathcal{A}(\sigma) = \Lambda_\sigma$. The conductivity σ is approximately recovered as $\mathcal{R}_\alpha(\Lambda_\sigma^\delta)$.

Definition 15.2.1. *Let $M > 0$ and $0 < \rho < 1$ and take the domain $\mathcal{D}(\mathcal{A})$ to be the set of functions $\sigma : \Omega \to \mathbb{R}$ satisfying*

$$\|\sigma\|_{C^2(\overline{\Omega})} \leq M, \tag{15.30}$$

$$\sigma(z) \geq M^{-1} \quad \text{for every } z \in \Omega, \tag{15.31}$$

$$\sigma(z) \equiv 1 \quad \text{for } \rho < |z| < 1. \tag{15.32}$$

The data space $Y \subset \mathcal{L}(H^{1/2}(\partial\Omega), H^{-1/2}(\partial\Omega))$ consists of bounded linear operators $\Lambda : H^{1/2}(\partial\Omega) \to H^{-1/2}(\partial\Omega)$ satisfying

$$\Lambda(1) = 0, \tag{15.33}$$

$$\int_{\partial\Omega} \Lambda(f)\,ds = 0 \quad \text{for every } f \in H^{1/2}(\partial\Omega). \tag{15.34}$$

We equip Y with the usual operator norm $\|\cdot\|_Y = \|\cdot\|_{H^{1/2}(\partial\Omega) \to H^{-1/2}(\partial\Omega)}$.

The constants M and ρ are a priori knowledge about the unknown conductivity. The one-dimensional conditions (15.33) and (15.34) can usually be easily enforced in practice; this is the role of the zero blocks in (13.11).

We adapt Definition 3.1 of [132] and Definitions 2.1 and 2.3 of [269] to the present nonlinear setting in Banach spaces as follows.

Definition 15.2.2. *A family of continuous mappings $\mathcal{R}_\alpha : Y \to L^\infty(\Omega)$ parameterized by $0 < \alpha < \infty$ is a* regularization strategy *for $\mathcal{A}$ if*

$$\lim_{\alpha \to 0} \|\mathcal{R}_\alpha(\Lambda_\sigma) - \sigma\|_{L^\infty(\Omega)} = 0 \tag{15.35}$$

for each fixed $\sigma \in \mathcal{D}(\mathcal{A})$. Further, a regularization strategy with a choice $\alpha = \alpha(\delta)$ of regularization parameter as a function of noise level is called admissible *if*

$$\alpha(\delta) \to 0 \quad \text{as } \delta \to 0, \tag{15.36}$$

and for any fixed $\sigma \in \mathcal{D}(\mathcal{A})$ the following holds:

$$\sup_{\Lambda_\sigma^\delta \in Y} \left\{ \|\mathcal{R}_{\alpha(\delta)}(\Lambda_\sigma^\delta) - \sigma\|_{L^\infty(\Omega)} : \|\Lambda_\sigma^\delta - \Lambda_\sigma\|_Y \leq \delta \right\} \to 0 \quad \text{as } \delta \to 0. \tag{15.37}$$

The definition of the regularization strategy $\mathcal{R}_\alpha : Y \to X$ for the D-bar method consists of four steps detailed in Figure 15.3. The truncation of the scattering transform at the cutoff frequency $R(\delta)$ can be viewed as a nonlinear low-pass filtering.

The following theorem is proved in [277], showing that $\mathcal{R}_{\alpha(\delta)}$ is a regularization strategy in the sense of Definition 15.2.2 when the noise amplitude $\delta > 0$ is small enough.

Theorem 15.2. *Let Ω be the unit disc. Assume that $M > 0$ and $0 < \rho < 1$ are given, and let $\mathcal{D}(\mathcal{A})$ be as in Definition 15.2.1. Then there exists a constant $\delta_0 > 0$, depending only on M and ρ, with the following properties. Let $\sigma \in \mathcal{D}(\mathcal{A})$ be arbitrary and let $\|\Lambda_\sigma^\delta - \Lambda_\sigma\|_Y \leq \delta < \delta_0$. Then $\mathcal{R}_{\alpha(\delta)}$, defined in the flowchart of Figure 15.3, is a well-defined regularization strategy and*

$$\alpha(\delta) = \frac{1}{R(\delta)}, \qquad R(\delta) = -\frac{1}{10}\log\delta,$$

15.2. Regularization via nonlinear low-pass filtering

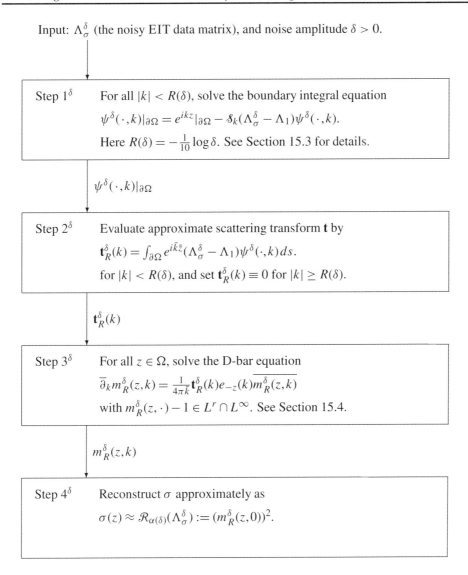

Figure 15.3. Flowchart summary of the regularized EIT reconstruction method introduced in [277]. The starting point of the process is the noisy measurement matrix Λ_σ^δ together with the a priori known noise level $\delta > 0$. Compare to Figure 15.1.

is an admissible choice of regularization parameter. Moreover, $\mathcal{R}_\alpha$ satisfies the estimate

$$\|\mathcal{R}_{\alpha(\delta)}(\Lambda_\sigma^\delta) - \sigma\|_{L^\infty(\Omega)} \leq C(-\log \delta)^{-1/14}. \qquad (15.38)$$

It is also shown in [277] that a spectral-theoretical argument can be used to extend Theorem 15.2 into arbitrarily large noise amplitudes.

15.3 Numerical solution of the boundary integral equation

We wish to solve numerically the boundary integral equation

$$\psi^\delta(z,k)|_{\partial\Omega} = e^{ikz}|_{\partial\Omega} - \mathcal{S}_k(\Lambda_\sigma^\delta - \Lambda_1)\psi^\delta(\cdot,k), \tag{15.39}$$

where k is a fixed, nonzero complex number. Furthermore,

$$\mathbf{t}_R^\delta(k) = \begin{cases} \int_{\partial\Omega} e^{i\bar{k}\bar{z}}(\Lambda_\sigma^\delta - \Lambda_1)\psi(\cdot,k)\,ds & \text{for } |k| < R, \\ 0 & \text{for } |k| \geq R. \end{cases} \tag{15.40}$$

Our strategy is to write (15.39) in a truncated Fourier basis and reduce the solution to matrix inversion. We assume that the noisy measurement Λ_σ^δ is given as a matrix $\mathbf{L}_\sigma^\delta$ acting on the Fourier basis; simulation of data in that form is explained in Section 13.2.

Choose $N > 0$. We represent a function $f \in H^s(\partial\Omega)$ approximately by the truncated Fourier series vector

$$\widehat{f} := \begin{bmatrix} \widehat{f}(-N) \\ \widehat{f}(-N+1) \\ \vdots \\ \widehat{f}(0) \\ \vdots \\ \widehat{f}(N-1) \\ \widehat{f}(N) \end{bmatrix},$$

where the Fourier coefficients are defined for $-N \leq n \leq N$ by

$$\widehat{f}(n) := \langle f, \varphi_n \rangle = \frac{1}{\sqrt{2\pi}} \int_0^{2\pi} f(\theta) e^{-in\theta}\,d\theta.$$

By standard Fourier series theory we get

$$f(\theta) \approx \sum_{n=-N}^{N} \widehat{f}(n)\varphi_n(\theta).$$

Let us approximate all linear operators appearing in (15.39) by matrices as explained in Section C.2.1. We know analytically that $\Lambda_1 \varphi_n = |n|\varphi_n$, so the $(2N+1) \times (2N+1)$ matrix representing the operator Λ_1 is

$$\mathbf{L}_1 := \operatorname{diag}[N, N-1, \ldots, 2, 1, 0, 1, 2, \ldots, N-1, N]. \tag{15.41}$$

As in Section 14.3.2, denote the standard Green's function for the Laplace operator by $G_0(z) = -(2\pi)^{-1}\log|z|$ and set $H_k = G_k - G_0 \in C^\infty(\mathbb{R}^2)$. Recall from (14.32) that

$$H_k(z) = H_1(kz) - \frac{1}{2\pi}\log|k|. \tag{15.42}$$

15.3. Numerical solution of the boundary integral equation

Use the above to decompose the single layer operator S_k as

$$\begin{aligned} S_k\phi(z) &= \int_{\partial\Omega} G_k(z-w)\phi(w)\,ds(w) \\ &= \int_{\partial\Omega} G_0(z-w)\phi(w)\,ds(w) + \int_{\partial\Omega} H_k(z-w)\phi(w)\,ds(w) \\ &= S_0\phi(z) + \mathcal{H}_k\phi(z) - \frac{\log|k|}{2\pi} \int_{\partial\Omega} \phi(w)\,ds(w), \end{aligned} \quad (15.43)$$

where the integral operator $\mathcal{H}_k$ is defined by

$$\mathcal{H}_k\phi(z) = \int_{\partial\Omega} H_1(k(z-y))\phi(y)\,ds(y). \quad (15.44)$$

Now the third term in (15.43) does not contribute to (15.39) at all since

$$\int_{\partial\Omega} (\Lambda_\sigma^\delta f)(y)\,ds(y) = 0 = \int_{\partial\Omega} (\Lambda_1 f)(y)\,ds(y),$$

the former equation by (15.34). In our case of Ω being the unit disc, the standard single-layer operator S_0 coincides with the ND map of the Laplace operator multiplied by $1/2$. Hence the matrix representation of S_0 is given by

$$\mathbf{S}_0 = \frac{1}{2}\operatorname{diag}\left[\frac{1}{N}, \frac{1}{N-1}, \ldots, \frac{1}{2}, 1, 0, 1, \frac{1}{2}, \ldots, \frac{1}{N-1}, \frac{1}{N}\right].$$

It remains to find a matrix $\mathbf{H}_k$ for the operator $\mathcal{H}_k$. We define the elements of $\mathbf{H}_k = [\mathbf{H}_k(m,n)]$ by

$$\mathbf{H}_k(m,n) := \langle \mathcal{H}_k \varphi_n, \varphi_m \rangle = \frac{1}{2\pi} \int_0^{2\pi} (\mathcal{H}_k e^{in\theta}) e^{-im\theta}\,d\theta. \quad (15.45)$$

Here $m \in \{-N,\ldots,N\}$ is the row index and $n \in \{-N,\ldots,N\}$ is the column index. Numerical evaluation of the integrals in (15.45) is reduced by the identity

$$H_1(z) = G_1(z) - G_0(z) = e^{iz}g_1(z) + \frac{1}{2\pi}\log|z| \quad (15.46)$$

to the algorithm for Faddeev's fundamental solution $g_1(z)$ given in Section 14.3.2.

We remark that possible numerical problems in the evaluation of H_1 using (15.46) related to the singularities of $g_1(z)$ and $\log|z|$ for z near zero can be avoided as follows:

- For $|z| \geq 0.7$, we are away from the singularities and can compute $H_1(z)$ directly using (15.46).

- For $|z| < 0.7$, we compute first $H_1(w)$ for w ranging on an equispaced angular grid on the unit disc. Then, because H_1 is a harmonic function, we can evaluate $H_1(z)$ using the classical Poisson kernel involving integration over the unit disc.

The approximate solution of (15.39) is now given in the truncated frequency domain by

$$\widehat{\psi|_\Omega} = [I + \mathbf{S}_k(\mathbf{L}_\sigma^\delta - \mathbf{L}_1)]^{-1}(\widehat{e^{ikz}|_{\partial\Omega}})$$
$$= [I + (\mathbf{S}_0 + \mathbf{H}_k)(\mathbf{L}_\sigma^\delta - \mathbf{L}_1)]^{-1}(\widehat{e^{ikz}|_{\partial\Omega}}), \qquad (15.47)$$

where we used the decomposition (15.43), and $\widehat{e^{ikz}|_{\partial\Omega}}$ stands for the Fourier expansion of e^{ikz}, calculated as follows. Write $z = e^{i\theta}$ and calculate as in [232, Section 2] to get

$$e^{ikz} = \sum_{n=-\infty}^{\infty} a_n(k) e^{in\theta} \quad \text{with} \quad a_n(k) = \begin{cases} \frac{(ik)^n}{n!}, & n \geq 0, \\ 0, & n < 0. \end{cases} \qquad (15.48)$$

The vector $\widehat{e^{ikz}|_{\partial\Omega}}$ thus takes the explicit form

$$\widehat{e^{ikz}|_{\partial\Omega}} = \sqrt{2\pi} \begin{bmatrix} 0 \\ 0 \\ \vdots \\ 0 \\ 1 \\ ik \\ -k^2/2 \\ \vdots \\ (ik)^N/N! \end{bmatrix}.$$

According to the theory, the inverse matrix $[I + (\mathbf{S}_0 + \mathbf{H}_k)(\mathbf{L}_\sigma^\delta - \mathbf{L}_1)]^{-1}$ exists for all k in the disc $D(0, -\frac{1}{10}\log\delta)$, at least for small enough δ and when the order N of trigonometric approximation is high enough.

Finally, we need to evaluate the approximate truncated scattering transform using a discrete version of formula (15.40). Once the Fourier coefficient vector $\widehat{\psi|_\Omega}$ is solved from (15.47), set

$$\widehat{u} = (\mathbf{L}_\sigma^\delta - \mathbf{L}_1)\widehat{\psi|_\Omega} \in \mathbb{C}^{2N+1} \qquad (15.49)$$

and define a function $u: \partial\Omega \to \mathbb{C}$ using the truncated Fourier series inversion:

$$u(\theta) = \sum_{n=-N}^{N} \widehat{u}(n)\varphi_n(\theta). \qquad (15.50)$$

Then the scattering transform can be computed for any $k \neq 0$ satisfying $|k| < -\frac{1}{10}\log\delta$ approximately using the formula

$$\mathbf{t}_R^\delta(k) \approx \int_0^{2\pi} e^{i\bar{k}\exp(-i\theta)} u(\theta)\, d\theta. \qquad (15.51)$$

The approximation in (15.51) is most accurate for k near zero.

15.4 Numerical solution of the D-bar equation

The first solution method published for the integral formulation (15.27) of the actual D-bar equation (15.24) was an adaptation of the Nystrom method; see [415]. The faster algorithm discussed here is a modification of the method introduced by Vainikko in [449] for the Lippmann–Schwinger equation related to the Helmholtz equation. A convergence proof of the method is found in [279]. See also [131] for a reformulated approach which is outside the scope of this book.

Assume we are given truncated scattering data $\mathbf{t}_R : \mathbb{C} \to \mathbb{C}$ satisfying $\mathrm{supp}(\mathbf{t}_R) \subset D(0, R)$. Define a pointwise multiplication operator T_R by

$$T_R f(k) = \frac{\mathbf{t}_R(k)}{4\pi \bar{k}} e^{-i(kz + \bar{k}\bar{z})} f(k). \tag{15.52}$$

Also, denote complex conjugation of a function as an operator $\rho(f) = \bar{f}$. Then the D-bar equation (15.24) with truncated data can be written as

$$\frac{\partial}{\partial \bar{k}} m_R(z, k) = T_R \rho(m_R(z, k)), \tag{15.53}$$

where $z \in \mathbb{R}^2$ is considered as a fixed parameter. Note that (15.53) is not complex-linear in m_R but only real-linear due to the complex conjugation ρ. We will return to this aspect in Section 15.4.2.

Let us comment on the solvability of the truncated-data D-bar equation (15.53) together with the asymptotic condition $m(z, k) \sim 1$ for fixed $z \in \mathbb{R}^2$ and large $|k|$. We know from [350, proof of Theorem 4] that $\mathbf{t}$ is continuous outside the origin. Further, by [415, Theorem 3.1] we know that $|\mathbf{t}(k)| \leq C|k|^2$ for k near zero. Thus the apparently singular function $\mathbf{t}_R(k)/\bar{k}$ is actually bounded inside the disc $|k| < R$. Furthermore, $\mathbf{t}_R(k)$ vanishes when $|k| \geq R$, and consequently $\mathbf{t}_R(k)/\bar{k} \in L^p(\mathbb{R}^2)$ for all $1 \leq p \leq \infty$. The following corollary is then a simple consequence of [350, Theorem 4.1].

Corollary 15.4.1. *Equation* (15.53) *has a unique solution satisfying* $m_R(z, \cdot) - 1 \in L^{p_0}(\mathbb{R}^2)$ *for any given and fixed exponent* $p_0 > 2$.

Recall that the solution of the D-bar equation (15.24) satisfies the integral equation (15.27). Similarly, the unique solution of (15.53) satisfies the integral equation

$$m_R(z, k) = 1 + \frac{1}{(2\pi)^2} \int_{D(0,R)} \frac{\mathbf{t}_R(k')}{(k - k')\bar{k}'} e_{-z}(k') \overline{m_R(z, k')} dk'_1 dk'_2 \tag{15.54}$$

that contains the asymptotic behavior explicitly as the constant function 1 appears in the right-hand side. Note that the domain of integration in the right-hand side of (15.54) is the disc $D(0, R)$ instead of the plane $\mathbb{R}^2$. This allows us to make the following key observation:

(KO) Let $z \in \mathbb{R}^2$ be fixed. If we know the restriction $m_R(z, k)$ to the disc $|k| < R$, then we know $m_R(z, k)$ for any $k \in \mathbb{C}$ by substituting $m_R(z, k)|_{D(0,R)}$ to the right-hand side of equation (15.54).

Let us write (15.54) in a more compact form. Recall the solid Cauchy transform P from (14.54); here we use slightly different notation adapted to the equation in the k-plane:

$$Pf(k) = -\frac{1}{\pi} \int_{\mathbb{R}^2} \frac{f(k')}{k' - k} dk'_1 dk'_2. \tag{15.55}$$

Then (15.54) takes the form
$$m_R = 1 + PT_R \rho(m_R). \tag{15.56}$$

Now (15.56) is formally similar to the generalized Lippmann–Schwinger equation (14.23), apart from the real-linear complex conjugation operator ρ. Consequently, our solution strategy for (15.56) parallels the strategy for (14.23).

15.4.1 Reduction to a periodic equation

This section is closely related to Section 14.3.3; the main differences are that here the multiplicator function is supported in a disc of radius $R > 0$ instead of radius 1 and that the Green's functions are different.

Take $\epsilon > 0$ and set $s = 2R + 3\epsilon$. Define $Q := [-s, s)^2$ and introduce a periodic version of (15.56) as follows. Choose an infinitely smooth cutoff function $\eta \in C_0^\infty(\mathbb{R}^2)$ satisfying

$$\eta(k) = \begin{cases} 1 & \text{for } 2R + \epsilon > |k|, \\ \text{smooth} & \text{for } 2R + \epsilon \le |k| < 2R + 2\epsilon, \\ 0 & \text{for } 2R + 2\epsilon \le |k|, \end{cases} \tag{15.57}$$

and $0 \le \eta(k) \le 1$ for all $k \in \mathbb{C}$.

Define a $2s$-periodic approximate Green's function $\widetilde{\beta}$ by setting $\widetilde{\beta}(k) := \eta(k)/(\pi k)$ for $k \in Q$ and extending periodically:

$$\widetilde{\beta}(k + 2j_1 s + i 2 j_2 s) = \frac{\eta(k)}{\pi k} \qquad \text{for } k \in Q \setminus 0, \quad j_1, j_2 \in \mathbb{Z}. \tag{15.58}$$

Define a periodic approximate solid Cauchy transform by

$$\widetilde{P} f(k) = (\widetilde{\beta} \, \widetilde{*} \, f)(k) = \int_Q \widetilde{\beta}(k - k') f(k') \, dk_1' dk_2', \tag{15.59}$$

where $\widetilde{*}$ denotes convolution on the torus; then $\widetilde{P}$ is a compact operator on $L^2(Q)$. We remark that (15.59) differs from (14.60) essentially only by the size of the square Q.

We interpret the pointwise multiplication operator T_R defined by (15.52) periodically in the obvious way of copying the function $\frac{\mathbf{t}_R(k)}{4\pi \bar{k}} e^{-i(kz + \bar{k}\bar{z})}$ to each tile; this is simple because the function is supported in the disc $D(0, R)$. Also, the complex conjugation operator ρ works the same way in the periodic framework.

Theorem 15.3. *Let $z \in \mathbb{R}^2$. There exists a unique $2s$-periodic solution to*

$$\widetilde{m}_R = 1 + \widetilde{P} T_R \rho(\widetilde{m}_R). \tag{15.60}$$

Furthermore, the solutions of (15.56) and (15.60) agree on the disc of radius R: $m_R(z, k) = \widetilde{m}_R(z, k)$ for $|k| < R$.

Proof. To show uniqueness, assume that $\widetilde{m}_R$ and $\widetilde{v}_R$ both satisfy (15.60). We will prove that $\widetilde{m}_R = \widetilde{v}_R$.

15.4. Numerical solution of the D-bar equation

Let φ be a function with $\mathrm{supp}(\varphi) \subset D(0,R)$, and denote by $\widetilde{\varphi}$ the periodic extension of φ. Note that the functions $(\pi k)^{-1}$ and $\widetilde{\beta}(k)$ coincide for $|k| < 2R + \epsilon$, as can be seen from equations (15.57) and (15.58). Hence the following identity holds for $|k| < R + \epsilon$:

$$(P\varphi)(k) = \frac{1}{\pi} \int_{\mathbb{C}} \frac{\varphi(k')}{k-k'} dk'_1 dk'_2 = \int_Q \widetilde{\beta}(k-k')\widetilde{\varphi}(k') dk'_1 dk'_2 = (\widetilde{P}\widetilde{\varphi})(k). \tag{15.61}$$

Define a nonperiodic function $m_R : \mathbb{R}^2 \to \mathbb{C}$ by the formulas

$$m_R(k) = \widetilde{m}_R(k) \qquad \text{for } |k| < R + \epsilon, \tag{15.62}$$
$$m_R(k) = 1 + PT_R \rho(\widetilde{m}_R) \quad \text{for } |k| > R. \tag{15.63}$$

Equation (15.61) and the fact that the function $T_R \rho(\widetilde{m}_R)$ is supported in $D(0,R)$ imply that (15.62) and (15.63) agree in the annulus $R < |z| < R + \epsilon$.

Further, define another nonperiodic function $\nu_R : \mathbb{R}^2 \to \mathbb{C}$ by replacing m by ν in formulas (15.62) and (15.63).

A combination of (15.61), (15.62), (15.63) and (15.60) shows that m_R and ν_R both satisfy (15.53). Moreover, formula (15.63) implies that $m_R(z,k) \sim 1/k$ and $\nu_R(z,k) \sim 1/k$ for fixed z and large $|k|$. Corollary 15.4.1 then implies that $m_R \equiv \nu_R$; in particular, by (15.62) we have $\widetilde{m}_R(z,\cdot)|_{D(0,R)} = \nu_R(z,\cdot)|_{D(0,R)}$. But then it follows from (15.60) that $\widetilde{m}_R \equiv \widetilde{\nu}_R$.

Existence follows from a Fredholm argument. Namely, a solution to (15.60) can be found as $\widetilde{m}_R = [I - \widetilde{P}T_R\rho]^{-1} 1$, provided that the inverse operator exists. Since $\widetilde{P}$ is compact, the operator $[I - \widetilde{P}T_R\rho]$ is of the form "identity + compact" and consequently Fredholm of index zero. The uniqueness proved above then provides existence. □

Our key observation (KO) now yields the following result.

Corollary 15.4.2. *The unique solution of (15.56) can be written in terms of the unique solution of the periodic equation (15.60) as*

$$m_R(z,k) = 1 + PT_R \rho(\widetilde{m}_R(z,\cdot)|_{D(0,R)}).$$

15.4.2 Numerical solution of the periodic equation

As in Section 15.4.1, take a square $Q := [-s,s]^2$ with some $s > 2$ as the basic tile of periodic tessellation of the plane. Similarly to Section 14.3.4, choose a positive integer m, denote $M = 2^m$, and set $h = 2s/M$. Define a grid $\mathcal{G}_m \subset Q$ by

$$\mathcal{G}_m = \{jh \mid j \in \mathbb{Z}_m^2\},$$

where

$$\mathbb{Z}_m^2 = \{(j_1, j_2) \in \mathbb{Z}^2 \mid -2^{m-1} \leq j_1 < 2^{m-1},\ -2^{m-1} \leq j_2 < 2^{m-1}\}.$$

Note that the number of points in $\mathcal{G}_m$ is M^2. See Figure 15.4(a) for a picture of the grid $\mathcal{G}_2$.

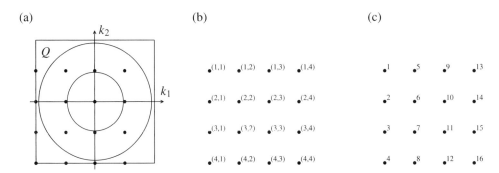

Figure 15.4. (a) Computational 4×4 grid $\mathcal{G}_2$, drawn as black dots, in the square $Q := [-s,s]^2$ with $s = 2R + 3\epsilon$. The two circles shown have radii R and $2R$. (b) Matrix-style numbering of the grid points. (c) Vector-style numbering of the grid points.

We define the grid approximation of a function $f : Q \to \mathbb{C}$ based on representing the values of f at the grid points as a complex $M \times M$ matrix. The matrix-style grid approximation $f_h^M \in \mathbb{C}^{M \times M}$ is defined by the formula

$$f_h^M(\ell_1, \ell_2) = f\left(h(-2^{m-1} - 1 + \ell_1), h(2^{m-1} - \ell_2)\right), \tag{15.64}$$

where $1 \leq \ell_1 \leq M$ is the row index and $1 \leq \ell_2 \leq M$ is the column index. See Figure 15.4(b) for an illustration of the matrix-style numbering of the grid points.

We use FFT to implement the periodic approximate Cauchy transform $\widetilde{P}$ defined in (15.59). The idea is to compute the periodic convolution on the torus as multiplication in the frequency domain. We represent the periodic approximate Green's function $\widetilde{\beta}$ defined in (15.58) using the matrix-style grid approximation (15.64) with the exception that we substitute zero as the element corresponding to the origin, where the Green's function is singular. Given a periodic function f, the transform $\widetilde{P}f$ can be approximately computed by

$$(\widetilde{P}f)_h^M \approx h^2\, \mathrm{IFFT}\bigl(\mathrm{FFT}(\widetilde{\beta}_h^M) \cdot \mathrm{FFT}(f_h^M)\bigr),$$

where FFT and IFFT stand for the two-dimensional direct and inverse fast Fourier transform, respectively.

We are interested in approximate numerical solution of the real-linear periodic equation (15.60). We write it in the form

$$[I - \widetilde{P}T_R \rho]\widetilde{m}_R = 1, \tag{15.65}$$

and solve it using the matrix-free iterative method GMRES as discussed in Appendix D. Equation (15.65) is not complex-linear but only real-linear, so the approach described in Section 14.3.4 will not work without modification. One possibility would be to use a dedicated iterative solver, such as introduced in [131]. Instead of that, we follow here the approach of keeping the real and imaginary parts separate so that GMRES is applied to a real-linear problem of dimension $2M^2$.

15.4. Numerical solution of the D-bar equation

We define another grid approximation of f, based on representing the values of f at the grid points as a real-valued vector of length $2M^2$. The matrix and vector representations are equivalent and related to each other via the two numberings illustrated in Figures 14.4(b) and 14.4(c). The vector-style grid approximation $\vec{f}_h \in \mathbb{R}^{2M^2}$ is defined by the formula

$$\vec{f}_h = \begin{bmatrix} \operatorname{Re} f(k^{(1)}) \\ \vdots \\ \operatorname{Re} f(k^{(M^2)}) \\ \operatorname{Im} f(k^{(1)}) \\ \vdots \\ \operatorname{Im} f(k^{(M^2)}) \end{bmatrix}. \qquad (15.66)$$

In (15.66) the grid points are denoted by $k^{(1)}, k^{(2)}, \ldots, k^{(M^2)} \in \mathcal{G}_m$, where the numbering follows the scheme illustrated by Figure 14.4(c).

Also, we introduce two functions for moving between the two discrete representations. The functions are called "matrl" and "vecrl" and they are closely related to the functions "mat" and "vec" discussed in Section 14.3.4.

Let $\mathbf{f} = [\mathbf{f}(1), \mathbf{f}(2), \mathbf{f}(3), \ldots, \mathbf{f}(2M^2)]^T \in \mathbb{R}^{2M^2}$ be an arbitrary vertical vector with real-valued elements. Define the function $\operatorname{matrl} : \mathbb{R}^{2M^2} \to \mathbb{C}^{M \times M}$ by the formula

$$(\operatorname{matrl}(\mathbf{f}))(\ell_1, \ell_2) = \mathbf{f}((\ell_2 - 1)M + \ell_1) + i\mathbf{f}(M^2 + (\ell_2 - 1)M + \ell_1).$$

Further, define the map $\operatorname{vecrl} : \mathbb{C}^{M \times M} \to \mathbb{R}^{2M^2}$ uniquely by the requirement $\operatorname{vecrl}(\operatorname{matrl}(\mathbf{f})) = \mathbf{f}$ for all vectors $\mathbf{f} \in \mathbb{R}^{2M^2}$.

Now we can compute $\widetilde{m}(z, \cdot)_h^M$ as the solution of (15.65) using the following algorithm:

1. Construct initial guess $\mathbf{m}_0 \in \mathbb{R}^{2M^2}$. A suitable choice is the vertical vector with the first M^2 elements equal to 1 and the last M^2 elements equal to 0.

2. Call GMRES with initial guess $\mathbf{m}_0$ and the linear operator in the left-hand side of (15.65) described by the routine

$$\mathbf{f} \mapsto \mathbf{f} - \operatorname{vecrl}\left(h^2 \operatorname{IFFT}\left(\operatorname{FFT}(\widetilde{\beta}_h^M) \cdot \operatorname{FFT}(T_h^M \cdot \overline{\operatorname{matrl}(\mathbf{f})})\right)\right),$$

where $T_h^M := \left(\frac{\mathbf{t}_R(k)}{4\pi \bar{k}} e^{-i(kz + \bar{k}\bar{z})}\right)_h^M$ and $\cdot$ denotes elementwise multiplication of matrices. Here $z \in \mathbb{R}^2$ is a fixed parameter. Denote the output vector of GMRES by $\mathbf{m} \in \mathbb{C}^{M^2}$.

3. Obtain solution of (15.65) by $\widetilde{m}(z, \cdot)_h^M = \operatorname{matrl}(\mathbf{m})$.

Finally we have

$$m(z, k_{(\ell_1, \ell_2)}) \approx \widetilde{m}(z, \cdot)_h^M(\ell_1, \ell_2) \qquad (15.67)$$

for all grid points $k_{(\ell_1, \ell_2)} \in \mathcal{G}_m$ satisfying $|k_{(\ell_1, \ell_2)}| < R$.

The above solution method has computational complexity $M^2 \log M$ when the solutions are computed on an $M \times M$ grid. The complexity of the first numerical D-bar solver described in [415] is M^4.

15.5 Regularized reconstructions

Let us illustrate the regularized D-bar method with a twice continuously differentiable conductivity $\sigma \in C^2(\Omega)$, which is a simulated smooth heart-and-lungs phantom. Here Ω is the unit disc, and $\sigma(z) \equiv 1$ for z in a neighborhood of $\partial\Omega$. Figure 15.8 shows the phantom in the bottom right plot.

15.5.1 Ground truth: Computing t directly from σ

When studying the quality and properties of the numerical reconstruction method, it is useful to be able to compare the scattering transform recovered from noisy data to the actual scattering transform. This is of course not possible in real-world measurement situations, but in the present simulated scenario we have the luxury of knowing the conductivity σ that is to be reconstructed from boundary measurements.

Define $q = \sigma^{-1/2} \Delta \sigma^{1/2}$. By zero extension we get $q \in C_0^0(\mathbb{R}^2)$. We choose a 128×128 Cartesian grid of points in the square $[-12, 12] \subset \mathbb{C}$ and compute $\mathbf{t}(k)$ numerically for each grid point satisfying $|k| < 12$. With a fixed grid point k we compute the function $m(z, k)$ by taking $m = 3$, $M = 8$, and $s = 2.1$ and using the method described in Section 14.3.4. The scattering transform can then be evaluated approximately as

$$\mathbf{t}(k) = \int_{\mathbb{R}^2} e^{-i(kz + \bar{k}\bar{z})} q(z) m(z, k) \, dx \, dy,$$

where the integral is simply approximated as a numerical quadrature with constant weight h^2 over the uniform z-grid of size 256×256.

See Figures 15.5 and 15.6 for the results.

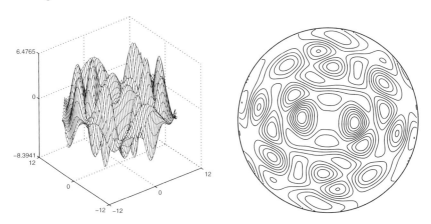

Figure 15.5. Real part of the nonlinear Fourier transform $\mathbf{t}(k)$ of the potential q shown in Figure 14.7. Here k ranges in the disc $|k| < 12$.

15.5. Regularized reconstructions

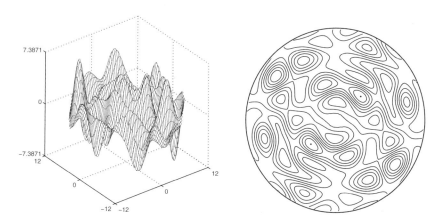

Figure 15.6. Imaginary part of the nonlinear Fourier transform $\mathbf{t}(k)$ of the potential q shown in Figure 14.7. Here k ranges in the disc $|k| < 12$.

15.5.2 Simulation of continuum model EIT data

We compute a matrix approximation R_σ to the ND map $\mathcal{R}_\sigma$ as explained in Section 13.2.2. The Fourier series are truncated by taking $N = 16$. We construct a finite element triangulation by successive refinements of an initial mesh; the resulting mesh has 131585 node points and 262144 triangles.

We construct a $(2N + 1) \times (2N + 1) = 33 \times 33$ matrix approximation $\mathbf{L}_\sigma^\delta$ to the DN map Λ_σ as explained in Section 13.2.3.

To estimate the size of δ, we repeat the above computation for the matrix approximation $\mathbf{L}_1^\delta$ to the DN map Λ_1. Since we know $\mathbf{L}_1$ analytically, we can estimate the size of the computational error in $\mathbf{L}_1^\delta$, and the result is

$$\|\mathbf{L}_1^\delta - \mathbf{L}_1\|_{H^{1/2}(\partial\Omega) \to H^{-1/2}(\partial\Omega)} \approx 10^{-6}.$$

Assuming that the finite element computation for σ has similar accuracy, we conclude that $\delta \approx 10^{-6}$ for $\mathbf{L}_\sigma^\delta$, too.

We construct data with various levels of noise as explained in Section 13.4 so that we can analyze the regularization properties of our method. In particular, we take δ to equal $10^{-6}, 10^{-5}, 10^{-4}, 10^{-3}$, and 10^{-2}.

15.5.3 Computation of the scattering transform from data

We choose a finite collection $\mathcal{K}$ of k-values by including all points $k \in \mathbb{C}$ satisfying $|k| < 10$ and being of the form $k = \frac{j_1}{5} + i\frac{j_2}{5}$ with some integers j_1, j_2.

For each nonzero $k \in \mathcal{K}$ we compute the matrix $\mathbf{H}_k$ using (15.45) and solve for $\widehat{\psi|_\Omega}$ from equation (15.47). We evaluate $\mathbf{t}^\delta(k)$ for each nonzero $k \in \mathcal{K}$ by (15.51), and we set $\mathbf{t}^\delta(0) = 0$ as this is known analytically to be the correct value.

We remark that it may be a good idea to store the $\mathbf{H}_k$ to separate data files for reuse as they do not depend on the conductivity at all.

Note that we have now evaluated $\mathbf{t}^\delta(k)$ in a grid inside the disc $|k| < 10$ regardless of the noise level δ. The values of $\mathbf{t}^\delta(k)$ for $|k|$ outside some radius depending on δ are numerical garbage, so we need to truncate $\mathbf{t}^\delta$. Theoretically it would be correct, at least for small enough δ, to choose the truncation radius as $R(\delta) = -\frac{1}{10}\log\delta$. However, we demonstrate the power of the nonlinear low-pass filter regularization by choosing the radius as follows:

$$R(\delta) = \sup\left\{0 < \rho < 10 : \frac{\|\mathbf{t}-\mathbf{t}^\delta\|_{L^\infty(D(0,\rho))}}{\|\mathbf{t}\|_{L^\infty(D(0,10))}} < 0.01\right\}, \qquad (15.68)$$

where $\mathbf{t}$ is the actual scattering transform computed in Section 15.5.1 and $\mathbf{t}^\delta$ is the approximate scattering transform computed from simulated noisy EIT data. Of course, the sup operation in (15.68) is implemented in a discrete fashion. See Figure 15.10 for a plot of the reconstructions from the resulting frequency cutoff radii $R(\delta)$.

Now in practice one cannot use the strategy (15.68) for the choice of $R(\delta)$ because $\mathbf{t}$ is unknown. However, the reason we demonstrate such a choice is that the "official" choice $R(\delta) = -\frac{1}{10}\log\delta$ is far too pessimistic, as seen in Figure 15.10. So while (15.68) cannot be used as such, in every practical application it is probably possible to find a method for choosing $R(\delta)$ larger than $-\frac{1}{10}\log\delta$ but still producing good results. Typically it is quite easy to see by visual inspection after what radius the quality of $\mathbf{t}^\delta(k)$ deteriorates. See Figure 15.7 for an illustration.

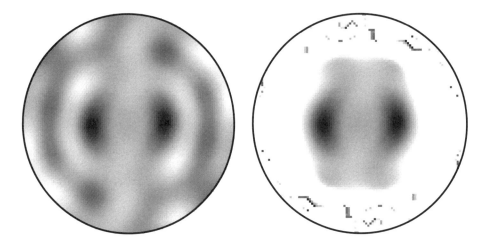

Figure 15.7. Effect of noise in the scattering transform. Left: real part of the scattering transform $\mathbf{t}(k)$ computed in the disc $|k| < 10$ using the knowledge of conductivity σ. This is the same function as is shown in Figure 15.5. Right: scattering transform computed from the simulated DN map with added noise of relative magnitude 0.01%. White areas in the right image denote function values larger than maximum value or smaller than minimum value in the left image. An image like this was the motivation for the nonlinear low-pass filtering regularization approach: the only good quality part of the transform computation from real data is located in a neighborhood of the origin.

15.5. Regularized reconstructions

We implement the nonlinear low-pass filtering with the formula

$$\mathbf{t}_R^\delta(k) := \begin{cases} \int_0^{2\pi} e^{i\bar{k}\exp(-i\theta)} u(\theta) d\theta & \text{for } k \in \mathcal{K}, |k| < R(\delta), \\ 0 & \text{for } k \in \mathcal{K}, |k| \geq R(\delta), \end{cases} \quad (15.69)$$

where u is defined by (15.49) and (15.50).

15.5.4 Solving the D-bar equation

We choose the reconstruction grid $\mathcal{Z}$ in the z-plane to be a 64×64 uniformly spaced Cartesian square grid with corners $\pm 1 \pm i$. The D-bar equation is solved for each $z \in \mathcal{Z}$ satisfying $|z| < 1$.

For a fixed $z \in \mathcal{Z}$ we proceed as follows. The computational parameters are

δ	R	M
10^{-2}	2.5	7
10^{-3}	3.5	8
10^{-4}	4.3	8
10^{-5}	5.9	8
10^{-6}	6.7	8

We solve the periodic equation (15.65) as explained in Section 15.4.2. Finally we use (15.67) to define the reconstructed conductivity as

$$\sigma(z) \approx (m(z,0))^2 \approx (\widetilde{m}_h^M(2^{M-1}+1, 2^{M-1}+1))^2.$$

Repeating the above computation for each $z \in \mathcal{Z}$ satisfying $|z| < 1$ produces pictures of the reconstructed conductivity as shown in Figures 15.8 and 15.9.

As a further example, we include a conductivity representing an idealized cross-section of a human chest with an abnormality in one lung, representing, for example, fluid. The conductive heart has a conductivity of 6 in this example, the lungs have a conductivity of 2, the fluid has a conductivity of 4, and the background has a conductivity of 1. Here the organ boundaries are smoothed, but the interface between the fluid and the lung is discontinuous. In Figure 15.11 a reconstruction is shown from data with 0.01% noise using a truncation radius of $R = 5.4$. The maximum reconstructed value is 5.9237, which occurs correctly in the heart region, and the minimum reconstructed value of 0.7974 is correctly found in the background, but is lower than the actual value of the background by 20%.

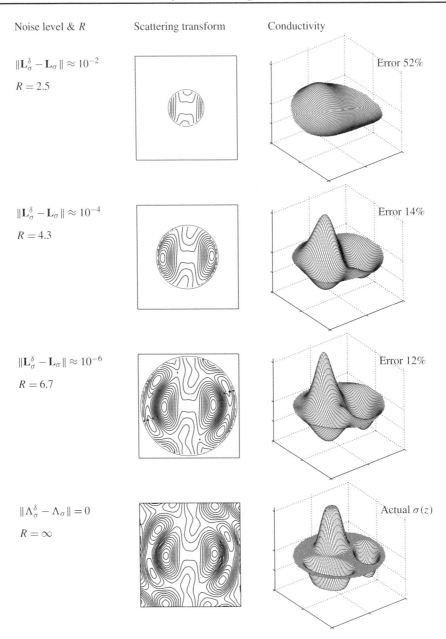

Figure 15.8. Regularized D-bar reconstructions. Left column: plots of the real part of the scattering transform, with indication of the truncation radius R. Right: corresponding reconstructions. The bottom row shows the ground truth for comparison. For a color version of these plots, see Figure 15.9.

15.5. Regularized reconstructions

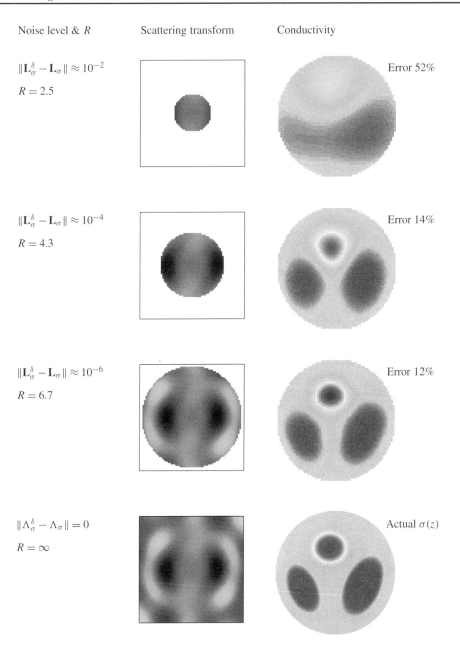

Figure 15.9. Regularized D-bar reconstructions. Left column: plots of the real part of the scattering transform, with indication of the truncation radius R. Right: corresponding reconstructions. The bottom row shows the ground truth for comparison.

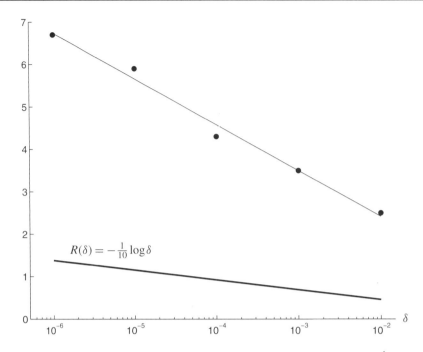

Figure 15.10. Comparison between the theoretical choice $R(\delta) = -\frac{1}{10}\log\delta$ of the cutoff frequency, and numerically observed maximal choices (black dots). The thin line is a least-squares linear fit to the observed radii. It is possible to use larger R in practice than what is assured to be safe theoretically.

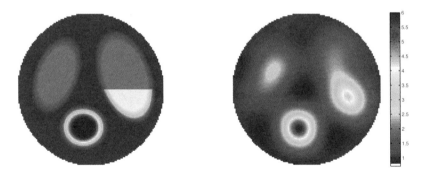

Figure 15.11. Left: Conductivity representing an idealized cross-section of a human chest with an abnormality in one lung, representing, for example, fluid. Right: reconstruction from data with 0.01% noise using a truncation radius of $R = 5.4$. Only one colorbar is included since the plots are on the same scale. The dynamic range is 102.5%

The dynamic range of the reconstruction is 102.5%. Further examples with discontinuous conductivities are provided in Chapter 16. Also, this same example, but with discontinuous organ boundaries, is reconstructed using the method of Astala and Päivärinta in Section 16.3.6.

Chapter 16
Other direct solution methods for EIT

The regularized D-bar method of Chapter 15 is arguably the most developed direct method for the solution of the inverse conductivity problem. However, it is theoretically justified only for twice differentiable conductivities, while practically interesting conductivities are typically piecewise smooth. We discuss several other methods in this chapter, all applicable to discontinuous conductivities.

We present in Section 16.1 a couple of modifications to the D-bar method of Chapter 15 which have been applied to experimental data.

The linearized method proposed by Calderón [64] is discussed in Section 16.2; it aims to find a small deviation from constant background conductivity and can be implemented as a practical algorithm. Moreover, we show how Calderón's method can be seen as an approximation to the D-bar method.

Section 16.3 shows how the uniqueness proof by Astala and Päivärinta [19] leads to a practical EIT algorithm, whose noise-robustness is based on a similar low-pass filtering than the one used in Chapter 15.

Section 16.4 deals with the *enclosure method* designed for recovering the convex hull of a set of inclusions in a known background.

16.1 D-bar methods with approximate scattering transforms

Methods that approximate the scattering transform with some degree of linearization can provide faster and simpler implementations that are also practical to use with experimental data. We will discuss two ways of computing approximations to the scattering transform, both of which involve computing lower order approximations to the CGO solutions ψ on the boundary.

16.1.1 The $\mathbf{t}^{\text{exp}}$ approximation

The first approximation we will consider is known as the $\mathbf{t}^{\text{exp}}$ approximation. The $\mathbf{t}^{\text{exp}}$ approximation was first introduced in [415] and has been used on experimental data in [234,

233, 345, 347, 346]. The idea is to approximate $\psi(z,k)|_{\partial\Omega}$ by its asymptotic behavior e^{ikz} in the computation of the scattering transform. This avoids the first step of the algorithm in Chapter 15—computing the traces of the CGO solutions ψ on the boundary—and can be viewed as a linearizing assumption since the approximation $\psi|_{\partial\Omega} \approx e^{ikx}$ is used instead of solving for ψ on the boundary using the nonlinear equation (15.22). As in Section 15.2 the scattering transform must be truncated since it grows in the presence of noise. We define

$$\mathbf{t}_R^{\exp}(k) := \begin{cases} \int_{\partial\Omega} e^{i\bar{k}\bar{z}}(\Lambda_\sigma - \Lambda_1)e^{ikz}ds & \text{for } |k| \leq R, \\ 0 & \text{for } |k| > R. \end{cases} \quad (16.1)$$

Denote the Fourier coefficients for e^{ikz} by $a_n(k)$, as in (15.48), and the continuous inner product by $\langle \cdot, \cdot \rangle$. Then

$$\mathbf{t}_R^{\exp}(k) = \sum_{m=0}^{\infty}\sum_{n=0}^{\infty} a_m(\bar{k})a_n(k)\langle e^{im\theta}, (\Lambda_\sigma - \Lambda_1)e^{in\theta}\rangle, \quad |k| \leq R. \quad (16.2)$$

It is left as an an exercise (Exercise 16.1.1) to find an expression for $\mathbf{t}_R^{\exp}$ involving these coefficients.

In [275] it was shown that the $\bar{\partial}$ equation with $\mathbf{t}_R^{\exp}$ has a unique solution which is smooth with respect to x, and the reconstruction using $\mathbf{t}_R^{\exp}$ is smooth and stable. Formula (16.1) allows the evaluation of $\mathbf{t}^{\exp}(k)$ for L^∞ conductivities, and the D-bar method is found to be effective even when the conductivity does not satisfy the assumptions of the original reconstruction theorem, as is the case with experimental data, which will be discussed below. Previous work [318, 28] shows that the exact D-bar reconstruction algorithm is stable in a restricted sense, namely, as a map defined on the range of the forward operator $\Lambda: \sigma \mapsto \Lambda_\sigma$. In [275] we show that the approximate reconstruction is continuously defined on the entire data space $\mathcal{L}(H^{1/2}(\partial\Omega), H^{-1/2}(\partial\Omega))$. As an application of the stability, we consider mollified versions γ_λ of a piecewise continuous conductivity distribution σ, and show that reconstructions of σ_λ converge to reconstructions of σ as $\lambda \to 0$. This means that no systematic artifacts are introduced when the reconstruction method is applied to conductivities outside the assumptions of the theory.

Write the corresponding D-bar equation with the scattering transform $\mathbf{t}(k)$ replaced by $\mathbf{t}_R^{\exp}$ as

$$\mu_R^{\exp}(z,s) = 1 + \frac{1}{(2\pi)^2}\int_{|k|\leq R} \frac{\mathbf{t}_R^{\exp}(k)}{(s-k)\bar{k}} e_{-z}(k)\overline{\mu_R^{\exp}(z,k)}dk_1 dk_2. \quad (16.3)$$

This defines a modified D-bar algorithm consisting of the following steps:

1. Compute $\mathbf{t}_R^{\exp}$ by (16.1).

2. Solve (16.3) for $\mu_R^{\exp}$.

3. Compute $\sigma_R^{\exp}(z) = \mu_R^{\exp}(z,0)^2$.

To establish the stability of this reconstruction algorithm, we start by formulating the reconstruction procedure as an operator. Let $L_c^p(\mathbb{R}^2)$ denote the space of $L^p(\mathbb{R}^2)$ functions with compact support, and define for $k \in \mathbb{C}$ the linear operator $\mathcal{T}_R^{\exp}: \mathcal{L}(H^{1/2}(\partial\Omega), H^{-1/2}(\partial\Omega)) \to L_c^\infty(\mathbb{R}^2)$ by

$$(\mathcal{T}_R^{\exp}L)(k) = \chi_{|k|<R}\frac{1}{4\pi\bar{k}}\int_{\partial\Omega}(e^{i\bar{k}\bar{z}}-1)L(e^{ikz}-1)ds(z). \quad (16.4)$$

16.1. D-bar methods with approximate scattering transforms

Define further for $p > 2$ the nonlinear operator

$$\mathcal{S} : L_c^p(\mathbb{R}^2) \to C^\infty(\overline{\Omega}), \qquad \phi \mapsto \mu(x,0),$$

where $\mu(x,\cdot)$ is the unique solution to (15.24) as established in [275, Lemma 3.4]. By composition we then define $\mathcal{M}_R^{\exp} : \mathcal{L}(H^{1/2}(\partial\Omega), H^{-1/2}(\partial\Omega)) \to C^\infty(\overline{\Omega})$ by

$$\mathcal{M}_R^{\exp} = \mathcal{S} \circ \mathcal{T}_R^{\exp}. \tag{16.5}$$

Using this notation it is clear that

$$(\sigma_R^{\exp}(z))^{1/2} = \mu_R^{\exp}(z,0) = \mathcal{M}_R^{\exp}(\Lambda_\sigma - \Lambda_1), \tag{16.6}$$

since $(\Lambda_\gamma - \Lambda_1)1 = 0$ and $\int_{\partial\Omega}(\Lambda_\gamma - \Lambda_1)f\,ds(x) = 0$ for all $f \in H^{1/2}(\partial\Omega)$. Thus $\mathcal{M}_R^{\exp}$ is an operator that implements the reconstruction algorithm based on the truncated approximate scattering data.

If $\mathcal{M}_R^{\exp}$ is continuous as an operator from $\mathcal{L}(H^{1/2}(\partial\Omega), H^{-1/2}(\partial\Omega))$ into $C^\infty(\overline{\Omega})$, then the reconstruction algorithm using $\mathbf{t}_R^{\exp}$ is stable. The boundedness of the operator $\mathcal{T}_R^{\exp}$ follows from the following lemma.

Lemma 16.1 (see [275, Lemma 4.1]). *The operator $\mathcal{T}_R^{\exp}$ is bounded from the space $\mathcal{L}(H^{1/2}(\partial\Omega), H^{-1/2}(\partial\Omega))$ into $L_c^\infty(\mathbb{R}^2)$ and satisfies*

$$\|\mathcal{T}_R^{\exp} L\|_{L^\infty(\mathbb{R}^2)} \leq C e^{2R} \|L\|_{\mathcal{L}(H^{1/2}(\partial\Omega), H^{-1/2}(\partial\Omega))}. \tag{16.7}$$

The continuity of the operator $\mathcal{S}$ follows from the following lemma.

Lemma 16.2 (see [275, Corollary 4.3]). *The operator $\mathcal{S}$ is bounded from $L_c^p(\mathbb{R}^2)$, $p > 2$, into $L^\infty(\Omega)$ and*

$$\|\mathcal{S}(\phi_1) - \mathcal{S}(\phi_2)\|_{L^\infty(\Omega)} \leq C \|\phi_1 - \phi_2\|_{L^p(\mathbb{R}^2)}, \tag{16.8}$$

where C depends on p, the support of ϕ_1, ϕ_2, and $\|\phi_1\|_{L^p(\mathbb{R}^2)}, \|\phi_2\|_{L^p(\mathbb{R}^2)}$.

Since the linear operator $\mathcal{T}_R^{\exp}$ is bounded and the operator $\mathcal{S}$ is continuous, this implies that $\mathcal{M}_R^{\exp} = \mathcal{S} \circ \mathcal{T}_R^{\exp}$ is continuous.

Reconstructions from several discontinuous conductivities such as the example in Section 12.5 were studied in [344] and [275]. Some of the results from these papers are included here. Reconstructions using $\mathbf{t}_R^{\exp}$ on nine discontinuous examples with varying contrast and support are found in Figure 16.1. Cross-sectional profiles of the conductivity and reconstructions are displayed to better illustrate the properties of the reconstructions. It clear that the support and contrast have a strong influence on the quality of the reconstruction using $\mathbf{t}_R^{\exp}$. Figure 16.2 illustrates the growth of the approximate scattering transform $\mathbf{t}^{\exp}$ even in the absence of noise for these nine examples. A study like this one is the topic of a project in Section 17.2.

The D-bar method with the $\mathbf{t}_R^{\exp}$ approximation was first implemented on noncircular domains in [346] where its robustness with respect to errors in input currents, output voltages, electrode placement, and domain shape modeling was studied using simulated data. The method was found to be quite robust in the presence of each kind of error. Results from

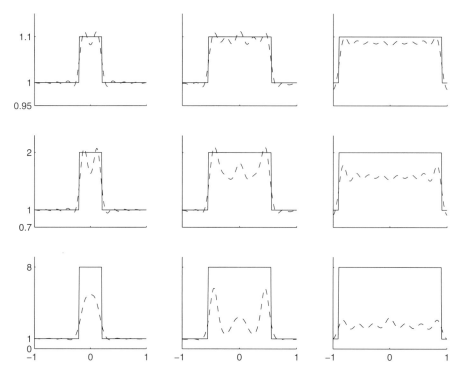

Figure 16.1. Cross-sectional profiles of the actual (solid) and reconstructed (dashed) radially symmetric discontinuous conductivity distributions. The contrast changes in each row, so that σ has a jump at r to $\sigma = 1.1$ in the top row, $\sigma = 2.0$ in the middle row, and $\sigma = 8.0$ in the bottom row. The radius at which the jump occurs varies according to column with $r = 0.2$ in the first column, $r = 0.55$ in the second column, and $r = 0.9$ in the third column. The truncation radius for $\mathbf{t}_R^{\exp}$ was taken to be $R = 15$ for the first two rows and $R = 12$ for the last row. Note that the vertical axis limits are the same in each row of plots [275].

this work on a simulated phantom chest are found in Figure 16.3. In the simulation, the chest has perimeter 900 mm, the electrodes have height and width 25.4 mm × 25.4 mm, and trigonometric current patterns (12.40) with amplitude 1 were applied in the FEM code described in Section 13.3.

Results on experimental data

To apply any of the D-bar methods we have presented here to experimental data, a matrix approximation to the DN map is first required. This is formed in the same manner as was used for the simulated voltage data computed by the FEM in Section 13.3.3. Once this map has been computed, we can proceed with the steps outlined above, computing $\mathbf{t}_R^{\exp}$ for our measured data set(s). As mentioned above, $\mathbf{t}_R^{\exp}$ will blow up in the presence of noise, as is the case with experimental data, and some experimentation is typically required to choose the truncation radius R.

16.1. D-bar methods with approximate scattering transforms   253

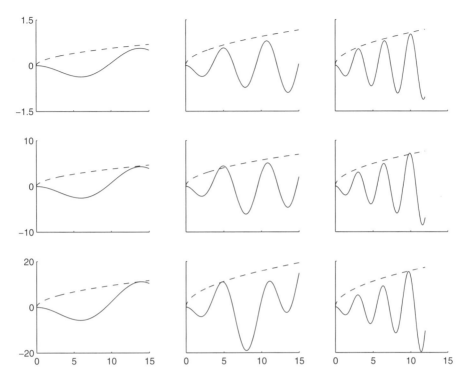

Figure 16.2. Profiles of approximate scattering transforms $\mathbf{t}_R^{\exp}$ (solid lines) for the discontinuous conductivity distributions in Figure 16.1 with constant multiples of $\sqrt{|k|}$ superimposed (dashed lines) to illustrate the growth of $\mathbf{t}^{\exp}$. Note that the vertical axis limits are the same in each row of plots [275].

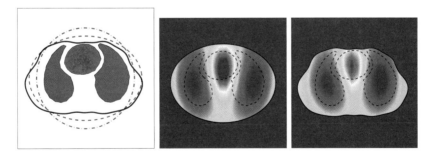

Figure 16.3. Left: an illustration of the phantom chest domain along with the simulated heart and lungs. The ellipse and circle boundaries used in [346] are included. Center: conductivity reconstruction modeling the domain as an ellipse. Right: conductivity reconstruction modeling the domain correctly as a cross-sectional chest. The conductivity was chosen to be 1 in the background, 1/3 in the lung region, and 2 in the heart region [346].

The D-bar method was first applied to experimental tank data in [234]. In this work, data was collected using the ACT3 system [130] at Rensselaer Polytechnic Institute[3] on a tank of radius 15 cm containing saline of conductivity 424 mS/m and agar of conductivity 750 mS/m simulating a heart and agar of conductivity 240 mS/m simulating lungs. Trigonometric current patterns (12.40) with amplitude 0.2 mA were applied on the 32 electrodes, which were 1.6 cm high and 2.5 cm wide. A photo of the configuration is found in Figure 16.4. Reconstructions of the tank are given in Figure 16.4 using three successive approximations to the scattering transform: first, $\mathbf{t}_R^{exp}$ with $R = 3.5$, second, an approximation introduced in the next section $\mathbf{t}_R^0$ with $R = 4$, and third, $\mathbf{t}_R^\delta(k)$ defined in Section 15.3 with $R = 3.6$, computed by Miguel Montoya. Difference images with this data set using $\mathbf{t}_R^{exp}$ can be found in [234].

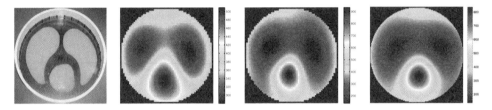

Figure 16.4. Left to right: experimental tank configuration, static reconstruction of the phantom chest using the $\mathbf{t}_R^{exp}$ with $R = 3.5$, static reconstruction of the phantom chest using the $\mathbf{t}^0$ approximation defined in Section 16.1.2 with $R = 4$ [114], static reconstruction of the phantom chest using the $\mathbf{t}_R^\delta(k)$ approximation defined in Section 15.3, computed by Miguel Montoya. The scale on the reconstructions is in mS/m.

In an experiment using data collected with an EIT system at the University of São Paulo, Brazil, a glass cup 65 mm in diameter was placed in three positions in a saline-filled circular tank 300 mm in diameter filled 14 mm high with 30 bar-shaped electrodes 10 mm wide. Currents of 4 mA were injected pairwise on the electrodes according to the "skip 1" pattern defined analogously to the "skip 3" pattern (12.43). The resistivity of the saline was approximately $1.285 \, \Omega - m$. Images of the *resistivity* (reciprocal of conductivity) are found in Figure 16.5. These were computed using $\mathbf{t}_R^{exp}$ with $R = 5.5$. Glass is a very highly resistive target and its location was determined with good precision in the reconstructions.

The D-bar method was first applied to human data in [233]. The data set was chosen to study the ability to detect conductivity changes caused by perfusion in the thorax in a healthy subject. Since the conductivity changes due to perfusion (blood flow between the heart and lungs) are smaller than changes due to ventilation, the data was collected during breath-holding. For this study, 100 frames of archival data measured by the ACT3 EIT system collected at 18 frames/sec during breath-holding were used. In the experiment the trigonometric current patterns (12.40) with an amplitude of 0.85 mA were applied on 32 electrodes 29 mm high by 24 mm wide located around the circumference of the male subject's chest. The chest had a circumference of 90 cm, and so its cross-section was modeled by a circle of radius 14.3 cm.

In [233], a new type of difference image was formed by using differences between the DN map for each frame of data and a reference data set in place of $\Lambda_\sigma - \Lambda_1$ in (16.1).

---

[3] ACT3 is a 32-electrode system operating at 28.8 kHz that applies currents and measures the real and quadrature components of the voltage on all 32 electrodes simultaneously.

16.1. D-bar methods with approximate scattering transforms

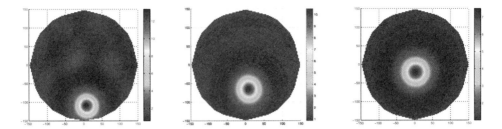

Figure 16.5. Left: static resistivity image of a glass cup near the boundary of a 300 mm diameter tank. Center: image of the cup nearer the center of the tank. Right: image of the cup in the center. The data for these images was collected on a pairwise current injection system at the University of São Paulo, Brazil.

The reference frame was chosen by hand to correspond to approximately the midpoint in the cardiac cycle. Denoting the conductivity of the reference data set by σ_{ref} and the conductivity of the jth frame by σ_j, define the truncated *differencing scattering transform* $\mathbf{t}_R^{\text{dif}}$ by

$$\mathbf{t}_R^{\text{dif}}(k;\sigma_j,\sigma_{\text{ref}}) \equiv \mathbf{t}_R^{\text{exp}}(k;\sigma_j) - \mathbf{t}_R^{\text{exp}}(k;\sigma_{\text{ref}})$$

$$= \int_{\partial\Omega} e^{i\bar{k}\bar{z}}(\Lambda_{\sigma_j}-\Lambda_1)e^{ikz}d\sigma(z) - \int_{\partial\Omega} e^{i\bar{k}\bar{z}}(\Lambda_{\sigma_{\text{ref}}}-\Lambda_1)e^{ikz}ds(z)$$

$$= \int_{\partial\Omega} e^{i\bar{k}\bar{z}}(\Lambda_{\gamma_j}-\Lambda_{\sigma_{\text{ref}}})e^{ikz}d\sigma(z). \quad (16.9)$$

Likewise, define $\mu_R^{\text{dif}}(z,s;\sigma_j,\sigma_{\text{ref}})$ by

$$\mu_R^{\text{dif}}(z,s;\sigma_j,\sigma_{\text{ref}}) = 1 + \frac{1}{(2\pi)^2}\int_{|k|\le R} \frac{\mathbf{t}_R^{\text{dif}}(k;\sigma_j,\sigma_{\text{ref}})}{(s-k)\bar{k}}e_{-z}(k)\overline{\mu_R^{\text{dif}}(z,s;\sigma_j,\sigma_{\text{ref}})}dk_1 dk_2. \quad (16.10)$$

Notice that due to the nonlinearity of the problem, this does not result in $\mu_R^{\text{dif}}(z,k;\sigma_j,\sigma_{\text{ref}}) = \mu_R^{\text{exp}}(z,k;\sigma_j) - \mu_R^{\text{exp}}(z,k;\sigma_{\text{ref}})$ and so the reconstruction is not a direct estimate of $\sigma_j - \sigma_{\text{ref}}$. However, if $\Lambda_{\sigma_{\text{ref}}} \approx \Lambda_1$, then $\mathbf{t}_R^{\text{exp}}(k;\sigma_{\text{ref}})$ is small, and $\mu_R^{\text{exp}}(z,k;\sigma_{\text{ref}}) \approx 1$ so that

$$\mu_R^{\text{exp}}(z,s;\sigma_j) \approx 1 + \frac{1}{(2\pi)^2}\int_{|k|\le R} \frac{\mathbf{t}_R^{\text{dif}}(k;\sigma_j,\sigma_{\text{ref}})}{(s-k)\bar{k}}e_{-z}(k)\overline{\mu_R^{\text{dif}}(z,s;\sigma_j,\sigma_{\text{ref}})}dk_1 dk_2$$

$$= \mu_R^{\text{dif}}(z,s;\sigma_j,\sigma_{\text{ref}}).$$

Furthermore, if the scattering transform without the Born approximation were used in the definition of $\mathbf{t}_R^{\text{dif}}$, this would not result in any meaningful scattering transform because each scattering transform requires knowledge of the CGO solution on the boundary for the non-homogeneous conductivity distribution.

The reconstructions were plotted using the ACT3 display system, which includes a low-pass filter. We selected a sequence of 24 consecutive images representing about 1.33 seconds, which was approximately one cardiac cycle at the low heart rate of this subject, 45 beats/min. In Figure 16.6 a sequence of 24 images representing approximately one cardiac cycle are displayed; dorsal is at the top, ventral is at the bottom of the images, and the

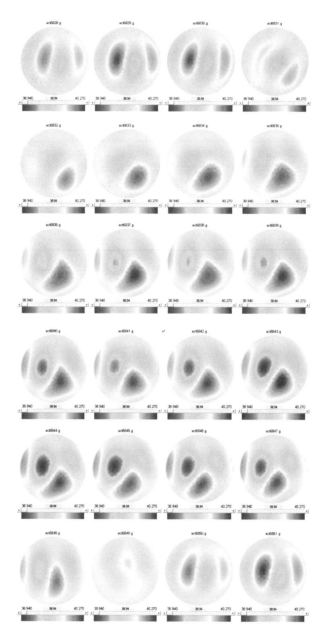

Figure 16.6. A sequence of 24 difference images from human data representing perfusion in approximately one cardiac cycle reconstructed by the D-bar method using the $\mathbf{t}_R^{\exp}$ approximation to the scattering transform. Dorsal is at the top, and the subject's left is at the viewer's right. Red represents regions of high conductivity relative to the reference image, which was in mid-contraction of the heart, and blue represents regions of low conductivity [233].

subject's left is on the viewer's right. A movie of the whole reconstructed sequence can be viewed at

http://www.math.colostate.edu/~mueller/cardiacsequence.

A thorough discussion of the images can be found in [233].

16.1.2 The t^0 approximation

More accurate approximations to the scattering transform can be computed by using the boundary integral equation (15.39). To compute an approximation to the operator S_k to first order, let $G_0(z) := -(2\pi)^{-1} \log|z|$. Then

$$I + S_k(\Lambda_\sigma - \Lambda_1) = I + S_0(\Lambda_\sigma - \Lambda_1) + \mathcal{H}_k(\Lambda_\sigma - \Lambda_1).$$

This approximation was first introduced in [344], further studied in [276], and applied to experimental data, both tank and human, in [114]. Define

$$\psi^0(z,k)|_{\partial\Omega} := e^{ikz}|_{\partial\Omega} - S_0(\Lambda_\sigma - \Lambda_1)\psi^0(\cdot,k),$$
$$\mathbf{t}^0(k) := \int_{\partial\Omega} e^{i\bar{k}\bar{z}}(\Lambda_\sigma - \Lambda_1)\psi^0(\cdot,k)ds(z).$$

For k near zero [344]

$$|\mathbf{t}(k) - \mathbf{t}^0(k)| \leq C|k|^3.$$

Let us study the functions $\mathbf{t}^{\text{exp}}$, $\mathbf{t}^0$, and $\mathbf{t}(k)$ corresponding to Example 1 in [344]. Both the conductivity σ and the potential $q = \sigma^{-1/2}\Delta\sigma^{1/2}$ are rotationally symmetric. See Figure 16.7. The scattering transform $\mathbf{t}(k)$ and Fourier transform $\widehat{q}(-2k_1, 2k_2)$ of q are plotted in Figure 16.8. The computation of $\mathbf{t}(k)$ relies on the numerical solution of the Lippmann–Schwinger equation (14.23). Notice how $\mathbf{t}(|k|) \approx \widehat{q}(2|k|)$ for large values of $|k|$.

For this radially symmetric example, we can simulate the diagonal DN map Λ_σ very precisely using Theorem 13.1 as explained in Section 13.1 and in [344]. This allows us to compute the functions $\mathbf{t}^{\text{exp}}$, $\mathbf{t}^0$, and $\mathbf{t}(k)$ very precisely; see the top plot in Figure 16.9. Since the example is a radially symmetric conductivity, the scattering transform $\mathbf{t}(k)$ and its approximations $\mathbf{t}^{\text{exp}}$ and $\mathbf{t}^0$ are real-valued and radially symmetric [415, 344]. Thus, we only plot profiles of the scattering transforms and reconstructions.

We add simulated noise of relative amplitude 0.0001 to the DN map and compute the functions $\mathbf{t}^{\text{exp}}$, $\mathbf{t}^0$, and $\mathbf{t}(k)$. As evidenced in the bottom plot in Figure 16.9, the numerical accuracy breaks down for $|k|$ exceeding roughly the threshold $|k| = 5$. Reconstructions from each of these scattering transforms are found in Figure 16.10.

Results on experimental data

In [114] the experimental tank data described in Section 16.1.1 was revisited and reconstructions were computed using the $\mathbf{t}^0$ approximation to the scattering transform. The result is shown in Figure 16.4. The improvement in spatial resolution is evident, as is an increase in dynamic range. In the same work the human chest data depicting perfusion described in Section 16.1.1 was also used to study the $\mathbf{t}^0$ approximation. Those results using a trunca-

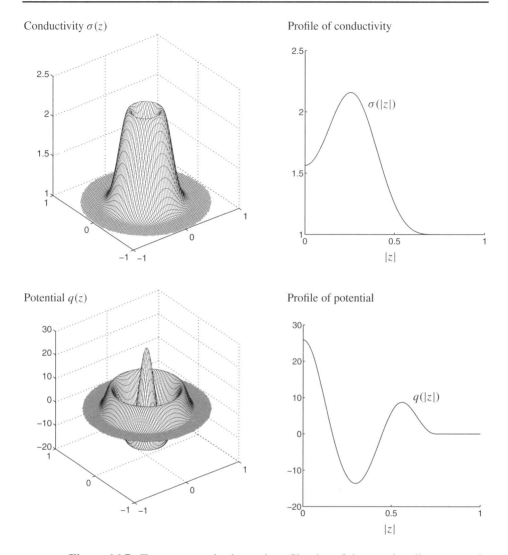

Figure 16.7. Top row: mesh plot and profile plot of the rotationally symmetric conductivity $\sigma(z) = \sigma(|z|)$. Bottom row: mesh plot and profile plot of the corresponding potential $q(z) = q(|z|)$.

tion radius of $R = 4.4$ are found in Figure 16.11. The images are not successive, but were instead chosen to give a more concise representation of the cardiac cycle. Following the reference data set (as set zero), we depict the second, fourth, sixth, ninth, tenth, sixteenth, eighteenth, twentieth, twenty-first, and twenty-second image in the sequence. The eleventh through fifteenth were omitted since that segment corresponds to diastole, and the heart region mainly becomes slowly redder (more conductive), and the lung regions become bluer (more resistive). Due to the fact that systole is a much more rapid process than diastole, more images are included during that portion of the cardiac cycle. Since exact information about the conductivity distribution inside the subject's chest is not available, the accuracy of

16.2. Calderón's method

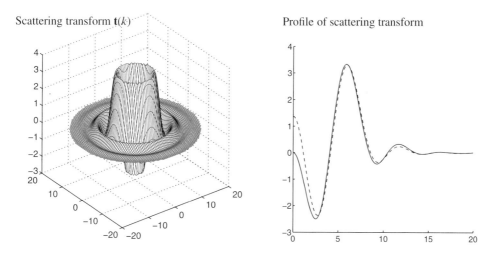

Figure 16.8. Left: mesh plot of the rotationally symmetric scattering transform $\mathbf{t}(k) = \mathbf{t}(|k|)$ corresponding to the potential q shown in Figure 16.7. Right: profile plots of the nonlinear Fourier transform $\mathbf{t}(|k|)$ (solid line) and the linear Fourier transform $\widehat{q}(2|k|)$ (dashed line). Note that the linear and nonlinear Fourier transforms are close to each other for large $|k|$; this is consistent with theory.

these images cannot be compared to those in [233]. However, the reconstructions seem to exhibit improved resolution of the heart and lungs. As in Figure 16.6, these are difference images with the same reference image as Figure 16.6.

The idea of replacing the Faddeev Green's function G_k by G_0 and computing an approximation to the CGO solutions was also employed in [189] in a D-bar algorithm for computing complex conductivities. (See Section 12.8.4 for a brief discussion.) One example from this work is included in Figure 16.12.

Exercise 16.1.1. *Use a Taylor series expansion and $z = e^{in\theta}$ to compute the Fourier coefficients of e^{ikz} and substitute them into the $\mathbf{t}^{\text{exp}}$ approximation to the scattering transform.*

Exercise 16.1.2. *Compute the scattering transform $\mathbf{t}^{\text{exp}}$ from (16.1) for a discontinuous conductivity of the form (12.25) using formula (16.2).*

16.2 Calderón's method

In addition to the proof that the linearized problem has a unique solution, the second significant result in [64] is a reconstruction method for approximating conductivities that are a small perturbation from a constant. As in Section 14.1 assume that $\sigma(z) = 1 + \delta(z)$, where the spatial variable $z = (x, y) \in \mathbb{R}^2$.

Consider the bilinear form

$$B(\Phi_1, \Phi_2) = \frac{1}{2}[Q_\sigma(w_1 + w_2) - Q_\sigma(w_1) - Q_\sigma(w_2)]. \tag{16.11}$$

Computation using ideal data

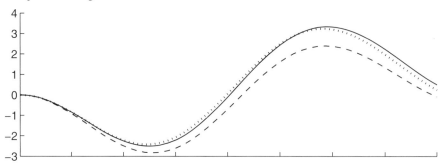

Computation using noisy data

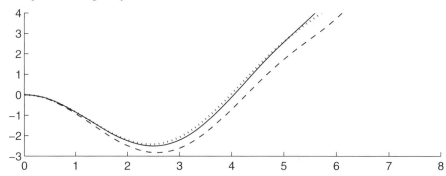

Figure 16.9. Profiles of the following radial and real-valued functions: scattering transform $\mathbf{t}^{\text{BIE}}(|k|)$ (solid line) computed by first computing ψ from (15.22), Born approximation $\mathbf{t}^{\text{exp}}(|k|)$ (dashed line) and the approximation $\mathbf{t}^0(|k|)$ (dotted line) based on the regular single-layer operator. Top plot: computations from unrealistically accurate DN map. Note that $\mathbf{t}^{\text{BIE}}(|k|)$ coincides with $\mathbf{t}(|k|)$ shown in Figure 16.8. Bottom plot: computations from DN matrix with 0.0001% added noise. Note how the accuracy is lost for $|k| > 5$ due to noise.

Then by (12.23) and choosing u_1 and u_2 as in (14.8) to be

$$u_1(z) = e^{i\pi(\xi \cdot z) + \pi(a \cdot z)}, \tag{16.12}$$

$$u_2(z) = e^{i\pi(\xi \cdot z) - \pi(a \cdot z)}, \tag{16.13}$$

where $a, \xi \in \mathbb{R}^2$ with $\xi \cdot a = 0$ and $|\xi| = |a|$,

$$B(\Phi_1, \Phi_2) = \int_D (1+\delta)(\nabla u_1 \cdot \nabla u_2) + \delta(\nabla u_1 \cdot \nabla v_2 + \nabla v_1 \cdot \nabla u_2) + (1+\delta)\nabla v_1 \cdot \nabla v_2 \, dz. \tag{16.14}$$

Note that $\nabla u_1 \cdot \nabla u_2 = -2\pi^2 |z|^2 e^{2\pi(\xi \cdot z)}$. Dividing (16.14) by $-2\pi^2 |\xi|^2$ results in

$$\frac{B(\Phi_1, \Phi_2)}{-2\pi^2 |\xi|^2} = \int_D (1+\delta) e^{2\pi(z \cdot x)} dz - R(\xi),$$

16.2. Calderón's method

Reconstructions from ideal data

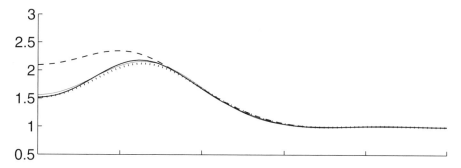

Reconstructions from noisy data

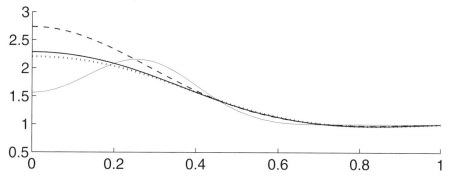

Figure 16.10. Profiles of the following radial and real-valued functions: the actual conductivity $\sigma(|z|)$ (thin solid line), reconstruction from the scattering transform $\mathbf{t}^{\text{BIE}}$ (thick solid line) computed by first computing ψ from (15.22), reconstruction from the Born approximation $\mathbf{t}^{\text{exp}}$ (dashed line) and reconstruction from the approximation $\mathbf{t}^0$ (dotted line). Top plot: computations from unrealistically accurate DN map, truncation radius $|k| = 10$. Bottom plot: computations from DN matrix with 0.0001% added noise. Truncation radius $|k| = 4$.

where

$$R(\xi) = \frac{1}{2\pi^2 |\xi|^2} \int_D \delta(\nabla u_1 \cdot \nabla v_2 + \nabla v_1 \cdot \nabla u_2) + (1+\delta)\nabla v_1 \cdot \nabla v_2 \, dz.$$

Since

$$\hat{\sigma}(\xi) = \int_D \sigma(z) e^{2\pi i (\xi \cdot z)} \, dz,$$

this results in an equation of the form

$$\hat{F}(\xi) = \hat{\sigma}(\xi) + R(\xi), \qquad (16.15)$$

where

$$\hat{F}(\xi) = -\frac{1}{2\pi^2 |\xi|^2} B(u_1, u_2).$$

Difference image with truncation radius R=4.4

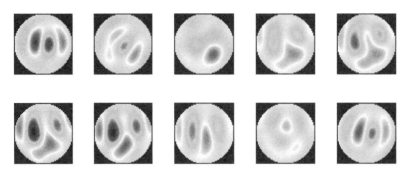

Figure 16.11. Selected reconstructions depicting perfusion in the human chest. Following the reference data set (as set zero), we display the second, fourth, sixth, ninth, tenth, sixteenth, eighteenth, twentieth, twenty-first, and twenty-second image in the sequence. Red represents regions of high conductivity relative to the reference image, and blue represents regions of low conductivity. These were computed using the $\mathbf{t}^0$ approximation to the scattering transform. Compare to Figure 16.6 [114].

If σ is extended to be zero outside D, then $\hat{\sigma}(\xi)$ can be interpreted as the Fourier transform of σ. $\hat{F}(\xi)$ can be determined from the measured data because it is given by the bilinear form (16.11), which is related to the DN data as follows:

$$B(\Phi_1, \Phi_2) = \int_{\partial\Omega} u_1 \Lambda_\sigma u_2 ds. \tag{16.16}$$

Calderón shows that the term $R(\xi)$ is bounded if the perturbation δ is small. He shows that in general

$$|R(\xi)| \leq C \|\delta\|_\infty^2 e^{2\pi|\xi|r},$$

where C is a constant and r is the radius of the smallest sphere containing the domain. However, this error can grow exponentially as ξ gets large. For the special case when

$$|\xi| \leq \frac{2-\alpha}{2\pi r} \log \frac{1}{\|\delta\|_\infty},$$

for α a constant between 1 and 2, one has

$$|R(\xi)| \leq C \|\delta\|_\infty^\alpha.$$

In light of this error bound, $\hat{F}(\xi)$ is a good approximation for $\hat{\sigma}$, provided δ is small. Note that if ξ is too large, the error will also increase, and in fact the reconstructions become highly oscillatory.

Since $\sigma \in L^\infty$ as Calderón suggests, it is a good idea to apply a mollifier η to avoid a Gibb's phenomenon in the inversion. The mollifier, η, is a compactly supported smooth function on $\mathbb{R}^n$ with $\int_{\mathbb{R}^n} \eta(y) dy = 1$. Since it must decay exponentially, the Fourier transform of the mollifier, $\hat{\eta}$, is a C^∞ function. Furthermore, the support of $\hat{\eta}$ is in $\{\xi : |\xi| \leq 1\}$

16.2. Calderón's method

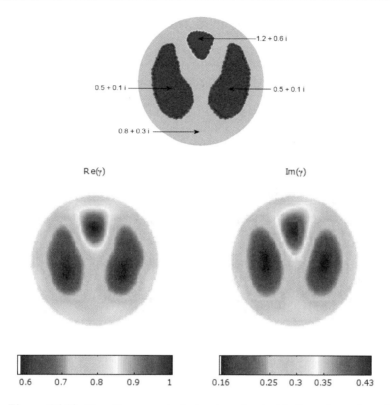

Figure 16.12. Top: the numerical phantom chest with the values of conductivity and permittivity indicated. Bottom: reconstructions of the real (left) and imaginary (right) parts of the conductivity from 0.01% added noise [189]. The cutoff frequency was $R = 4.3$. The dynamic range is 60% for the conductivity and 56% for the permittivity [189].

and $\hat{\eta}(0) = 1$. The usual properties of the Fourier transform imply that $\hat{\eta}\left(\frac{\xi}{\beta}\right) = \beta^n \eta(\beta z) =: \eta_\beta(z)$. Equation (16.15) becomes

$$\hat{F}(\xi)\hat{\eta}\left(\frac{\xi}{\beta}\right) = \hat{\sigma}(\xi)\hat{\eta}\left(\frac{\xi}{\beta}\right) + R(\xi)\hat{\eta}\left(\frac{\xi}{\beta}\right). \qquad (16.17)$$

Taking the inverse Fourier transform results in

$$(\sigma * \eta_\beta)(z) = (F * \eta_\beta)(z) + \rho(z), \qquad (16.18)$$

where $*$ denotes convolution. Calderón establishes an error bound on $\rho(z)$ that goes to zero if $\delta \to 0$. This results in the following formula to approximate σ:

$$\sigma(z) \approx (\sigma * \eta_\beta)(z) \approx (F * \eta_\beta)(z). \qquad (16.19)$$

For further details the reader is referred to the original paper [64], or to [48], which includes an implementation with experimental data. The results in the latter work suggest that the use of a mollifier as discussed by Calderón in [64] result in more accurate reconstructions.

16.2.1 Calderón's method as an approximation of the D-bar method

The connection of Calderón's method to the D-bar method described in Section 16.1.1 was derived in [275]. Following the analysis in [275], we write Calderón's method in the context of the approximate scattering transform $\mathbf{t}^{\exp}$ defined in (16.1) and compare it to the D-bar method.

Integrating by parts in (16.1) gives

$$\mathbf{t}^{\exp}(k) = \int_\Omega (\sigma - 1)\nabla u(z,k) \cdot \nabla(e^{i\overline{k}\bar{z}}) dz, \qquad (16.20)$$

where

$$\nabla \cdot (\sigma - 1)\nabla u = 0 \text{ in } \Omega, \quad u|_{\partial\Omega} = e^{ikz}.$$

When $\|\sigma - 1\|_{L^\infty(\Omega)}$ is small then u is close to e^{ikz} inside Ω. Indeed, if we write $u = e^{ikz} + \delta u$ for $\delta u \in H_0^1(\Omega)$ satisfying $\nabla \cdot \sigma \nabla \delta u = -\nabla \cdot (\sigma - 1)\nabla(e^{ikz})$, we have the estimate

$$\|\delta u\|_{H^1(\Omega)} \leq C\|\sigma - 1\|_{L^\infty(\Omega)} e^{|k|r}, \qquad (16.21)$$

where r is the radius of the smallest ball containing Ω. Substituting $u = e^{ikz} + \delta u$ into (16.20) and dividing by $-2|k|^2$ we obtain

$$-\frac{\mathbf{t}^{\exp}(k)}{2|k|^2} = -\frac{1}{2|k|^2} \int_\Omega (\sigma - 1)\nabla(e^{ikz} + \delta u) \cdot \nabla(e^{i\overline{k}\bar{z}}) dx$$

$$= \int_\Omega (\sigma - 1) e^{i(kz + \bar{k}\bar{z})} dz + R(k)$$

$$= 2\pi \mathcal{F}(\chi_\Omega(\sigma - 1))(-2k_1, 2k_2) + R(k), \qquad (16.22)$$

where $\mathcal{F}$ denotes the Fourier transform and

$$R(k) = -\frac{1}{2|k|^2} \int_\Omega (\sigma - 1)\nabla \delta u \cdot \nabla(e^{i\overline{k}\bar{z}}) dz.$$

Using (16.21), it is not difficult to obtain

$$|R(k)| \leq C\|\sigma - 1\|_{L^\infty(\Omega)}^2 e^{2|k|r}. \qquad (16.23)$$

Recall that the idea behind Calderón's method is to multiply (16.22) by a smooth cutoff function and then apply the inverse Fourier transform. Let $\hat{\eta} \in C_0^\infty(\mathbb{R}^2)$ be a nonnegative function supported in the unit ball with $\hat{\eta} = 1$ near the origin, and let β be a positive parameter determining the cutoff radius. Then from (16.22) we obtain

$$\mathcal{F}(\chi_\Omega(\sigma - 1))(-2k_1, 2k_2)\hat{\eta}(k/\beta) = -\frac{\mathbf{t}^{\exp}(k)}{4\pi|k|^2}\hat{\eta}(k/\beta) - R(k)\hat{\eta}(k/\beta).$$

Changing variables $s = (s_1, s_2) = 2(-k_1, k_2)$ gives

$$\mathcal{F}(\chi_\Omega(\sigma - 1))(s_1, s_2)\hat{\eta}\left(\frac{(-s_1, s_2)}{2\beta}\right)$$
$$= -\frac{\mathbf{t}^{\exp}((-s_1, s_2)/2)}{\pi|s|^2}\hat{\eta}\left(\frac{(-s_1, s_2)}{2\beta}\right) - R\left(\frac{(-s_1, s_2)}{2}\right)\hat{\eta}\left(\frac{(-s_1, s_2)}{2\beta}\right). \qquad (16.24)$$

16.2. Calderón's method

Inverting $\mathcal{F}$ and neglecting the second term in (16.24) yields an approximation to σ:

$$\sigma^{\text{app}}(z) - 1 = -\frac{1}{2\pi} \int_{\mathbb{R}^2} e^{iz \cdot s} \frac{\mathbf{t}^{\exp}((-s_1, s_2)/2)}{\pi |s|^2} \hat{\eta}\left(\frac{(-s_1, s_2)}{2\beta}\right) ds_1 ds_2.$$

Changing back the variables in the integral to $(k_1, k_2) = (-s_1, s_2)/2$ yields the formula

$$\begin{aligned}
\sigma^{\text{app}}(z) - 1 &= -\frac{1}{2\pi^2} \int_{\mathbb{R}^2} e^{2i(-xk_1 + yk_2)} \frac{\mathbf{t}^{\exp}(k)}{4|k|^2} \hat{\eta}(k/\beta) 4 dk_1 dk_2 \\
&= -\frac{2}{(2\pi)^2} \int_{\mathbb{R}^2} e^{-i(zk + \bar{z}\bar{k})} \frac{\mathbf{t}^{\exp}(k)}{|k|^2} \hat{\eta}(k/\beta) dk_1 dk_2.
\end{aligned} \quad (16.25)$$

The reconstruction σ^{app} is an approximation of a low-pass filtered version of σ. In light of (16.23), choosing the parameter β as in [64] with $0 < \alpha < 1$ to be

$$\beta = \frac{1-\alpha}{2r} \log \frac{1}{\|\sigma - 1\|_{L^\infty(\Omega)}} \quad (16.26)$$

results in the error estimate

$$\begin{aligned}
\|\sigma^{\text{app}}(z) - \eta_\beta * \sigma\|_{L^\infty} &\leq \|R(k)\hat{\eta}(k/\beta)\|_{L^1(\mathbb{R}^2)} \\
&\leq C\|\sigma - 1\|_{L^\infty(\Omega)}^{1+\alpha} (\log(\|\sigma - 1\|_{L^\infty(\Omega)}))^2.
\end{aligned}$$

Note that when $\|\sigma - 1\|_{L^\infty(\Omega)}$ is sufficiently small this error is much smaller than $\|\sigma - 1\|_{L^\infty(\Omega)}$. Since in most applications the approximate magnitudes of the conductivities comprising $\sigma(z)$ are known, an estimate to σ in (16.26) can be computed.

In summary, the algorithm proposed in [64] is tantamount to the following:

1. Compute $\mathbf{t}^{\exp}(k)$ by (16.1).

2. Construct a low-pass filter $\hat{\eta}(k/\beta)$.

3. Compute the approximation σ^{app} by (16.25).

Calderón's method using (16.25) can be seen as a three-step approximation of the D-bar method:

1. In (15.27), $\mathbf{t}(k)$ is approximated by $\mathbf{t}^{\exp}(k)\hat{\eta}(k/\beta)$, where $\hat{\eta}(k/\beta)$ is a smooth cutoff function.

2. The function μ in the integral in the right-hand side of (15.27) is approximated by its asymptotic value $\mu \sim 1$.

3. The square function in $\sigma(z) \approx (m(z, 0))^2$ is linearized: $(1 - h)^2 \sim 1 - 2h$.

In contrast, the D-bar method using $\mathbf{t}_R^{\exp}$ makes only the first approximation (with sharp cutoff).

Exercise 16.2.1. *Prove formula (16.16), relating the bilinear form (16.11) to the DN data.*

Exercise 16.2.2. *Find the Fourier expansions of the CGO solutions u_1 and u_2 defined by (16.12) and (16.13), respectively.*

Exercise 16.2.3. *Prove the estimate (16.23).*

16.3 The Astala–Päivärinta method

The constructive uniqueness proof given by Astala and Päivärinta in [19, 18] answered the long-standing uniqueness question in two dimensions for conductivities in L^∞. A computational reconstruction method based on that proof was introduced in [16]. The method is noniterative, provides a noise-robust solution of the full nonlinear EIT problem, and assumes no regularity in the conductivity distribution, contrary to previous direct EIT algorithms in the literature. In particular, the method presented here and published in [16] is applicable to the class of piecewise smooth conductivities. This class is important since it describes the conductivity distribution in the human body, as well as those arising in process tomography, such as stratified media in a pipeline.

The reconstruction procedure consists of three steps:

(i) Recover traces of CGO solutions on the boundary $\partial\Omega$ from the DN map by solving a boundary integral equation. See Section 16.3.3 for details.

(ii) Compute approximate values of CGO solutions inside the unit disc using the *low-pass transport matrix*. See Section 16.3.4.

(iii) Reconstruct the conductivity from the recovered values of the CGO solutions inside Ω using differentiation and simple algebra. See Section 16.3.5.

Computationally the most demanding part of the reconstruction method is the evaluation of the low-pass transport matrix in Step (ii). This step is also the main difference between this direct reconstruction procedure and others based on D-bar equations.

16.3.1 CGO solutions

Let $\Omega \subset \mathbb{R}^2$ be the unit disc and let $\sigma : \Omega \to (0, \infty)$ be an essentially bounded measurable function satisfying $\sigma(z) \geq c > 0$ for almost every $z \in \Omega$. For simplicity, assume that $\sigma \equiv 1$ outside Ω.

Let $k \in \mathbb{C}$. We wish to construct a unique solution u_1 to

$$\nabla \cdot (\sigma(z)\nabla u_1(z,k)) = 0, \quad u_1(z,k) \sim e^{ikz}, \quad \text{as } |z| \to \infty, \tag{16.27}$$

and a unique solution u_2 to

$$\nabla \cdot \left(\frac{1}{\sigma(z)}\nabla u_2(z,k)\right) = 0, \quad u_2(z,k) \sim ie^{ikz}, \quad \text{as } |z| \to \infty. \tag{16.28}$$

The exponential behavior of the CGO solutions is used for constructing a nonlinear Fourier analysis for the inverse conductivity problem, and k can be thought of as a frequency-domain variable.

Solutions for (16.27) and (16.28) can be constructed by defining

$$\mu(z) = \frac{1-\sigma(z)}{1+\sigma(z)} \tag{16.29}$$

and considering solutions to the Beltrami equation

$$\overline{\partial}_z f_\mu = \mu \overline{\partial_z f_\mu}. \tag{16.30}$$

It was shown in Section 14.4 how to construct solutions to (16.30) with the following

16.3. The Astala–Päivärinta method

asymptotic behavior:

$$f_\mu(z,k) = e^{ikz}(1+\omega(z,k)), \quad \text{with} \quad \omega(z,k) = \mathcal{O}\left(\frac{1}{z}\right) \text{ as } |z| \to \infty. \quad (16.31)$$

Now the following theorem from [19] gives the connection between equations (16.27), (16.28), and (16.30).

Theorem 16.3. *Assume $u \in H^1(D)$ is real-valued and satisfies (16.27). Then there exists $v \in H^1(D)$, unique up to a constant, such that $f = u + iv$ satisfies (16.30), where μ is defined by (16.29). Conversely, if $f \in H^1(D)$ satisfies (16.30) with a real-valued μ, then $u = \text{Re}(f)$ and $v = \text{Im}(f)$ satisfy*

$$\nabla \cdot (\sigma \nabla u) = 0 \quad \text{and} \quad \nabla \cdot \left(\frac{1}{\sigma} \nabla v\right) = 0,$$

respectively, where $\sigma = (1-\mu)/(1+\mu)$.

Proof. Let w be the vector $w^T = [-\sigma \partial_2 u, \sigma \partial_1 u]$, where $\partial_1 = \frac{\partial}{\partial x}, \partial_2 = \frac{\partial}{\partial y}$. Then by (16.30), $\partial_2 w_1 = \partial_1 w_2$, since

$$\partial_2 w_1 - \partial_1 w_2 = -\partial_2(\sigma \partial_2 u) - \partial_1(\sigma \partial_1 u)$$
$$= -\nabla \cdot (\sigma u)$$
$$= 0.$$

Thus, there exists $v \in H^1(D)$, unique up to a constant such that

$$\partial_1 v = -\sigma \partial_2 u \quad \text{and} \quad \partial_2 v = \sigma \partial_1 u.$$

Let $f = u + iv$. Then f satisfies (16.30).

The proof of the converse is an exercise. $\square$

Note that the function v defined in Theorem 16.3 is only defined up to a constant. We enforce uniqueness through the normalization

$$\int_{\partial \Omega} v \, ds = 0.$$

16.3.2 The transport equation

Transporting the values of CGO solutions from $\mathbb{R}^2 \setminus \overline{\Omega}$ into Ω is a crucial technique in the uniqueness proof in [19].

According to [19, formula (8.3)], the solutions u_1 and u_2 of equations (16.27) and (16.28) are real-linearly independent:

$$u_2(z,k) \neq 0, \quad \frac{u_1(z,k)}{u_2(z,k)} \notin \mathbb{R},$$

where $z \in \mathbb{R}^2$ and $k \in \mathbb{C}$ are arbitrary. Thus we can write, for any $z_0 \in \mathbb{R}^2 \setminus \overline{\Omega}$,

$$u_1(z,k) = a_1 u_1(z_0,k) + a_2 u_2(z_0,k), \quad (16.32)$$
$$u_2(z,k) = b_1 u_1(z_0,k) + b_2 u_2(z_0,k), \quad (16.33)$$

where $a_j = a_j(z, z_0, k)$ and $b_j = b_j(z, z_0, k)$ are real-valued functions. We define the *transport matrix* by

$$T := \begin{bmatrix} a_1 & a_2 \\ b_1 & b_2 \end{bmatrix}. \tag{16.34}$$

Of course, the pair (16.32), (16.33) of equations is not useful for transporting values of solutions from z_0 into z unless we know a_j and b_j. But they can be determined from boundary measurements as follows.

Define

$$h_+ = \frac{1}{2}(f_\mu + f_{-\mu}), \qquad h_- = \frac{i}{2}(\overline{f_\mu} - \overline{f_{-\mu}}). \tag{16.35}$$

It is shown in [19, Section 5] that the following D-bar equations in the k variable hold for h_+ and h_-:

$$\overline{\partial}_k h_+ = \tau_\mu \overline{h_-}, \quad \overline{\partial}_k h_- = \tau_\mu \overline{h_+}.$$

It is then an exercise to show that both u_1 and u_2 satisfy the $\overline{\partial}_k$ equation

$$\overline{\partial}_k u_j(z, k) = -i\tau_\mu(k)\overline{u_j(z,k)}, \tag{16.36}$$

with the normalizations $u_1(z, 0) \equiv 1$ and $u_2(z, 0) \equiv i$. The coefficient τ_μ in (16.36) is defined by

$$\tau_\mu(k) = \frac{1}{2}(t_\mu(k) - t_{-\mu}(k)),$$

where

$$t_\mu(k) = \frac{1}{\pi} \int_\Omega \mu(z) \partial(e_k(z) M_\mu(z, k)) dz.$$

It is shown in [19, Section 8] that computing the $\overline{\partial}_k$ derivative of (16.32) and denoting $\alpha = a_1 + ia_2$ leads to

$$\overline{\partial}_k \alpha(z, z_0, k) = \nu_{z_0}(k) \overline{\partial_k \alpha(z, z_0, k)}, \tag{16.37}$$

where

$$\nu_{z_0}(k) := i \frac{h_-(z_0, k)}{h_+(z_0, k)}. \tag{16.38}$$

It also turns out that $\beta_\mu = i\alpha_{-\mu}$.

The reconstruction method for infinite-precision data described in [19] is based on the fact that the values of the CGO solutions at z_0 can be recovered from the knowledge of Λ_σ. Then ν_{z_0} can be evaluated using (16.38) and α and β found by solving the Beltrami equation (16.37) with certain asymptotics.

16.3.3 The boundary integral equation

With ω defined by (14.48), let

$$M_\mu(z, k) = 1 + \omega(z, k). \tag{16.39}$$

Then the following boundary integral equation holds [18]:

$$M_\mu(\cdot, k)|_{\partial\Omega} + 1 = (\mathcal{P}_\mu^k + \mathcal{P}_0) M_\mu(\cdot, k)|_{\partial\Omega}, \tag{16.40}$$

where $\mathcal{P}_\mu^k$ and $\mathcal{P}_0$ are projection operators to be discussed in more detail below.

16.3. The Astala–Päivärinta method

The numerical solution of (16.40) is done by

- writing real and imaginary parts separately;
- replacing all the operators by their $(4N+2) \times (4N+2)$ matrix approximations; and
- solving the resulting finite linear system for $|k| \leq R$, where $R > 0$ depends on the noise level.

There is an important connection between the projection operators in (16.40) and the DN map Λ_σ. This connection requires an understanding of the μ-Hilbert transform $\mathcal{H}_\mu : H^{1/2}(\partial\Omega) \to H^{1/2}(\partial\Omega)$ defined by mapping

$$\mathcal{H}_\mu : u_1|_{\partial\Omega} \longrightarrow u_2|_{\partial\Omega}.$$

However, this definition only holds for real-valued functions u. To extend this to complex-valued functions in $H^{1/2}(\partial\Omega)$, define

$$\mathcal{H}_\mu(iu) = i\mathcal{H}_{-\mu}(u).$$

Theorem 16.4. *The DN map Λ_σ uniquely determines $\mathcal{H}_\mu$, $\mathcal{H}_{-\mu}$, and $\Lambda_{\sigma^{-1}}$.*

Proof. Note first since v is the real part of $g = -if$, and $\overline{\partial}g \equiv -\mu\overline{\partial g}$,

$$\mathcal{H}_\mu \circ \mathcal{H}_{-\mu} u = \mathcal{H}_{-\mu} \circ \mathcal{H}_\mu u = -u + \frac{1}{2\pi}\int_{\partial\Omega} u\,ds.$$

So $\mathcal{H}_\mu$ uniquely determines $\mathcal{H}_{-\mu}$.

Choose a counterclockwise orientation of $\partial\partial\Omega$, and let ∂_T denote the tangential derivative on $\partial\partial\Omega$. Then $\partial_T = -\partial_1 + \partial_2$. By definition of Λ_σ, for $\phi \in C^\infty(\bar{\Omega})$,

$$\int_{\partial\Omega} \phi \Lambda_\sigma u\, ds = \int_\Omega \nabla\phi \cdot \sigma \nabla u\, dx$$
$$= \int_\Omega ((\partial_1\phi)\sigma\partial_1 u + (\partial_2\phi)\sigma\partial_2 u)\,dx$$
$$= \int_\Omega ((\partial_1\phi)\partial_2 v - (\partial_2\phi)\partial_1 v))\,dx$$
$$= \int_{\partial\Omega} (v\partial_1\phi - v\partial_2\phi)\,ds - \int_\Omega (v\partial_2\partial_1\phi - v\partial_1\partial_2\phi)\,dx$$
$$= -\int_{\partial\Omega} v\partial_T\phi\,ds.$$

So in the weak sense, $\partial_T \mathcal{H}_\mu(u) = \Lambda_\sigma u$. Note

$$-\mu = \frac{1 - \frac{1}{\sigma}}{1 + \frac{1}{\sigma}},$$

so $\Lambda_{\frac{1}{\sigma}}(u) = \partial_T \mathcal{H}_{-\mu}(u)$. □

From the proof, we see that in the weak sense for real-valued functions $g \in H^{1/2}(\partial\Omega)$,

$$\partial_T \mathcal{H}_\mu g = \Lambda_\sigma g, \qquad (16.41)$$

where ∂_T is the tangential derivative map along the boundary.

We will consider the trigonometric basis functions

$$\phi_n(\theta) = \begin{cases} \pi^{-1/2}\cos((n+1)\theta/2) & \text{for odd } n, \\ \pi^{-1/2}\sin(n\theta/2) & \text{for even } n. \end{cases} \qquad (16.42)$$

The DN map Λ_σ is approximately represented by the matrix $\mathbf{L}_\sigma$ defined by

$$(\mathbf{L}_\sigma)_{m,n} := \langle \Lambda_\sigma \phi_n, \phi_m \rangle.$$

The tangential derivative map ∂_T can be approximated in the basis (16.42) by the matrix $\mathbf{D}_T$ (see Exercise 16.3.2):

$$\mathbf{D}_T = \begin{bmatrix} 0 & 1 & & & & & \\ -1 & 0 & & & & & \\ & & 0 & 2 & & & \\ & & -2 & 0 & & & \\ & & & & \ddots & & \\ & & & & & 0 & N \\ & & & & & -N & 0 \end{bmatrix}. \qquad (16.43)$$

Define an averaging operator

$$\mathcal{L}\phi := |\partial\Omega|^{-1} \int_{\partial\Omega} \phi\, ds.$$

The real-linear operator $\mathcal{P}_\mu : H^{1/2}(\partial\Omega) \to H^{1/2}(\partial\Omega)$ is defined by

$$\mathcal{P}_\mu g = \frac{1}{2}(I + i\mathcal{H}_\mu)g + \frac{1}{2}\mathcal{L}g, \qquad (16.44)$$

where g may be complex-valued. Further, denote

$$\mathcal{P}_\mu^k := e^{-ikz}\mathcal{P}_\mu e^{ikz}. \qquad (16.45)$$

By equations (13.42), (16.41), and (16.43) we can approximate $\mathcal{H}_\mu$ acting on real-valued, zero-mean functions expanded in the basis (16.42) by

$$\widetilde{\mathbf{H}}_\mu := \mathbf{D}_T^{-1}\mathbf{L}_\sigma. \qquad (16.46)$$

In general, the traces of the CGO solutions at $\partial\Omega$ do not have mean zero, and so we append the basis function $\phi_0 = (2\pi)^{-1/2}$ to (16.42). This leads to the following $(2N+1) \times (2N+1)$ matrix approximation to the μ-Hilbert transform $\mathcal{H}_\mu$:

$$\mathbf{H}_\mu := \begin{bmatrix} 0 & 0 \\ 0 & \widetilde{\mathbf{H}}_\mu \end{bmatrix}. \qquad (16.47)$$

16.3. The Astala–Päivärinta method

Furthermore, we can approximate $\mathcal{H}_{-\mu}$ when we have $\mathbf{H}_\mu$ available. We have the identity

$$\mathcal{H}_\mu \circ (-\mathcal{H}_{-\mu})u = (-\mathcal{H}_{-\mu}) \circ \mathcal{H}_\mu u = u - \mathcal{L}u, \tag{16.48}$$

so $-\mathcal{H}_{-\mu}$ is the inverse operator of $\mathcal{H}_\mu$ in the subspace of zero-mean functions. Thus we may define a $(2N+1) \times (2N+1)$ matrix approximation to $\mathcal{H}_{-\mu}$:

$$\mathbf{H}_{-\mu} := \begin{bmatrix} 0 & 0 \\ 0 & -\widetilde{\mathbf{H}}_\mu^{-1} \end{bmatrix}. \tag{16.49}$$

Summarizing, once we have measured the DN matrix (13.42), we can approximate the μ-Hilbert transforms $\mathcal{H}_{\pm\mu}$ by matrices $\mathbf{H}_{\pm\mu}$ acting on the basis (16.42) augmented by the constant basis function $\phi_0 \equiv (2\pi)^{-1/2}$.

We represent complex-valued functions $g \in H^{1/2}(\partial\Omega)$ by expanding the real and imaginary parts separately and organizing the coefficients as the following vertical vector in $\mathbb{R}^{4N+2}$:

$$\widehat{g} = \begin{bmatrix} \langle \operatorname{Re} g, \phi_0 \rangle \\ \langle \operatorname{Re} g, \phi_1 \rangle \\ \vdots \\ \langle \operatorname{Re} g, \phi_{2N} \rangle \\ \langle \operatorname{Im} g, \phi_0 \rangle \\ \langle \operatorname{Im} g, \phi_1 \rangle \\ \vdots \\ \langle \operatorname{Im} g, \phi_{2N} \rangle \end{bmatrix} \in \mathbb{R}^{4N+2}.$$

Denote the direct and inverse transform by $\widehat{g} = \widetilde{\mathcal{F}} g$ and $g = \widetilde{\mathcal{F}}^{-1} \widehat{g}$. Note that the $(2N+1) \times (2N+1)$ matrix representation of the averaging operator $\mathcal{L}$ is given by

$$\mathbf{L} = \frac{1}{\sqrt{2\pi}} \operatorname{diag}[1, 0, \ldots, 0, 1, 0, \ldots, 0],$$

where the ones are located at elements $(1,1)$ and $(2N+2, 2N+2)$.

Now we can solve the boundary integral equation (16.40) approximately by solving the following equation for the transform coefficients of M_μ:

$$(I - \mathbf{P}_\mu^k - \mathbf{P}_0)\widehat{M}_\mu(\cdot, k)|_{\partial\Omega} = -\widetilde{\mathcal{F}}(1). \tag{16.50}$$

Here $\mathbf{P}_\mu^k$ and $\mathbf{P}_0$ stand for approximate implementations of the actions of the operators $\mathcal{P}_\mu^k$ and $\mathcal{P}_0$, respectively. The action of $\mathcal{P}_\mu$ (defined in (16.44)) in the transform domain is

$$\mathbf{P}_\mu \widehat{g} = \frac{1}{2}\left(I + i\begin{bmatrix} \mathbf{H}_\mu & 0 \\ 0 & \mathbf{H}_{-\mu} \end{bmatrix}\right)\widehat{g} + \frac{1}{2}\mathbf{L}\widehat{g},$$

where $\mathbf{H}_\mu$ and $\mathbf{H}_{-\mu}$ are given by (16.47) and (16.49), respectively. The action of $\mathcal{P}_\mu^k$ (defined in (16.45)) in the transform domain is

$$\mathbf{P}_\mu^k \widehat{g} = \widetilde{\mathcal{F}}\left(e^{-ikz} \cdot \widetilde{\mathcal{F}}^{-1}\left(\mathbf{P}_\mu \widetilde{\mathcal{F}}\left(e^{ikz} \cdot (\widetilde{\mathcal{F}}^{-1}\widehat{g})\right)\right)\right). \tag{16.51}$$

Now one can solve (16.50) for $\widehat{M}_\mu$ iteratively using GMRES.

Because of the conjugation with exponential functions in (16.51), the errors in $\mathbf{H}_\mu$ and $\mathbf{H}_{-\mu}$ get multiplied with numbers exponentially large in k, and so we can only reliably numerically solve the boundary integral equation (16.40) for k ranging in a disc $D(0, R)$, where the radius $R > 0$ depends on the noise level. This is where the exponential ill-posedness of the inverse conductivity problem shows up in the reconstruction method.

16.3.4 The low-pass transport matrix

Once M_μ is known on the boundary, so is $f_\mu(\cdot, k)|_{\partial\Omega}$ by (16.31) and (16.39). Equation (14.47) and the fact that μ is supported in Ω implies that $f_\mu(\cdot, k)$ is harmonic outside the unit disc. Thus, the coefficients in the Fourier series for the trace of f_μ on $\partial\Omega$ can also be used to expand f_μ as a power series outside Ω.

Choose then a point $z_0 \in \mathbb{R}^2 \setminus \overline{\Omega}$. As explained above, we know $f_\mu(z_0, k)$ and $f_{-\mu}(z_0, k)$ for any $|k| < R$. Use (16.35) to construct the function

$$v_{z_0}^{(R)}(k) := \begin{cases} i\dfrac{h_-(z_0,k)}{h_+(z_0,k)} & \text{for } |k| < R, \\ 0 & \text{for } |k| \geq R. \end{cases} \tag{16.52}$$

We next solve the truncated Beltrami equations

$$\overline{\partial}_k \alpha^{(R)} = v_{z_0}^{(R)}(k) \overline{\partial_k \alpha^{(R)}}, \tag{16.53}$$

$$\overline{\partial}_k \beta^{(R)} = v_{z_0}^{(R)}(k) \overline{\partial_k \beta^{(R)}}, \tag{16.54}$$

with solutions represented in the form

$$\alpha^{(R)}(z, z_0, k) = \exp(ik(z - z_0) + \varepsilon(k)), \tag{16.55}$$

$$\beta^{(R)}(z, z_0, k) = i\exp(ik(z - z_0) + \widetilde{\varepsilon}(k)), \tag{16.56}$$

where $\varepsilon(k)/k \to 0$ and $\widetilde{\varepsilon}(k)/k \to 0$ as $k \to \infty$. Requiring

$$\alpha^{(R)}(z, z_0, 0) = 1 \quad \text{and} \quad \beta^{(R)}(z, z_0, 0) = i \tag{16.57}$$

fixes the solutions uniquely.

Fix any nonzero $k_0 \in \mathbb{C}$ and choose any point z inside the unit disc. Denote $\alpha^{(R)} = a_1^{(R)} + ia_2^{(R)}$ and $\beta^{(R)} = b_1^{(R)} + ib_2^{(R)}$. We can now use the truncated transport matrix

$$T^{(R)} = T_{z,z_0,k_0}^{(R)} := \begin{pmatrix} a_1^{(R)} & a_2^{(R)} \\ b_1^{(R)} & b_2^{(R)} \end{pmatrix} \tag{16.58}$$

to compute

$$u_1^{(R)}(z, k_0) = a_1^{(R)} u_1(z_0, k_0) + a_2^{(R)} u_2(z_0, k_0), \tag{16.59}$$
$$u_2^{(R)}(z, k_0) = b_1^{(R)} u_1(z_0, k_0) + b_2^{(R)} u_2(z_0, k_0).$$

The truncation in (16.52) can be interpreted as a nonlinear low-pass filter in the k-plane. This is where the term *low-pass transport matrix* originates.

16.3. The Astala–Päivärinta method

To construct the truncated transport matrix we first find solutions η_1 and η_2 to the equation

$$\overline{\partial}_k \eta_j = \nu_{z_0}^{(R)}(k) \overline{\partial_k \eta_j}, \qquad (16.60)$$

with asymptotics

$$\eta_1(k) = e^{ik(z-z_0)}(1 + \mathcal{O}(1/k)), \qquad (16.61)$$
$$\eta_2(k) = i\, e^{ik(z-z_0)}(1 + \mathcal{O}(1/k)), \qquad (16.62)$$

respectively, as $|k| \to \infty$. Such solutions exist and are unique by [19, Theorem 4.2].

The solutions η_j are complex valued, but pointwise real-linearly independent by [14, Theorem 18.4.1]. Hence there are constants $A, B \in \mathbb{R}$ such that

$$A\,\eta_1(0) + B\,\eta_2(0) = 1.$$

We now set

$$\alpha^{(R)}(z, z_0, k) = A\,\eta_1(k) + B\,\eta_2(k).$$

Then $\alpha^{(R)}(z, z_0, k)$ satisfies (16.53) and the appropriate asymptotic conditions.

For computational construction of the frequency domain CGO solutions η_j satisfying (16.60)–(16.62), see [16, Section 5.2] and [217].

16.3.5 Final reconstruction steps

We know the approximate solutions $u_j^{(R)}(z, k_0)$ for $z \in \Omega$ and one fixed k_0. We use (16.35) and (14.51) to connect $u_1^{(R)}, u_2^{(R)}$ with $f_\mu^{(R)}, f_{-\mu}^{(R)}$. Define

$$\mu^{(R)}(z) = \frac{\overline{\partial} f_\mu^{(R)}(z, k_0)}{\partial f_\mu^{(R)}(z, k_0)}. \qquad (16.63)$$

Because of the truncation, the function $f_\mu^{(R)}(\cdot, k_0)$ is smooth and there are no difficulties in the numerical differentiation in (16.63). Also, from theoretical considerations we know that there is no division by zero in (16.63).

Finally we reconstruct the conductivity σ approximately as

$$\sigma^{(R)} = \frac{1 - \mu^{(R)}}{1 + \mu^{(R)}}. \qquad (16.64)$$

16.3.6 Numerical results

We illustrate the use of the method on four example conductivities studied in [16], as well as a conductivity representing an idealized cross-section of a human chest with an abnormality in one lung with the same values as in the example of Section 15.5.4. Of the conductivities studied in [16], the first is a simple heart-and-lungs phantom σ_1. The second and third are similar heart-and-lungs phantoms σ_2 and σ_3, both featuring a low-conductivity spine with σ_2 containing an additional tumor-like inhomogeneity inside one lung. The fourth example is a conductivity cross-section σ_4 of a stratified medium that could arise, for example, in an oil pipeline or other industrial process monitoring applications.

All these conductivities are discontinuous and therefore violate the assumptions of the D-bar method of Chapter 15.

The conductivity values for the first three examples are

Tissue type	σ_1	σ_2	σ_3
Background	1.0	1.0	1.0
Lungs	0.7	0.7	0.7
Spine	–	0.2	0.2
Heart (red)	2.0	2.0	2.0
Tumor (red)	–	2.0	–

The boundary measurements were simulated for $N = 16$ using FEM. Noise with a level of 0.01% was added to the DN map as described in Section 13.4. The boundary integral equation was solved as described above, and the cutoff frequency R was taken as large as possible without too much numerical error, resulting in $R = 6$ for the data with no added noise and $R = 5.5$ for the data with 0.01% noise in the first example. The cutoff $R = 5.5$ was used for σ_2 and σ_3 as well. For reconstructions of σ_1, the choice $z_0 = 1.4$ in (16.52) results in a function v_{z_0} with $R = 6$, as shown in Figure 16.13. The point k_0 was chosen to be the nearest k-grid point to 1.

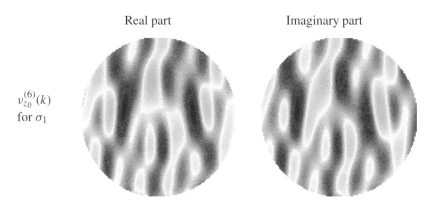

$v_{z_0}^{(6)}(k)$ for σ_1

Figure 16.13. The real (left) and imaginary (right) parts of the function $v_{z_0}(k)$ corresponding to the heart-and-lungs phantom σ_1 for $|k| < 6$. Numerical values of the real and imaginary parts range between -0.08 and 0.08 [16].

Figure 16.14 shows the reconstructions of σ_1 with relative errors given by

$$\frac{\|\sigma - \sigma^{(R)}\|_{L^2(\Omega)}}{\|\sigma\|_{L^2(\Omega)}} \cdot 100\%. \qquad (16.65)$$

The relative errors of the reconstructions are 11.6% for the noise-free data and 12.7% for the noisy data.

Further, we define the dynamic range of a reconstruction to be 100% multiplied by the ratio of the difference of the maximum and minimum values in the reconstruction to the difference of maximum and minimum values of the true conductivity. The minimum

16.3. The Astala–Päivärinta method

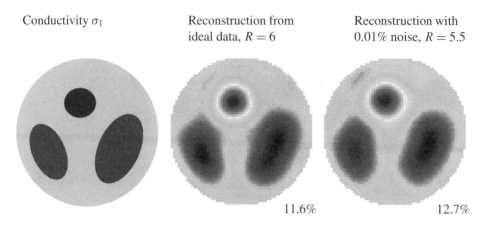

Figure 16.14. Heart-and-lungs conductivity phantom σ_1 and two different reconstructions. The colors (conductivity values) in the three images are directly comparable. The relative error percentages of the two reconstructions are computed using formula (16.65). Left: true conductivity σ_1. Center: reconstruction from noise-free data with $z_0 = 1.4$ and cutoff frequency $R = 6$. The minimum conductivity in the reconstruction is 0.637 (true value 0.7); the maximum conductivity is 1.997 (true value 2). Right: reconstruction from data with added noise of relative amplitude $\frac{\|E\|_{2,2}}{\|L_\sigma\|_{2,2}} \approx 10^{-4}$. Here $z_0 = 1.4$ and the cutoff frequency is $R = 5.5$. The minimum conductivity in the reconstruction is 0.637 and the maximum 1.870 [16].

conductivity in the noise-free reconstruction is 0.637 (compared to a true value of 0.7), and the maximum conductivity is 1.997 (compared to a true value of 2). This results in a dynamic range of 105%. The minimum conductivity in the reconstruction from data with 0.01% noise is 0.637 and the maximum 1.870, resulting in a dynamic range of 95%.

Figure 16.15 shows the reconstructions of σ_2 and σ_3, with relative errors 16.7% and 16.3% and dynamic ranges 109% and 106%, respectively. The latter are computed using the differences between the heart and spine since the maximum and minimum conductivity values occur there.

Figure 16.16 contains the reconstruction of the idealized cross-section of the human chest with fluid in one lung. As in Section 15.5.4, the heart has a conductivity of 6, the lungs have a conductivity of 2, the fluid has a conductivity of 4, and the background has a conductivity of 1. The organ boundaries are discontinuous, and the data, simulated by the continuum FEM, include 0.01% noise. The maximum reconstructed values is 5.9949, which occurs in the heart region as it should, and the minimum reconstructed value is 0.7855, which is found in the background, as it should be, but is lower than the actual value of the background by 21%. The dynamic range of the reconstruction is 104%.

Exercise 16.3.1. *Prove the converse of Theorem* 16.3.

Exercise 16.3.2. *Derive the matrix* (16.43) *as an approximation to the tangential derivative map in the basis* (16.42).

Exercise 16.3.3. *Derive equation* (16.36).

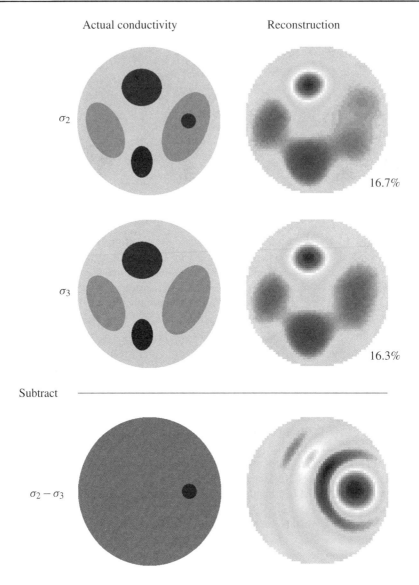

Figure 16.15. True conductivities σ_2 and σ_3 (left) and reconstructions from noise-free data (right). In each case, the cutoff frequency is $R = 5.5$ and $z_0 = 1.4$. The relative error percentages of the two reconstructions are computed using formula (16.65). The bottom row represents a difference image; that is, the reconstruction in the second row is subtracted from the reconstruction in the top row to form the image in the bottom row. In rows 1 and 2, the colormaps in the left and right image are the same. The images in the bottom row are each plotted on their own scale. The true maximum conductivity in the first and second rows is 2.0, and the true minimum conductivity is 0.2. The max and min in row one are 2.3319 and 0.3781, respectively. The max and min in row two are 2.2732 and 0.3728, respectively. The true contrast in the difference image is 1.3, while the contrast achieved in the difference image is 0.2073 [16].

16.4. The enclosure method of Ikehata 277

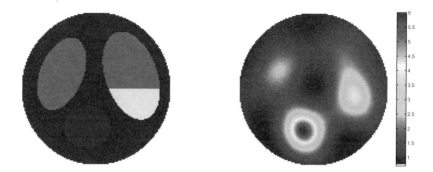

Figure 16.16. Left: Discontinuous conductivity representing an idealized cross-section of a human chest with an abnormality in one lung, representing, for example, fluid. Compare to Figure 15.11 in Chapter 15, which is a smoothed version of this example. Right: reconstruction using the method of Section 16.3 from data with 0.01% noise using a truncation radius of $R = 4.7$. Only one colorbar is included since the plots are on the same scale. The dynamic range of the reconstruction is 104%.

16.4 The enclosure method of Ikehata

The goal of the enclosure method is to find information about jump discontinuities or other anomalies in electric conductivity inside an unknown body. More precisely, the idea is to extract the convex hull of a set D containing the anomalies by approaching the set from all directions with half-spaces and detect from EIT measurement data when the half-space intersects D.

The enclosure method for EIT was introduced by Masaru Ikehata in [221]. It was first implemented numerically (simultaneously and independently) in [227] and [60]. Later it has been generalized and implemented for two- and three-dimensional EIT in [225, 228, 220, 226, 447, 219] and for inverse scattering in [224]. There is also recent progress in the case of Maxwell's equations; see [495].

The enclosure method is studied here in a two-dimensional setting. Choose Ω to be the unit disc: $\Omega := B(0, 1) \subset \mathbb{R}^2$. For simplicity we study conductivities of the form

$$\sigma(z) = 1 + c\chi_D(z), \tag{16.66}$$

where $c > 0$ and χ_D denotes the characteristic function of a subset $D \subset \Omega$. Assume that $D \subset B(0, \rho)$ with $\rho < 1$. Then there is a neighborhood of the boundary $\partial\Omega$, where $\sigma \equiv 1$.

Define the *support function* $h_D(\omega)$ of the set D by the formula

$$h_D(\omega) = \sup_{z \in D} \omega \cdot z, \tag{16.67}$$

where $\omega \in \mathbb{R}^2$ is a unit vector: $|\omega| = 1$. Note that if we know the support function of D for all directions ω, then we know the convex hull of D. See Figure 16.17. The enclosure method is designed for recovering $h_D(\omega)$ from the knowledge of Λ_σ.

The crucial object is the *indicator function* defined as follows. Given a direction $\omega = (\omega_1, \omega_2)$, define an orthogonal direction $\omega^\perp := (\omega_2, -\omega_1)$. We will use as Dirichlet data the following function:

$$f_\omega(z; \tau) := e^{\tau z \cdot \omega + i\tau z \cdot \omega^\perp}. \tag{16.68}$$

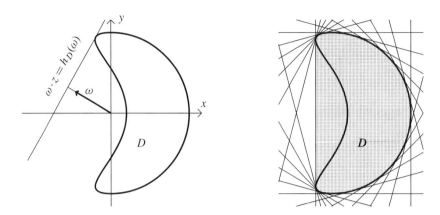

Figure 16.17. Left: inclusion D, direction ω and support function $h_D(\omega)$. Right: approximate convex hull (shaded area) given by the knowledge of the support function corresponding to 24 directions.

The indicator function is defined by

$$I_\omega(\tau) := \int_{\partial\Omega} f_\omega(z;\tau)\Big((\Lambda_\sigma - \Lambda_1)\overline{f_\omega(\cdot;\tau)}\Big) dS(z). \qquad (16.69)$$

Then, according to the proof in [221], asymptotically for large τ we have

$$\frac{1}{2}\log|I_\omega(\tau)| \approx h_D(\omega)\tau, \qquad (16.70)$$

see [227, formula (12)]. The right-hand side of (16.70) is linear in τ, and the slope gives the sought-for support function. The left-hand side of (16.70) can be recovered from measured data and becomes asymptotically linear in τ when $\tau \to \infty$.

In practice we cannot use arbitrarily large values of τ because the Dirichlet function (16.68) oscillates with frequency directly proportional to τ, and grows exponentially in τ whenever $z \cdot \omega > 0$. Applying such voltage potentials at boundaries of physical bodies becomes impossible at some point when τ is increased.

The practical method is based on the approximate idea that linearity holds already with small τ, and with roughly the correct slope. Using small values of τ only is a kind of regularization; an asymptotic analysis of this nonlinear regularization method is available in [222]. These are the steps:

1. Choose a collection of directions $\omega_1, \omega_2, \ldots, \omega_M \in S^1$.

2. Choose a collection of values for τ, say, $0 < \tau_1 < \tau_2 < \cdots < \tau_N < \infty$.

3. For all $1 \leq j \leq M$ and $1 \leq \ell \leq N$, apply the voltage potential $\overline{f_{\omega_j}(z;\tau_\ell)}$ for $z \in \partial\Omega$, and measure the corresponding current $\Lambda_\sigma\big(\overline{f_\omega(\cdot;\tau)}\big)$ through the boundary. Compute $\Lambda_1\big(\overline{f_\omega(\cdot;\tau)}\big)$ as well.

16.4. The enclosure method of Ikehata

4. Fix a direction ω_j. Evaluate computationally the following numbers:

$$q_\ell := \frac{1}{2}\log|I_{\omega_j}(\tau_\ell)| = \frac{1}{2}\log\left|\int_{\partial\Omega} f_{\omega_j}(z;\tau_\ell)\left((\Lambda_\sigma - \Lambda_1)\overline{f_{\omega_j}(z;\tau_\ell)}\right)dS(z)\right|$$

for all $1 \leq \ell \leq N$.

5. Fit a first-order polynomial of the form $y = ax + b$ to the data points (τ_ℓ, q_ℓ) in the sense of least squares. Then the slope of the fitted linear function gives an approximation to the support function: $a \approx h_D(\omega_j)$.

6. Repeat Steps 4 and 5 for all directions ω_j with $1 \leq j \leq M$. Approximate the convex hull of D by the polygon defined by the recovered support function values.

Let us illustrate the above method numerically. Consider the following simple conductivity distribution:

$$\sigma(z) = \begin{cases} 2 & \text{for } |z| \leq 0.5, \\ 1 & \text{for } 0.5 < |z| \leq 1. \end{cases} \quad (16.71)$$

In other words, $\sigma = 1 + \chi_D$ for $D = B(0, 1/2)$. Then the support function $h_D(\omega) \equiv 1/2$ for all $\omega \in S^1$. See the left image in Figure 16.18.

Choose $\tau_\ell := \ell$ for $\ell = 2, 3, \ldots, 6$. We take $\omega = [1, 0]^T$; in this symmetric situation any unit vector would do. We use the finite element method to simulate $\Lambda_\sigma\left(f_\omega(\cdot;\tau)\right)$ and $\Lambda_1\left(f_\omega(\cdot;\tau)\right)$, and we integrate numerically over the boundary to evaluate the numbers q_ℓ defined in Step 4 above. The result is shown as black dots in the right plot in Figure 16.18. The dots are arranged roughly in a linear fashion, and we fit a line through the data points in the least-squares sense. The resulting slope is 0.587, while the theoretically expected value is $h_D(\omega) = 0.500$. The error comes from the fact that formula (16.70) holds only asymptotically when τ is large.

Further studies of the enclosure method are left as a project work. See Section 17.1 for details.

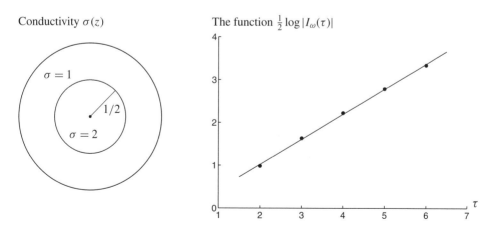

Figure 16.18. Numerical illustration of the enclosure method of Ikehata. Left: the conductivity distribution $\sigma = 1 + \chi_{B(0,1/2)}$. Right: the data points (τ_ℓ, q_ℓ) for $\ell = 2, 3, \ldots, 6$ are shown as black dots. The first-order least squares fit is shown as a thin line; the slope is 0.587, while $h_D(\omega) = 0.500$.

Chapter 17
Projects

In this chapter we include projects suitable for coursework or independent study on several topics from Part II of this book. These projects all involve programming and are designed to give the reader some intuition and familiarity with these methods. If you have not already, please read the general instructions about developing mathematical algorithms in the beginning of Chapter 10.

17.1 Enclosure method for EIT

The goal of this project is to implement the enclosure method as described in Section 16.4 and study its noise-robustness properties.

Let the domain Ω be the unit disc for simplicity. Also, let the conductivity be of the form $\sigma(z) = 1 + c\chi_D(z)$ with some constant $c > 0$.

17.1.1 Simulation of EIT data

Choose a suitable inclusion D. It's a good idea to start with something simple, such as a disc $D = B(0, \rho)$ with radius $0 < \rho < 1$. Once everything works with that case, you can move on to single discs with centers away from origin, collections of several discs, and to nonconvex inclusion components.

Given D, you need to compute numerically the following currents through the boundary: $\Lambda_\sigma\bigl(\overline{f_\omega(\cdot;\tau)}\bigr)$ and $\Lambda_1\bigl(\overline{f_\omega(\cdot;\tau)}\bigr)$. This can be done using the FEM as discussed in Section 13.2.1. You just need to specify the Dirichlet boundary condition $\overline{f_\omega(\cdot;\tau)}|_{\partial\Omega}$ instead of the Fourier basis functions considered in Section 13.2.1.

17.1.2 Calculation of the indicator function

You need to evaluate integrals coming from the definition (16.69) of the indicator function $I_\omega(\tau)$:

$$I_\omega(\tau) = \int_{\partial\Omega} f_\omega(z;\tau)\Big((\Lambda_\sigma - \Lambda_1)\overline{f_\omega(\cdot;\tau)}\Big)dS(z)$$

$$= \int_0^{2\pi} f_\omega(e^{i\pi\theta};\tau)\big(\Lambda_\sigma\,\overline{f_\omega(\cdot;\tau)}\big)(\theta)d\theta \qquad (17.1)$$

$$- \int_0^{2\pi} f_\omega(e^{i\pi\theta};\tau)\big(\Lambda_1\overline{f_\omega(\cdot;\tau)}\big)(\theta)d\theta. \qquad (17.2)$$

One practical way of doing this is to denote the angular parameter of the points shown in Figure 13.1(c) by $\theta_1,\ldots,\theta_M$, the (constant) angular displacement between the points by $\Delta\theta_j := \theta_{j+1} - \theta_j$, and applying the simple quadrature

$$\int_0^{2\pi} g(\theta)d\theta \approx \sum_{j=1}^M \Delta\theta_j\, g(\theta_j)$$

to the integrals (17.1) and (17.2).

We remark that in the solution of the Dirichlet problem one evaluates $\overline{f_\omega}$ at the points shown in Figure 13.1(b), whereas in the calculation of the integrals (17.1) and (17.2) one evaluates f_ω at the points shown in Figure 13.1(c). Being careful about this can save significant debugging time.

17.1.3 Experiments with noise-free data

Now you are all set for recovering some support function values from simulated EIT data.

Fix an inclusion D and a direction ω and calculate the true value of $h_D(\omega)$ analytically for comparison. Choose a collection of values for τ, say, $0 < \tau_1 < \tau_2 < \cdots < \tau_6 < \infty$. To begin with, you can take $\tau_\ell = \ell$ and modify them as you go along.

Evaluate computationally the following numbers:

$$q_\ell = \frac{1}{2}\log\left|\int_{\partial\Omega} f_{\omega_j}(z;\tau_\ell)\Big((\Lambda_\sigma - \Lambda_1)\overline{f_{\omega_j}(z;\tau_\ell)}\Big)dS(z)\right|.$$

Now fit a line in the least-squares sense to these three points:

$$(\tau_1,q_1),\quad (\tau_2,q_2),\quad (\tau_3,q_3).$$

Denote the slope of the fitted line by s. Further, fit a line in the least-squares sense to these three points as well:

$$(\tau_4,q_4),\quad (\tau_5,q_5),\quad (\tau_6,q_6).$$

Denote the slope of the second fitted line by $\tilde{s}$.

Repeat the above calculations for several directions ω. Which number is usually closer to $h_D(\omega)$, the first slope s or the second slope $\tilde{s}$? Why?

17.1.4 Experiments with noisy data

Repeat the experiments of Section 17.1.3 with noisy data. That is, add random numbers to the computed derivatives $\nu \cdot \nabla u|_{\partial\Omega}$, evaluated at centers of edge segments as shown in Figure 13.1(c). A suitable MATLAB command for this is `v = v+a*randn(size(v))`, where `v` is the vector containing the noise-free values of $\nu \cdot \nabla u|_{\partial\Omega}$ at the appropriate points, `randn` produces normally distributed random numbers with mean zero and standard deviation 1, and `a` stands for a positive constant $a > 0$ used to adjust the amount of noise.

Compute a few examples and vary the noise level $a > 0$. Which number is more sensitive to simulated noise, the first slope s or the second slope $\tilde{s}$? Why?

17.1.5 Visualization of reconstructions

Design three different inclusions D, each having some interesting properties. Some of them might have several components, and one of them can be a nonconvex inclusion such as the smooth "kite" shown in Figure 16.17. Compute approximate reconstructions from noisy and noise-free data for all three cases. Find a way to report your results visually so that each of these three quantities is visible:

- The original inclusion D.

- The true convex hull of D.

- The recovered approximate convex hull of D.

17.2 The D-bar method with Born approximation

We saw in Section 16.1.1 that taking a Born approximation to the CGO solutions to the Schrödinger equation results in an approximation to the scattering transform which we denote by $\mathbf{t}^{\text{exp}}$. There are several topics for a project which stem from this approach, all of which require its implementation.

Since the Born approximation involves setting the CGO solution $\psi(z,k)$ for $z \in \partial\Omega$ equal to its large $|z|$ asymptotic behavior, one may wish to study the validity of this approximation. Clearly, the closer $\psi(z,k)$ is to its asymptotic behavior, the better this approximation will be. Since e^{ikz} is a solution to Laplace's equation, this approximation will be most accurate when $q(z) = \Delta\sqrt{\sigma}/\sqrt{\sigma} \approx 0$, or when the conductivity is nearly constant. One project is to study the nature of the scattering transform and reconstructions as σ deviates from a constant in magnitude and in its spatial properties.

Another project is to compute the DN map with noisy data and study the effects of changing the truncation radius R, which serves as a regularization parameter, on $\mathbf{t}_R^{\text{exp}}$ and on the reconstruction. One can study the effects of noise on R, or the effects of the choice of R on the reconstruction.

We suggest working on the unit disc Ω for simplicity and with a conductivity σ that is 1 in a neighborhood of the boundary $\partial\Omega$.

As in Chapter 12, let L denote the number of electrodes and θ_ℓ the angle of the center of the ℓth electrode, where the electrodes are equally spaced with angle $\Delta\theta$ between their centers.

17.2.1 Simulation of data

Use the methods described in Section 13.2 to simulate data in the Fourier basis. Depending upon your goals, you may choose to consider one of the following kinds of conductivities:

(i) Rotationally symmetric σ with a jump on the circle $|z| = r_0$. In this case, the basis functions (13.1) are eigenfunctions for the DN map: $\Lambda_\sigma \varphi_n = \lambda_n \varphi_n$. The real-valued eigenvalues λ_n can be computed explicitly by hand, as shown in Section 13.1.

(ii) If you are adept with the FEM, or wish to implement the method described in Section 13.2, you can define a more general conductivity distribution σ and compute a matrix approximation to Λ_σ as explained in Section 13.2.3.

In a smaller project it is enough to work with case (i). If case (ii) is considered, it's a good idea to compare the results of (i) and (ii) for a rotationally symmetric case as a debugging step. In any case, for a project, it is recommended that you decide on a set of examples that you wish to study before you begin your computations. The addition of noise to the data is described in Section 13.1 for case (i) and Section 13.2 for case (ii).

17.2.2 Computation of the scattering transform

This is the first step of the algorithm since you will apply the Born approximation and replace $\psi(z,k)|_{\partial \Omega}$ by $e^{ikz}|_{\partial \Omega}$. Recall that we set

$$\mathbf{t}_R^{\exp}(k) := \begin{cases} \int_{\partial \Omega} e^{i\bar{k}\bar{z}}(\Lambda_\sigma - \Lambda_1)e^{ikz} ds & \text{for } |k| \leq R, \\ 0 & \text{for } |k| > R, \end{cases} \quad (17.3)$$

and we will choose R empirically. Denoting the Fourier coefficients for e^{ikz} by $a_n(k)$, we found

$$\mathbf{t}_R^{\exp}(k) = \sum_{m=0}^{\infty} \sum_{n=0}^{\infty} a_m(\bar{k}) a_n(k) \langle e^{im\theta}, (\Lambda_\sigma - \Lambda_1)e^{in\theta} \rangle, \quad |k| \leq R.$$

Replacing the DN maps by their matrix approximations and approximating the continuous inner product by a discrete inner product results in a computationally practical formula for $\mathbf{t}_R^{\exp}$. Note that in practice you only have the voltages on L electrodes, and so the continuous inner product can be approximated by a discrete inner product as follows:

$$\langle f, g \rangle \equiv \int_0^{2\pi} \overline{f(\theta)} g(\theta) d\theta \approx \Delta \theta \sum_{\ell=1}^{L} \overline{f(\theta_\ell)} g(\theta_\ell).$$

In the case (i) of a radially symmetric conductivity distribution, the approximation to $\mathbf{t}_R^{\exp}$ can be written in a nice series formulation in terms of the eigenvalues λ_n of the DN map; this was the topic of Exercise 16.1.2. Also, note that in case (i), the computed scattering transform should also be radially symmetric. This property is very valuable for debugging.

Now you are ready to compute $\mathbf{t}_R^{\exp}$ for the example or examples you chose for a variety of noise levels and truncation radii R. Since the scattering transform will blow up if R is chosen too large, plot your results on an appropriate grid and scale so that you can compare the effect of your choices on the blowup of $\mathbf{t}_R^{\exp}$. Note that if your conductivity is not rotationally symmetric, $\mathbf{t}_R^{\exp}$ will have a real and imaginary part.

17.2.3 Solution of the D-bar equation

The next step in the algorithm is to solve the D-bar equation for $k' \in \mathbb{C}$, $|k'| \leq R$,

$$\mu_R^{\exp}(z,k) = 1 + \frac{1}{(2\pi)^2} \int_{|k'| \leq R} \frac{\mathbf{t}_R^{\exp}(k')}{(k-k')\overline{k'}} e_{-z}(k') \overline{\mu_R^{\exp}(z,k')} dk_1' dk_2'. \qquad (17.4)$$

The one-grid method for solving equation (17.4) described in Section 15.4 is available from the authors as a package for solving D-bar equations of this form at the website www.inverseproblemsbook.org. Alternatively, you can program it yourself! Recall that the equation is solved for each z independently, and so this step of the algorithm can be trivially parallelized. You will need to define your choices of z, typically a grid on $[-1,1]^2$, with values of $|z| > 1$ not computed since we have chosen to work on the unit disc.

Several important things to note are that the k-grid for the solution of (17.4) by this method is of size $M = 2^m$ and may not coincide with the grid you chose for the computation of the scattering transform. You may need to use interpolation, such as MATLAB's `interp2` command, to evaluate the scattering transform on the appropriate grid for the D-bar equation solver. If the resulting linear system is solved by GMRES, one must first split the system into its real and imaginary components and solve the stacked system by GMRES. Finally, note that the solver will not converge properly if a reasonable truncation radius is not chosen.

17.2.4 Computation and visualization of reconstructions

Recall that the approximation to σ is computed by simply evaluating $\sigma_R^{\exp}(z) = \mu_R^{\exp}(z,0)^2$. A typical way of plotting the reconstruction is by using the `imagesc` command in MATLAB. For a three-dimensional plot, the MATLAB command `mesh` does the job, but often one wants to rotate the plot to get a good view of the reconstruction, which doesn't lend itself well to preparing a report. Plot your results on the same scale as the true conductivity distribution. In case (i), you may wish to plot cross-sections of your reconstructions, but keep in mind that it is a good debugging technique to see if your reconstructions are indeed radially symmetric (or very close to it!).

17.3 Calderón's method

The linearization-based method introduced by Calderón in [64] and described in Section 16.2 is a direct method tantamount to a further linearization of the D-bar method, as described in Section 16.2.1. Thus, it is an interesting exercise to compare the results of this project, which is to implement Calderón's method, to the results of the project on the D-bar method with Born approximation from Section 17.2. Like the D-bar method, there are several aspects of Calderón's method that can be studied, such as the effects of the contrast of the conductivity on the reconstruction, the use of a mollifying function, and the truncation radius in the computation of the inverse Fourier transform. All can be studied for noisy or noise-free data. We suggest choosing test problems in the same manner as in Section 17.2, that is, either rotationally symmetric σ with a jump on the circle $|z| = r_0$ or a more general conductivity distribution with data computed by the FEM.

17.3.1 Computational aspects

Recall from Section 16.2 that the conductivity is obtained by computing an inverse Fourier transform of a function $\hat{F}$ related to the measured data and the exponentially growing solutions u_1 and u_2 to Laplace's equation. That is,

$$\sigma(z) \approx \mathcal{F}^{-1}(\hat{F}(z)).$$

As in previous sections, we denote the spatial variable by z, but note that here $z = (x, y) \in \mathbb{R}^2$, not in $\mathbb{C}$. The Fourier transform variable is denoted by $\xi \in \mathbb{R}^2$. Recall that the exponentially growing solutions are defined by

$$u_1(z) = e^{i\pi(\xi \cdot z) + \pi(a \cdot z)},$$
$$u_2(z) = e^{i\pi(\xi \cdot z) - \pi(a \cdot z)},$$

where $a, \xi \in \mathbb{R}^2$ with $\xi \cdot a = 0$ and $|\xi| = |a|$. From Section 16.2

$$\hat{F}(\xi) = -\frac{1}{2\pi^2 |\xi|^2} B\left(e^{i\pi(\xi \cdot z) + \pi(a \cdot z)}, e^{i\pi(\xi \cdot z) - \pi(a \cdot z)}\right) \tag{17.5}$$

and

$$B(u_1, u_2) = \int_{\partial \Omega} u_1 \Lambda_\sigma u_2 \, ds. \tag{17.6}$$

Since you will compute data using one of the methods in Section 13.2, you are working with the Fourier basis functions defined by (13.1). In Exercise 16.2.2 you were asked to find Fourier expansions of u_1 and u_2. Denoting the Fourier coefficients of u_1 by $a_j(\xi)$ and those of u_2 by $b_j(\xi)$ and substituting these expansions into (17.5) and (17.6) results in a series formulation for $\hat{F}$:

$$\hat{F}(\xi) = \frac{-1}{2\pi^2 |\xi|^2} \sum_{j=0}^{\infty} \sum_{k=0}^{\infty} a_j(\xi) b_k(\xi) \int_0^{2\pi} e^{\mp ij\theta} \Lambda_\sigma e^{\pm ik\theta} d\theta.$$

Replacing the DN map Λ_σ by its matrix approximation and approximating the continuous inner product by a discrete inner product results in a computationally practical formula for $\hat{F}$.

Having computed $\hat{F}$, you have the option of either applying a mollifying function as suggested in [64] and then taking the inverse Fourier transform, or simply taking the inverse Fourier transform. If you choose to apply a mollifying function, one option is $\eta_s(z) = g_t(z) = te^{-\pi|z|^2/t}$, where $s = 1/\sqrt{t}$. The Fourier transform of η_s is $\hat{\eta}_s(\xi/s) = \hat{g}_t(\xi) = e^{-\pi t |\xi|^2}$. In either case you will need to restrict $\hat{F}(\xi)$ to a finite domain before computing the inverse Fourier transform, which can be done in a number of ways. A simple method is a straightforward Simpson's rule.

17.4 Inverse obstacle scattering

This project work will familiarize you with *inverse obstacle scattering* in the case of sound-hard obstacles. We work in a two-dimensional setting that arises from the following three-dimensional situation. Consider an infinitely long cylinder $D \times \mathbb{R} \subset \mathbb{R}^3$, where $D \subset \mathbb{R}^2$ is

17.4. Inverse obstacle scattering

a bounded domain of unknown shape and location. Assume that $\mathbb{R}^2 \setminus \overline{D}$ is connected and that the boundary ∂D is a smooth Jordan curve.

We probe the cylinder with acoustic waves sent from far away and modeled as plane waves. The direct problem is to compute the total field for all incoming directions from the knowledge of D, and then determine the far field pattern. The inverse problem is: given the far field pattern, reconstruct the set D. In Section 17.4.1 we explain how to compute numerically far field patterns for sound-hard obstacles.

The project is to first choose some suitable obstacles, for example, one or more disjoint discs, then compute the corresponding far field patterns with k being roughly equal to the diameter of the discs (you can experiment with larger and smaller k). Recover the obstacles approximately using one or more of the following methods:

- Linear sampling method [90, 98, 63].

- Factorization method [270, 269].

- Enclosure method [224].

- Any method in the book [381].

You can use the following quantity for measuring the relative error of the reconstructed obstacles:

$$\frac{\mu(D \setminus D_{\text{rec}}) + \mu(D_{\text{rec}} \setminus D)}{\mu(D)} \cdot 100\%,$$

where the reconstructed obstacle is denoted by D_{rec} and μ stands for area (Lebesgue measure in $\mathbb{R}^2$).

17.4.1 Numerical computation of far field patterns

Fix $d \in S^1$ and $k > 0$. We wish to solve numerically the following exterior Neumann problem:

$$\Delta w + k^2 w = 0 \quad \text{in } \mathbb{R}^2 \setminus \overline{D}, \quad (17.7)$$

$$\left.\frac{\partial w}{\partial \nu}\right|_{\partial D} = g,$$

$$\lim_{r \to \infty} \sqrt{r} \left(\frac{\partial w}{\partial r} - ikw\right) = 0.$$

In particular, we will take $g = -(\partial/\partial \nu)e^{ikz \cdot d}$.

We seek the solution as a single-layer potential representation

$$w(z) = \int_{\partial D} \Phi(z - y) f(y) ds(y), \quad z \in \mathbb{R}^2 \setminus \overline{D}, \quad (17.8)$$

with continuous density f. The fundamental solution Φ appearing in (17.8) is defined by

$$\Phi(z) = \frac{i}{4} H_0^{(1)}(k|z|), \quad (17.9)$$

where $H_0^{(1)}$ is the Hankel function of the first kind and order zero. Let f be a solution (not necessarily unique) of the integral equation

$$\left(\frac{1}{2}I - A\right)f = -g, \tag{17.10}$$

where the operator A is given by

$$(Af)(z) = \int_{\partial D} \frac{\partial \Phi(z-y)}{\partial \nu(z)} f(y) ds(y), \qquad z \in \partial D. \tag{17.11}$$

Note that A is the adjoint of the acoustic double-layer potential. Then w given by (17.8) solves the exterior Neumann problem (17.7); see, e.g., [404, Theorem 3.3.8] or [91].

We remark that Hankel functions satisfy

$$\frac{d}{dr} H_0^{(1)}(r) = -H_1^{(1)}(r) \tag{17.12}$$

for all $r > 0$.

Next we discuss the numerical solution of (17.10). Our approach makes use of [382]. Take a 1-periodic parameterization of the curve ∂D as $z = z(t) = (z_1(t), z_2(t))$ such that $|z'(t)| > 0$ for all $t \in \mathbb{R}$. (If D consists of several components, the derivative is considered piecewise in the obvious manner.) Choose the parameterization in such a way that the unit outward normal vector ν is given by $\nu(z(t)) = |z'(t)|^{-1}(z_2'(t), -z_1'(t))$. Denoting $f(t) = f(z(t))$ and $g(t) = g(z(t))$, (17.10) is equivalent to

$$\frac{1}{2}f(t) - (Af)(t) = -g(t), \tag{17.13}$$

where A defined by (17.11) can be written in the following form using (17.12):

$$(Af)(t) = \frac{i}{4} \int_{-1/2}^{1/2} \frac{\partial H_0^{(1)}(k|z(t)-y(s)|)}{\partial \nu(z(t))} f(s) |y'(s)| ds.$$

Applying the chain rule, we obtain

$$(Af)(t) = -\frac{ik}{4} \int_{-1/2}^{1/2} H_1^{(1)}(k|z(t)-y(s)|) \frac{\nu(z(t)) \cdot (z(t)-y(s))}{|z(t)-y(s)|} f(s) |y'(s)| ds. \tag{17.14}$$

We discretize the interval $[-1/2, 1/2]$ by choosing $N = 2^n$ for some $n > 1$ and setting

$$s_j = jh, \qquad j = -N/2, -N/2+1, \ldots, N/2-1, \qquad h = 1/N.$$

We then approximate the integral in (17.14) by the following sum:

$$\sum_{j=-N/2}^{N/2-1} H_1^{(1)}(k|z(t)-y(s_j)|) \frac{\nu(z(t)) \cdot (z(t)-y(s_j))}{|z(t)-y(s_j)|} f(s_j) |y'(s_j)|. \tag{17.15}$$

In the case that $t = s_j$ for some j in (17.15) there is a problem with singularity of $H_1^{(1)}(z)$ at $z = 0$. One can overcome this problem by omitting the term corresponding to $t = s_j$ in the

17.4. Inverse obstacle scattering

sum (17.15). Because of the weakly integrable nature of the singularity, the error caused by the omission becomes arbitrarily small when N grows.

Now that we have an approximate numerical implementation of the operator A, we can use some iterative solver (e.g., GMRES) to solve the Fredholm equation (17.13) numerically.

Note that with some unfortunate choices of obstacle D and wave number k equation (17.13) may not have a unique solution. Such problems can be overcome either by using a more general solution method, such as [292], or by just modifying the example so that the solution becomes unique. The latter may involve some experimentation with several D and k.

You can verify your solver by choosing an obstacle that contains the origin and considering the field $w(z) = \Phi(z)$ for $z \in \mathbb{R}^2 \setminus D$. Then w and g in (17.7) are explicitly known and you can compare the numerical results with the ground truth.

Appendix A

Banach spaces and Hilbert spaces

This appendix consists of definitions and essential theorems on Banach and Hilbert spaces and norms. The organization loosely follows the chapters on Banach and Hilbert spaces in Renardy and Rogers [396], and we recommend that text to the reader for a more thorough treatment and for the proofs we have omitted for brevity. We assume some background in analysis at the level of Rudin and familiarity with the L^p spaces as well.

Definition A.0.1. *A normed linear space that is complete in its metric is called a* Banach space. *That is, X is a Banach space if there exists a sequence $\{x_n\} \subset X$ with $\|x_n - x_m\| \to 0$ as $m, n \to \infty$; then $\{x_n\}$ converges in X.*

Recall that such a sequence is called a *Cauchy sequence*, and so a Banach space is a normed linear space in which all Cauchy sequences converge to an element in the space.

Examples of Banach spaces include the following:

- Bounded continuous functions on the closure of an open set $\Omega \subset \mathbb{R}^m$, denoted by $C_b(\bar{\Omega})$, equipped with norm $\|u\| = \sup_{x \in \bar{\Omega}} |u(x)|$.

- Bounded analytic functions on an open set $\Omega \subset \mathbb{C}$ with norm $\|u\| = \sup_{x \in \Omega} |u(z)|$.

- The L^p spaces, $1 \leq p < \infty$.

Definition A.0.2. *Let X be a vector space over $\mathbb{K} = \mathbb{R}$ or $\mathbb{C}$. An* inner product *on X is a function $\langle \cdot, \cdot \rangle : X \times X \to \mathbb{K}$ that, for all $x, y, z \in X$ and for all $\alpha \in \mathbb{K}$, satisfies*

1. $\langle x, y \rangle = \overline{\langle y, x \rangle}$,

2. $\langle \alpha x, y \rangle = \alpha \langle x, y \rangle$,

3. $\langle x + y, z \rangle = \langle x, z \rangle + \langle y, z \rangle$,

4. $\langle x, x \rangle \geq 0$, *and* $\langle x, x \rangle = 0$ *if and only if* $x = 0$.

The space $(X, \langle \cdot, \cdot \rangle)$ is called an inner product space.

Definition A.0.3. *An inner product space that is complete in its metric is called a* Hilbert space.

Definition A.0.4. *Two norms $\|\cdot\|_1$ and $\|\cdot\|_2$ are* equivalent *if there exist constants $m, M > 0$ such that*
$$m\|x\|_1 \leq \|x\|_2 \leq M\|x\|_1 \quad \text{for all } x \in X.$$

Definition A.0.5. *Elements x, y of a Hilbert space H are orthogonal if $\langle x, y \rangle = 0$.*

Definition A.0.6. *Let M be a linear subspace of a Hilbert space H. We define the* orthogonal complement of M by $M^\perp$ *by*
$$M^\perp = \{x \in H : \langle x, y \rangle = 0 \text{ for all } y \in M\}.$$

Note that the orthogonal complement is always a closed subspace.

Theorem A.1 (projection theorem). *Let M be a closed linear subspace of a Hilbert space H. Then every $x \in H$ has the unique decomposition $x = y + z$ where $y \in M$ and $z \in M^\perp$.*

Appendix B

Mappings and compact operators

Theorem B.1. *Let L be a linear mapping from a normed linear space X to a normed linear space Y. Then the following are equivalent:*

1. *L is continuous at 0.*

2. *L is continuous.*

3. *There exists $c \geq 0$ such that $\|Lx\| \leq c\|x\|$ for all $x \in X$ (boundedness).*

Item 3 above is the definition of a bounded linear mapping. If L is a bounded linear mapping from X to Y, we define its norm by

$$\|L\| = \sup_{\|x\|=1} \|Lx\|. \tag{B.1}$$

Definition B.0.7. *Let $\mathcal{L}(X,Y)$ denote the space of bounded linear mappings from X to Y.*

Note that if X and Y are Banach spaces, then $\mathcal{L}(X,Y)$ equipped with the norm (B.1) is also a Banach space.

Theorem B.2 (Riesz representation theorem). *Let $L : X \to \mathbb{C}$ be a bounded linear functional. Then there exists a unique $x_0 \in X$ such that $Lx = \langle x, x_0 \rangle$ for all $x \in X$. Moreover,*

$$\|L\| = \|x_0\|.$$

Definition B.0.8. *A* linear operator *from a Banach space X to a Banach space Y is the pair $(D(A), A)$, where $D(A) \subset X$ is called the domain of A and A is a linear transformation from $D(A)$ to Y. The* range *of A is the subspace*

$$\text{range}(A) = \{y \in Y : y = Ax \text{ for some } x \in D(A)\}.$$

The nullspace *of A is the subspace*

$$\mathcal{N}(A) = \{x \in X : Ax = 0\}.$$

Theorem B.3. *Let A be a linear operator from a Banach space X to a Banach space Y with domain D(A) and range* range(A). *Then the following two properties hold:*

1. *The inverse operator* A^{-1} : range(A) $\to$ X *exists if and only if* $\mathcal{N}(A) = \{0\}$.

2. *If the inverse operator exists, it is linear.*

Compact operators play an important role in inverse problems. We include several essential definitions and theorems. Most proofs are omitted for brevity, and the reader is again referred to [396] for a more complete exposition. The following theorem provides an important characterization of compact operators.

Theorem B.4. *The operator* $K : D(K) \subset X \to Y$ *is compact if and only if for every bounded sequence* $\{x_n\}$ *in* $D(K)$, $\{Kx_n\}$ *has a convergent subsequence.*

Theorem B.5. *Every compact operator is bounded.*

Proof. Assume K is an unbounded operator. Then there exists $\{u_n\} \subset D(K)$ with $\|u_n\| = 1$ such that $\lim_{n\to\infty} \|Ku\| = \infty$. Choose elements of $\{u_n\}$ to create a subsequence $\{\tilde{u}_n\}$ such that $\{K\tilde{u}_n\}$ is (strictly) monotone and $\lim_{n\to\infty} \|K\tilde{u}_n\| = \infty$. Note that since K is compact and $\|\tilde{u}_n\| = 1$, $\{K\tilde{u}_n\}$ must contain a convergent subsequence. This is a contradiction and proves that K is bounded. □

Note: Let H be a Hilbert space. If $K : D(K) \subset H \to H$ is compact, then K is bounded on $D(K)$ and there exists an extension $\tilde{K} \subset \mathcal{L}(H)$ such that $\|\tilde{K}\| = \|K\|$ and $\tilde{K} = K$ on $D(K)$. It can be shown that $\tilde{K}$ is compact. Thus if K is compact, we can take $D(K) = H$.

Note: If $K \in \mathcal{L}(H, H)$, K is not necessarily compact. As an example, consider the identity operator I on an infinite-dimensional Hilbert space H. The identity operator I is not compact.

Proof. Suppose I is compact. Let $\{e_n\}$ be an orthonormal basis of H. Since $\|e_n\| = 1$, $\{Ie_n\}$ must have a convergence subsequence $\{e'_n\}$. But this is impossible since $\|e'_n - e_n\|^2 = \langle e'_n - e_n, e'_n - e_n \rangle = \|e'_n\|^2 + \|e_n\|^2 - 2\Re\langle e'_n, e_n \rangle = 2$. Thus, no subsequence can be Cauchy. So I is not compact. □

Theorem B.6. *Let* $A : D(A) \subset X \to Y$ *be a linear operator.*

1. *If A is bounded and* range(A) *is finite-dimensional, then A is compact.*

2. *If* $D(A)$ *is finite-dimensional, then A is compact.*

   Proof.

1. Let $\{x_n\} \subset D(A)$ with $\|x_n\| \leq M$ for all n. Then $\{Ax_n\} \subset$ range(A) and $\|Ax_n\| \leq \|A\|M \leq \infty$. Since range(A) is a finite-dimensional Hilbert space, by the Bolzano–Weierstrass theorem, $\{Ax_n\}$ has a convergent subsequence. So A is compact.

2. Let $\{e_n\}_{n=1}^{N}$ be an orthonormal basis for $D(A)$, ($N < \infty$). Then

$$\text{linear span}\{Ae_n\}_{n=1}^{N} = \text{range}(A).$$

Thus, $\dim \mathcal{R}(A) \leq \dim D(A)$. So $A : D(A) \to \text{range}(A)$ which has dimension $N' \leq N$. Thus, A is a bounded operator. So A is compact. □

Recall Bessel's inequality: If $\{e_n\}$ is a orthonormal set in H, then for all $u \in H$, $\sum_{n=1}^{\infty} |\langle u, e_n \rangle|^2 \leq \|u\|^2$. Thus, whenever $\{e_n\}$ is an orthonormal set in H, then $\lim_{n \to \infty} \langle u, e_n \rangle = 0$ for all $u \in H$ since the above sum converges.

Definition B.0.9. *A sequence $\{f_n\}$ is said to converge weakly to $f \in H$ if $\langle f_n, u \rangle \to \langle f, u \rangle$ as $n \to \infty$ for all $u \in H$. We write this as $f_n \overset{wk}{\to} f$.*

Theorem B.7. *If $f_n \to f$ in H (i.e., $\|f_n - f\| \to 0$), then $(f_n \overset{wk}{\to} f)$ in H.*

Proof. Let $u \in H$. Then $|\langle f_n, u \rangle - \langle f, u \rangle| = |\langle f_n - f, u \rangle| \leq \|f_n - f\| \|u\| \to 0$ as $n \to \infty$. This completes the proof. □

Appendix C

Fourier transform and Sobolev spaces

We include here some essential definitions and theorems on Sobolev spaces and the Fourier transform. Additional reading on Sobolev spaces includes [4, 56, 136, 488, 394, 400].

C.1 Sobolev spaces on domains $\Omega \subset \mathbb{R}^n$

Definition C.1.1. *The* Fourier Transform *of a function defined on $\mathbb{R}^n$ is given by*

$$\mathcal{F}(x)(\xi) = \widehat{f}(\xi) = \frac{1}{(2\pi)^n} \int_{\mathbb{R}^n} f(x) e^{-ix \cdot \xi} dx.$$

A sufficient condition for the existence of the Fourier transform is $f \in L^1(\mathbb{R})$, in which case $\widehat{f} \in C_0(\mathbb{R})$, but may not belong to $L^1(\mathbb{R})$.

Theorem C.1. *If the Fourier transform of both $f(x)$ and $f'(x)$ exist, then*

$$\mathcal{F}(f'(x))(\xi) = i\xi \widehat{f}(\xi).$$

More generally, if $f(x)$ and all its derivatives up to order m have Fourier transforms, then

$$\mathcal{F}(f^{(k)}(x))(\xi) = (i\xi)^k \widehat{f}(\xi), \quad k = 1,\ldots,m.$$

If f and $\widehat{f} \in L^1(\mathbb{R})$, then the inverse Fourier transform is defined by

$$\mathcal{F}^{-1}(f)(x) = \int_{\mathbb{R}} \widehat{f}(\xi) e^{i\xi x} d\xi.$$

Definition C.1.2. *Define a multi-index α, $\alpha = (\alpha_1,\ldots,\alpha_n)$ and the partial derivative operator D^α by*

$$D^\alpha = \left(\frac{\partial}{\partial x_1}\right)^{\alpha_1} \cdots \left(\frac{\partial}{\partial x_n}\right)^{\alpha_n}.$$

For $x \in \mathbb{R}^n$, define x^α by $x^\alpha = \Pi_{k=1}^n x_k^{\alpha_k}$. Define $|\alpha| = \alpha_1 + \cdots + \alpha_n$ and $\alpha! = (\alpha_1!)\ldots(\alpha_n!)$.

Theorem C.2. *If $u, D^\alpha u \in L^2(\mathbb{R}^n)$, then $\mathcal{F}(D^\alpha u) = (i\xi)^\alpha \widehat{u}(\xi) \in L^2(\mathbb{R}^n)$. If $u, x^\alpha D^\alpha u \in L^2(\mathbb{R}^n)$, then $\mathcal{F}(x^\alpha D^\alpha u) = (i D_\xi)^\alpha \widehat{u}(\xi) \in L^2(\mathbb{R}^n)$.*

Note that the first part of the theorem implies that if u is smooth, then $\widehat{u}$ decays rapidly at ∞, and the second part of the theorem implies that if u decays rapidly at ∞, then $\widehat{u}$ is smooth.

Definition C.1.3. *A function $\phi(x)$ defined on $\mathbb{R}$ is a test function if ϕ is in $C^\infty(\mathbb{R})$ and has compact support. We write $\phi \in C_0^\infty(\mathbb{R})$.*

Theorem C.3. *$C_0^\infty(\mathbb{R})$ is a dense subspace of $L^p(\mathbb{R})$, $1 \leq p \leq \infty$.*

This theorem allows us to extend the Fourier transform to L^2 functions as follows:

Theorem C.4. *For every $f \in L^2(\mathbb{R})$ there exists a unique $\widehat{f} \in L^2(\mathbb{R})$ such that $\widehat{f} = \mathcal{F}(f)$ in the sense that if $\{\phi_n(x)\} \in C_0^\infty(\mathbb{R})$ is such that $\|\phi_n - f\|_2 \to 0$ as $n \to \infty$, then $\widehat{\phi}_n = \mathcal{F}(\phi_n) \in L^2(\mathbb{R})$ is such that $\|\widehat{\phi}_n - \widehat{f}\|_2 \to 0$ as $n \to \infty$,*

In the case that the weak derivative of a function lies in L^2, the Fourier transform can be a useful tool to compute the weak derivative. Such a weak derivative is also called a *derivative in the L^2 sense*.

Definition C.1.4. *Let α be a multi-index. The function $D^\alpha f$ (which is actually an equivalence class of functions) is called the αth weak derivative of f if for any test function $\phi \in C_0^\infty(\Omega)$*

$$\langle D^\alpha f, \phi \rangle \equiv \int_\Omega D^\alpha f(x) \phi(x) dx = (-1)^{|\alpha|} \langle f, D^\alpha \phi \rangle.$$

Example 1:

$$g(x) = \begin{cases} 1 - |x| & \text{if } |x| < 1 \\ 0 & \text{if } |x| > 1 \end{cases}$$

$g(x) \in L^2(\mathbb{R})$, is continuous and has compact support but is not differentiable. Since $g \in L^2(\mathbb{R})$, $\widehat{g}$ exists by the previous theorem, and one can show

$$\widehat{g}(\xi) = \frac{2}{\xi^2}(1 - \cos\xi).$$

Note that $i\xi \widehat{g}(\xi) = \frac{2i}{\xi}(1 - \cos\xi) \in L^2(\mathbb{R})$. Thus, $\mathcal{F}(g'(x)) = i\xi \widehat{g}(\xi)$ and we can define the derivative of g in the L^2 sense by

$$g'(x) = \mathcal{F}^{-1}(i\xi \widehat{g}(\xi)) = \begin{cases} -\text{sgn}(x) & \text{if } |x| < 1, \\ 0 & \text{if } |x| > 1. \end{cases}$$

Example 2:

$$p(x) = \begin{cases} |x| & \text{if } |x| < 1, \\ 0 & \text{if } |x| > 1, \end{cases}$$

$p(x) \in L^2(\mathbb{R})$, has compact support but is not continuous and is not differentiable. Since $p \in L^2(\mathbb{R})$, $\widehat{p}$ exists by the previous theorem, and one can show

$$\widehat{p}(\xi) = \frac{2}{\xi^2}(\xi \sin\xi - (1-\cos\xi)).$$

So $i\xi \widehat{p}(\xi) = 2i\sin\xi - \frac{2i}{\xi}(1-\cos\xi) \notin L^2(R)$. So p does not have a derivative in the sense of L^2. However, one can show that $p(x)$ has a weak derivative (in the sense of generalized functions).

Theorem C.5 (Sobolev embedding theorem in $\mathbb{R}$). *Suppose $f \in L^2(\mathbb{R})$ and $\xi^p \widehat{f}(\xi) \in L^2(\mathbb{R})$ for $p \leq M$. Then f has derivatives in the L^2-sense of all orders less than or equal to M and these L^2 derivatives belong to C_0 for $p \leq M-1$.*

Proof. $\xi^p \widehat{f}(\xi) \in L^2(\mathbb{R})$ for $p \leq M$ implies $f^{(p)} = \mathcal{F}^{-1}((i\xi)^p \widehat{f}(\xi))$ in the L^2-sense and $(1+\xi^2)^{M/2}\widehat{f}(\xi) \in L^2(\mathbb{R})$. We must show that $\xi^p \widehat{f}(\xi) \in L^1(\mathbb{R})$ for $q \leq M-1$ since this implies

$$\mathcal{F}^{-1}((i\xi)^q \widehat{f}(\xi)) \in C_0(\mathbb{R}) \quad \text{and}$$
$$\mathcal{F}^{-1}((i\xi)^q \widehat{f}(\xi)) = f^{(q)}(x) \text{ a.e.}$$

By Hölder's inequality

$$\int_{\mathbb{R}} |\xi|^q |\widehat{f}(\xi)| d\xi = = \int_{\mathbb{R}} \frac{|\xi|^q}{(1+\xi^2)^{M/2}} (1+\xi^2)^{M/2} |\widehat{f}(\xi)| d\xi$$
$$\leq \left(\int_{\mathbb{R}} \frac{|\xi|^{2q}}{(1+\xi^2)^M} d\xi \right)^{1/2} \left(\int_{\mathbb{R}} (1+\xi^2)^M |\widehat{f}(\xi)|^2 d\xi \right)^{1/2}$$
$$\leq C \left(\int_{\mathbb{R}} \frac{|\xi|^{2q}}{(1+\xi^2)^M} d\xi \right)^{1/2}$$
$$< \infty \text{ provided } q \leq M-1. \quad \square$$

Definition C.1.5. *Define, for $s \geq 0$, the Sobolev spaces*

$$H^s(\mathbb{R}^n) = \{u \in L^2(\mathbb{R}^n) : (1+|\xi|^2)^{s/2}\widehat{u}(\xi) \in L^2(\mathbb{R}^n)\}$$

with

$$\|u\|_s^2 = \int_{\mathbb{R}^n} (1+|\xi|^2)^s \widehat{u}(\xi)^2 d\xi$$

and

$$(u,v)_s = \int_{\mathbb{R}^n} (1+|\xi|^2)^s \widehat{u}(\xi)^2 \widehat{v}(\xi) d\xi.$$

An equivalent norm on H^s is defined by

$$\|u\|_{H^s}^2 = \sum_{|\alpha| \leq s} \|D^\alpha u\|_{L^2}^2.$$

Note: (1) $H^0 = L^2$. (2) $s \geq t \geq 0$ implies $H^s \subset H^t \subset H^0$.

Theorem C.6. *Any differential operator of order m is a bounded linear mapping from H^s into H^{s-m} for $s > m$.*

Example 3: Consider $u(x)$ on $\Omega = (0,2)$ defined by

$$u(x) = \begin{cases} x^2, & 0 < x \leq 1, \\ 2x^2 - 2x + 1, & 1 < x < 2. \end{cases}$$

Note $u \in C^1(\Omega)$ and

$$u'(x) = \begin{cases} 2x, & 0 < x \leq 1, \\ 4x - 2, & 1 < x < 2. \end{cases}$$

However, u'' does not exist in the classical sense at 1. The weak derivative of u' is

$$u''(x) = \begin{cases} 2, & 0 < x \leq 1, \\ 4, & 1 < x < 2 \end{cases}$$

and $u, u', u'' \in L^2(0,2)$. However, the distributional derivative of u'' is $2\delta(x-1) \notin L^2(0,2)$. Thus,

$$u \in H^2(0,2), \quad \|u\|_{H^2}^2 = \int_0^2 (u^2 + (u')^2 + (u'')^2) dx = 71.37,$$

$$u \in H^1(0,2), \quad \|u\|_{H^1}^2 = \int_0^2 ((u')^2 + (u'')^2) dx = 39,$$

$$u \in H^0(0,2), \quad \|u\|_{H^0}^2 = \int_0^2 (u'')^2 dx = 20.$$

Example 4: Consider u defined on $\Omega = (-1,1) \times (-1,1)$ defined by

$$u(x,y) = \begin{cases} x, & x > 0, \\ 0, & x \leq 0. \end{cases}$$

Let $\phi \in C_0^\infty(\Omega)$. Then

$$\int_\Omega \frac{\partial u}{\partial y} \phi(x,y) dx dy = -\int_\Omega u \frac{\partial \phi}{\partial y} dx dy$$

$$= -\int_0^1 \int_{-1}^1 x \frac{\partial \phi}{\partial y} dy dx$$

$$= -\int_0^1 x(\phi(x,1) - \phi(x,-1)) dx$$

$$= 0 \quad \text{since } \phi \in C_0^\infty(\Omega).$$

Thus, $\frac{\partial u}{\partial y} = 0$.

$$\int_\Omega \frac{\partial u}{\partial x}\phi(x,y)dxdy = -\int_\Omega u\frac{\partial \phi}{\partial}dxdy$$
$$= -\int_{-1}^1 \int_0^1 x\frac{\partial \phi}{\partial x}dxdy$$
$$= -\int_{-1}^1 \left(x(\phi(x,y)|_{x=0}^1 - \int_0^1 \phi(x,y)dx \right) dy$$
$$= -\int_{-1}^1 \left((\phi(1,y)-0) - \int_0^1 \phi(x,y)dx \right) dy$$
$$= \int_{-1}^1 \int_0^1 \phi(x,y)dxdy$$
$$= \int_\Omega H(x)\phi(x,y)dxdy.$$

Thus, $\frac{\partial u}{\partial x} = H(x)$. Since $\frac{\partial^2 u}{\partial x^2} = \delta_x$, $\frac{\partial^2 u}{\partial x^2} \notin L^2(\Omega)$. So $U \in H^1(\Omega)$.

Theorem C.7 (Sobolev embedding). *Let Ω be a bounded domain in $\mathbb{R}^n$ with Lipschitz boundary Γ. If $m - k > \frac{n}{2}$, then every function $H^m(\Omega)$ belongs to $C^k(\bar\Omega)$. Furthermore, the embedding $H^m(\Omega) \subset C^k(\bar\Omega)$ is continuous.*

C.2 Fourier series and spaces $H^s(\partial\Omega)$

We restrict ourselves here to $\Omega \subset \mathbb{R}^2$ being the unit disc. Parameterize the boundary (unit circle) as
$$\partial\Omega = \{(\cos\theta, \sin\theta) \mid 0 \le \theta < 2\pi\}.$$

Denote by $C^\infty(\partial\Omega)$ the space of smooth test functions defined on $\partial\Omega$. We define Fourier coefficients for $f \in C^\infty(\partial\Omega)$ by the formula

$$\widehat{f}(n) := \frac{1}{\sqrt{2\pi}} \int_0^{2\pi} f(\theta) e^{-in\theta} d\theta, \qquad n \in \mathbb{Z}. \tag{C.1}$$

Definition (C.1) extends to the space $C^\infty(\partial\Omega)'$ of distributions in the standard way.

The Fourier coefficients (C.1) offer a particularly transparent way of defining Sobolev spaces $H^s(\partial\Omega)$ consisting of "s times weakly differentiable functions" on $\partial\Omega$. This works for all $s \in \mathbb{R}$, not only for positive integers s. We will make use of the following fact:

$$f \in L^2(\partial\Omega) \quad \text{if and only if} \quad \sum_{n=-\infty}^\infty |\widehat{f}(n)|^2 < \infty. \tag{C.2}$$

Let us first discuss the construction of the space $H^1(\partial\Omega)$. We could use the straightforward definition

$$H^1(\partial\Omega) := \{f \in L^2(\partial\Omega) : f'(\theta) \in L^2(\partial\Omega)\}, \tag{C.3}$$

where the derivative f' is taken in the sense of distributions, but we prefer an equivalent approach based on Fourier coefficients. Use integration by parts to compute

$$\begin{aligned}
\widehat{f'}(n) &= \frac{1}{\sqrt{2\pi}} \int_0^{2\pi} \frac{df}{d\theta}(\theta) e^{-in\theta} \, d\theta \\
&= -\frac{1}{\sqrt{2\pi}} \int_0^{2\pi} f(\theta) \frac{d}{d\theta}(e^{-in\theta}) \, d\theta \\
&= \frac{in}{\sqrt{2\pi}} \int_0^{2\pi} f(\theta) e^{-in\theta} \, d\theta \\
&= in \widehat{f}(n).
\end{aligned}$$

Combining this with (C.3), we can give an alternative but equivalent definition of the Sobolev space $H^1(\partial\Omega)$:

$$H^1(\partial\Omega) := \left\{ f \in L^2(\partial\Omega) : \sum_{n=-\infty}^{\infty} (1+|n|^2)|\widehat{f}(n)|^2 < \infty \right\}. \tag{C.4}$$

The reason for using $1+|n|^2$ in (C.4) instead of just $|n|^2$ is to avoid having zero weight on the coefficient $\widehat{f}(0)$. The crucial point is that the expressions $1+|n|^2$ and $|n|^2$ behave asymptotically similarly when $|n| \to \infty$.

We see from (C.4) that weak differentiability of a function $f \in L^2(\partial\Omega)$ is related to the decay rate of $|\widehat{f}(n)|$ as $n \to \infty$. This idea can be taken further by defining for all $s \geq 0$

$$H^s(\partial\Omega) := \left\{ f \in L^2(\partial\Omega) : \sum_{n=-\infty}^{\infty} (1+|n|^2)^s |\widehat{f}(n)|^2 < \infty \right\}. \tag{C.5}$$

Note that inserting $s = 0$ into (C.5) gives $H^0(\partial\Omega) = L^2(\partial\Omega)$ and that with $s = 1$ formula (C.5) coincides with (C.4).

We proceed to give explicit definitions of Hilbert space inner products and norms for the Sobolev spaces $H^s(\partial\Omega)$ with any real number $s \geq 0$. Define

$$\varphi_n^s(\theta) = \frac{w_{-s}(n)}{\sqrt{2\pi}} e^{in\theta}, \tag{C.6}$$

where $n \in \mathbb{Z}$ and the multiplier function is defined by

$$w_s(n) := (1+|n|^2)^{s/2}. \tag{C.7}$$

Define an inner product for the space $H^s(\partial\Omega)$ by

$$\langle f, g \rangle_s := \langle f, g \rangle_{H^s(\partial\Omega)} = \sum_{n=-\infty}^{\infty} w_s(n) \widehat{f}(n) \overline{w_s(n) \widehat{g}(n)}. \tag{C.8}$$

We write

$$\|f\|_s = \|f\|_{H^s(\partial\Omega)} = \langle f, f \rangle_s^{1/2}. \tag{C.9}$$

Note that $\langle \cdot, \cdot \rangle_0$ is the usual inner product of $L^2(\partial\Omega)$.

C.2. Fourier series and spaces $H^s(\partial\Omega)$

Compute the Fourier coefficients of the functions φ_m^s by (C.1):

$$\widehat{\varphi_m^s}(n) = \frac{w_{-s}(m)}{2\pi} \int_0^{2\pi} e^{i(m-n)\theta} d\theta = w_{-s}(m)\delta_m(n), \qquad (C.10)$$

where $\delta_m(n) = 1$ when $m = n$ and zero otherwise. Now substituting (C.10) into (C.8) shows that (C.6) is an orthonormal basis for the Hilbert space $H^s(\partial\Omega)$:

$$\langle \varphi_n^s, \varphi_\ell^s \rangle_s = \delta_n(\ell). \qquad (C.11)$$

The definition of spaces $H^s(\partial\Omega)$ for negative $s \in \mathbb{R}$ can be done by duality construction: $H^{-s}(\partial\Omega) = \bigl(H^s(\partial\Omega)\bigr)'$. Informally, one can always check whether a function f belongs to $H^s(\partial\Omega)$ by studying the finiteness of the expression $\sum_{n=-\infty}^{\infty}(1+|n|^2)^s|\widehat{f}(n)|^2$.

The above definitions can be generalized to more general domains Ω than the unit disc, but that is outside the scope of this book. The interested reader can learn about such theory in [4].

C.2.1 Matrix approximation of operators

Assume there exists a bounded linear operator $\mathcal{A} : H^s(\partial\Omega) \to H^r(\partial\Omega)$. We wish to construct a matrix approximation to $\mathcal{A}$.

Denote $f \in H^s(\partial\Omega)$ and $g = \mathcal{A}f \in H^r(\partial\Omega)$. Then if the Fourier series have the correct convergence properties,

$$g(\theta) = \mathcal{A}f(\theta) = \mathcal{A} \sum_{n=-\infty}^{\infty} \widehat{f}(n)\varphi_n(\theta) = \sum_{n=-\infty}^{\infty} \widehat{f}(n)\mathcal{A}\varphi_n(\theta), \qquad (C.12)$$

where $\widehat{f}(n) = \langle f, \varphi_n \rangle$ and the Fourier basis functions φ_n are as defined in (13.1). Taking the inner product with a basis function φ_m in (C.12) results in

$$\widehat{g}(m) := \langle \varphi_m, g \rangle = \left\langle \varphi_m, \sum_{n=-\infty}^{\infty} \widehat{f}(n)\mathcal{A}\varphi_n(z) \right\rangle = \sum_{n=-\infty}^{\infty} \widehat{f}(n)\langle \varphi_m, \mathcal{A}\varphi_n(z) \rangle,$$

and we see that knowledge of $\langle \varphi_m, \mathcal{A}\varphi_n \rangle$ completely determines $g = \mathcal{A}f$ for known f (or, equivalently, for known $\widehat{f}(n)$).

Given $N > 0$, we approximate functions $f \in H^s(\partial\Omega)$ and $g \in H^r(\partial\Omega)$ using the following truncated sums:

$$f(\theta) \approx \sum_{n=-N}^{N} \widehat{f}(n)\varphi_n(\theta), \qquad g(\theta) \approx \sum_{n=-N}^{N} \widehat{g}(n)\varphi_n(\theta). \qquad (C.13)$$

We construct a matrix approximation $A : \mathbb{C}^{2N+1} \to \mathbb{C}^{2N+1}$ to the operator $\mathcal{A}$ by setting $A := [A_{m,n}]$ and defining the matrix elements $A_{m,n}$ as follows:

$$A_{m,n} := \langle \mathcal{A}\varphi_n, \varphi_m \rangle = \frac{1}{2\pi} \int_0^{2\pi} (\mathcal{A}e^{in\theta})e^{-im\theta} d\theta. \qquad (C.14)$$

In definition (C.14) we use the following nonstandard but convenient indexing: $m \in \{-N, \ldots, N\}$ for the rows and $n \in \{-N, \ldots, N\}$ for the columns.

To see how the matrix approximation (C.14) can be used computationally, denote the vectors $\widehat{f}, \widehat{g} \in \mathbb{C}^{2N+1}$ of Fourier coefficients of f and g by

$$\widehat{f} := \begin{bmatrix} \widehat{f}(-N) \\ \widehat{f}(-N+1) \\ \vdots \\ \widehat{f}(0) \\ \vdots \\ \widehat{f}(N-1) \\ \widehat{f}(N) \end{bmatrix}, \qquad \widehat{g} := \begin{bmatrix} \widehat{g}(-N) \\ \widehat{g}(-N+1) \\ \vdots \\ \widehat{g}(0) \\ \vdots \\ \widehat{g}(N-1) \\ \widehat{g}(N) \end{bmatrix}.$$

The idea is that the equality $\widehat{g} = A\widehat{f}$ should hold. Let us check this for the first component of the vector $A\widehat{f}$; the rest of the components can be studied similarly.

$$\begin{aligned}
(A\widehat{f})(-N) &= \sum_{n=-N}^{N} A_{-N,n} \widehat{f}(n) \\
&= \sum_{n=-N}^{N} \langle \mathcal{A}\varphi_n, \varphi_{-N} \rangle \langle f, \varphi_n \rangle \\
&= \left\langle \left(\sum_{n=-N}^{N} \langle f, \varphi_n \rangle \mathcal{A}\varphi_n \right), \varphi_{-N} \right\rangle \\
&= \left\langle \mathcal{A} \left(\sum_{n=-N}^{N} \langle f, \varphi_n \rangle \varphi_n \right), \varphi_{-N} \right\rangle \\
&= \langle \mathcal{A}f, \varphi_{-N} \rangle \\
&= \widehat{g}(-N).
\end{aligned}$$

Next we discuss the approximate evaluation of the operator norm of $\mathcal{A} : H^s(\partial\Omega) \to H^r(\partial\Omega)$ from the matrix A. The norm is defined by

$$\|\mathcal{A}\|_{\mathcal{L}(H^s(\partial\Omega), H^r(\partial\Omega))} = \sup_{f \in H^s(\partial\Omega)} \frac{\|\mathcal{A}f\|_r}{\|f\|_s}. \tag{C.15}$$

Using (C.8), (C.9), and (C.11) we see that the norm $\|f\|_s$ can be approximated by

$$\|f\|_s \approx \left(\sum_{n=-N}^{N} |\langle f, \varphi_n^s \rangle_s|^2 \right)^{1/2}. \tag{C.16}$$

C.2. Fourier series and spaces $H^s(\partial\Omega)$

Define vectors $\vec{f} \in \mathbb{C}^{2N+1}$ and $\vec{g} \in \mathbb{C}^{2N+1}$ as follows:

$$\vec{f} = \begin{bmatrix} \langle f, \varphi^s_{-N} \rangle_s \\ \vdots \\ \langle f, \varphi^s_0 \rangle_s \\ \vdots \\ \langle f, \varphi^s_N \rangle_s \end{bmatrix}, \qquad \vec{g} = \begin{bmatrix} \langle \mathcal{A}f, \varphi^r_{-N} \rangle_r \\ \vdots \\ \langle \mathcal{A}f, \varphi^r_0 \rangle_r \\ \vdots \\ \langle \mathcal{A}f, \varphi^r_N \rangle_r \end{bmatrix}.$$

Then we can use Euclidean norms according to (C.16):

$$\|f\|_s \approx \|\vec{f}\|_{\mathbb{C}^{2N+1}}, \qquad \|\mathcal{A}f\|_r \approx \|\vec{g}\|_{\mathbb{C}^{2N+1}}. \tag{C.17}$$

Let us compute

$$\langle \mathcal{A}f, \varphi^r_m \rangle_r = \left\langle \mathcal{A}\left(\sum_{n=-N}^{N} \langle f, \varphi^s_n \rangle_s \varphi^s_n(\theta)\right), \varphi^r_m \right\rangle_r$$

$$= \sum_{n=-N}^{N} \langle f, \varphi^s_n \rangle_s \langle \mathcal{A}\varphi^s_n, \varphi^r_m \rangle_r,$$

so defining a $(2N+1) \times (2N+1)$ matrix $B = [B_{m,n}] := [\langle \mathcal{A}\varphi^s_n, \varphi^r_m \rangle_r]$ yields

$$\vec{g} = B\vec{f}. \tag{C.18}$$

Now combining (C.15), (C.17), and (C.18) yields

$$\|\mathcal{A}\|_{\mathcal{L}(H^s, H^r)} \approx \sup_{\vec{f} \in \mathbb{C}^{2N+1}} \frac{\|B\vec{f}\|_{\mathbb{C}^{2N+1}}}{\|\vec{f}\|_{\mathbb{C}^{2N+1}}} = \|B\|_{\mathcal{L}(\mathbb{C}^{2N+1})}, \tag{C.19}$$

where $\|B\|_{\mathcal{L}(\mathbb{C}^{2N+1})}$ is the standard matrix operator norm.

Let us determine the matrix element $B_{m,n} = \langle \mathcal{A}\varphi^s_n, \varphi^r_m \rangle_r$ in terms of the matrix approximation A defined by (C.14) for $\mathcal{A}$. Truncate formula (C.8) and apply (C.10) and (C.1) to get

$$\langle \mathcal{A}\varphi^s_n, \varphi^r_m \rangle_r \approx \sum_{\ell=-N}^{N} w_r(\ell) \widehat{\mathcal{A}\varphi^s_n}(\ell) \overline{w_r(\ell) \widehat{\varphi^r_m}(\ell)}$$

$$= w_r(m) \widehat{\mathcal{A}\varphi^s_n}(m)$$

$$= w_r(m) w_{-s}(n) \frac{1}{2\pi} \int_0^{2\pi} (\mathcal{A}e^{in\theta}) e^{-im\theta} d\theta$$

$$= w_r(m) w_{-s}(n) A_{m,n}.$$

Finally, (C.19) takes the form

$$\|\mathcal{A}\|_{\mathcal{L}(H^s, H^r)} \approx \|[w_r(m) w_{-s}(n) A_{m,n}]\|_{\mathcal{L}(\mathbb{C}^{2N+1})}. \tag{C.20}$$

C.3 Traces of functions in $H^m(\Omega)$

Let Ω be an open domain and suppose u is continuous on $\bar{\Omega} = \Omega \cup \Gamma$. Then we can find u on the boundary Γ simply by evaluating u at those points. Define the *trace operator* T to be the linear operator on a continuous function $u \in C(\bar{\Omega})$, restricting u to the boundary $\partial \Omega$. That is, $T : C(\bar{\Omega}) \to C(\partial \Omega)$ or $T(u) = u|_{\partial \Omega}$.

What if $u \in H^m(\Omega)$? Then u is not necessarily continuous, and the trace cannot be defined in this manner. Also note that functions in H^m are equivalence classes of functions, defined up to a set of measure zero, and $\partial \Omega$ is a set of measure 0.

Theorem C.8. *Let Ω be a domain with Lipschitz boundary Γ. Then for all*

$$u \in C^1(\bar{\Omega}) \to C(\Gamma), \quad Tu = u|_\Gamma$$

satisfies

$$\|Tu\|_{L^2(\Gamma)} \leq C \|u\|_{H^1(\Omega)}.$$

Theorem C.9 (trace theorem). *Let Ω be a bounded domain in $\mathbb{R}^n$ with Lipschitz boundary Γ. Then*

(i) *there exists a unique bounded linear operator $T : H^1(\Omega) \to L^2(\Gamma)$ (i.e., $\|Tu\|_{L^2(\Gamma)} \leq C \|u\|_{H^1(\Omega)}$) with the property that if $u \in C^1(\bar{\Omega})$, then $Tu = u|_\Gamma$ in the conventional sense;*

(ii) *the range of T is dense in $L^2(\Gamma)$.*

More generally, if $u \in H^m(\Omega)$ and $T_j u \equiv \frac{\partial^j u}{\partial \nu}$, $0 \leq j \leq m-1$, then $T_j : H^m(\Omega) \to H^{m-j-1/2}(\Gamma)$ is a continuous, linear surjection.

Now if $u \in H^1(\Omega)$, we can unambiguously define boundary values of u on Γ. $u = u_0$ means $u = u_0$ a.e. on Γ.

Lastly, recall the theorem from last time: Any differential operator of order m is a bounded linear mapping from H^s into H^{s-m} for $s > m$. Then, for example, (as in EIT), if $u \in H^1(\Omega)$, then $Tu = u|_\Gamma \in L^2(\Gamma)$. In fact, since $j = 0$, $Tu \in H^{1-0-1/2}(\Gamma) = H^{1/2}(\Gamma)$ and $\frac{\partial u}{\partial \nu}|_\Gamma \in H^{-1/2}(\Omega)$.

Definition C.3.1. $H^{-m}(\Omega)$ *is the dual space of $H^m(\Omega) = \{$bounded linear functionals on $H^m(\Omega)\}$.*

Appendix D

Iterative solution of linear equations

Suppose we want to solve the system $Ax = b$ of linear equations, where $b \in \mathbb{R}^n$ and A is an $n \times n$ invertible matrix. Of course, the straightforward solution method is writing $x = A^{-1}b$, in MATLAB notation x=inv(A)*b, but when n is large this may be computationally infeasible.

One possibility is to use an iterative solution method, such as GMRES, based on the use of a sequence of Krylov subspaces spanned by sets

$$\{b\}, \{b, Ab\}, \{b, Ab, A^2b\}, \ldots \{b, Ab, A^2b, \ldots, A^mb\}, \ldots.$$

One advantage of these methods is that the inverse matrix A^{-1} need not be constructed. Another benefit of GMRES is that in certain situations we need not construct the matrix A at all; it is enough to provide GMRES with a computational routine returning the vector $Av \in \mathbb{R}^n$ for any given vector $v \in \mathbb{R}^n$.

We will begin with motivation for the definition of the Krylov subspaces. Suppose we wish to solve the linear system $Ax = b$, and the matrix A may be poorly conditioned. Given a preconditioner K that is somehow simpler than A, we will let x_0 denote the solution to

$$Kx_0 = b$$

and consider $A(x_0 + z) = b$, where z is then a correction vector satisfying $Az = b - Ax_0$. We can form a new approximation to x by constructing $x_1 = x_0 + z_0$, where z_0 solves

$$Kz_0 = b - Ax_0.$$

This can be repeated, giving successive approximations to approximate x. Namely, $x_2 = x_0 + z_0 + z_1$, where z_1 satisfies

$$Kz_1 = b - Ax_1,$$

and so on. In some methods, K is updated as well, but we will consider a static K. At the $(i+1)$st step we have

$$x_{i+1} = x_i + z_i = x_i + K^{-1}(b - Ax_i) = x_i + \widetilde{b} - \widetilde{A}x_i. \qquad (D.1)$$

Note that if K is the identity, we arrive at the *Richardson iteration*:

$$x_{i+1} = b + (I-A)x_i = x_i + r_i,$$

where r_i is the residual $b - Ax_i$. In general, the matrix $\widetilde{A}$ in (D.1) represents a preconditioned matrix, and in the following material we will drop the tildes and regard A as such.

Note that

$$x_{i+1} = x_i + r_i,$$
$$b - Ax_{i+1} = b - Ax_i - Ar_i,$$
$$r_{i+1} = r_i - Ar_i = (I-A)r_i,$$

so that $\|r_{i+1}\| \leq \|I-A\| \|r_i\|$. Thus, the iterations converge for all x_0 if $\|I-A\| < 1$. This puts some restrictions on the preconditioner.

We can define the space in which successive approximations to x_0 are located as follows:

$$\begin{aligned} x_{i+1} &= x_i + r_i \\ &= x_{i-1} + r_{i-1} + r_i \\ &= \cdots = r_0 + r_1 + \cdots + r_i \\ &= \sum_{j=0}^{i} (I-A)^j r_0 \\ &\in \text{span}\{r_0, Ar_0, \ldots, A^i r_0\} \\ &\equiv K^{i+1}(A; r_0). \end{aligned}$$

Definition D.0.2. *The m-dimensional space spanned by a given vector v and increasing powers of A applied to v up to the $(m-1)$st power is called the m-dimensional Krylov subspace generated by A and v and is denoted $K^m(A; v)$.*

Methods that attempt to generate approximations from the Krylov subspace are called *Krylov subspace methods*. There are four classes of such methods:

1. Ritz–Galerkin methods (the conjugate gradient method is an example).

2. Minimum norm residual approaches (GMRES is an example).

3. Petrov–Galerkin methods.

4. Minimum norm error methods.

For a symmetric positive definite matrix, the algorithm of choice is the conjugate gradient method. For the minimum norm residual approach, we need some further notation. Denote the basis vectors of $K^{i+1}(A, r_0)$ by $u_j = A^{j-1} r_0$ and U_j the $n \times j$ matrix with columns $u_1, \ldots, u_j$. One can show

$$AU_i = U_i B_i + u_{i+1} e_i^T, \tag{D.2}$$

where B_i is an $i \times i$ matrix with $b_{j+1,j} = 1$ and all other elements zero. Decomposing U_i as $U_i = Q_i R_i$ with $Q_i^T Q_i = I$ and R_i upper triangular, we obtain from (D.2),

$$AQ_i R_i = Q_i R_i B_i + u_{i+1} e_i^T,$$
$$AQ_i = Q_i R_i B_i R_i^{-1} + u_{i+1} e_i^T R_i^{-1}.$$

Let $\widetilde{H}_i = R_i B_i R_i^{-1}$. This is useful for deriving an orthonormal basis of $K^i(A; r_0)$, which we will denote $v_1, \ldots, v_i$. Denoting the matrix V_i to be the matrix with columns $v_1, \ldots, v_i$, one can show

$$AV_i = V_{i+1} H_{i+1,i}. \tag{D.3}$$

There are computational algorithms for constructing H and V that we will not discuss here.

D.0.1 The minimum norm residual approach

We seek $x_i \in K^i(A; r_0)$, $x_i = V_i y$ for which $\|b - Ax_i\|$ is minimal. Assume without loss of generality that our initial $x_0 = 0$. Defining $\rho = \|r_0\|_2$, from (D.3)

$$\|b - Ax_i\|_2 = \|b - AV_i y\|_2 = \|b - V_{i+1} H_{i+1,i} y\|_2$$
$$= \|\rho V_{i+1} e_1 - V_{i+1} H_{i+1,i} y\|_2$$
$$= \|V_{i+1}\|_2 \|\rho e_1 - H_{i+1,i} y\|_2.$$

Since $\|V_{i+1}\|_2 = 1$, we have $\|b - Ax_i\|_2 = \|\rho e_1 - H_{i+1,i} y\|_2$. This is zero when $\rho e_1 = H_{i+1,i} y$. Thus, we wish to solve the least-squares minimization problem

$$H_{i+1,i} y = \|r_0\|_2 e_1.$$

GMRES is a way of solving this minimization problem. In particular, it uses Givens rotations to annihilate the subdiagonal elements of $H_{i+1,i}$. The resulting upper triangular matrix is denoted by $R_{i,i}$:

$$H_{i+1,i} = Q_{i+1,i} R_{i,i},$$

where $Q_{i+1,i}$ is the product of the successive Givens eliminations of $H_{j+1,j}$ for $j = 1, \ldots, i$. After Givens transformations, the least-squares solution minimizes

$$\|H_{i+1,i} y - \rho e_1\|_2 = \|Q_{i+1,i} R_{i,i} y - \rho e_1\|_2$$
$$= \|Q_{i+1,i}^T Q_{i+1,i} R_{i,i} y - Q_{i+1,i}^T \rho e_1\|_2.$$

Since $Q_{i+1,i}^T Q_{i+1,i} = I$, $y = R_{i,i}^{-1} Q_{i+1,i}^T \rho e_1$, and now

$$x_i = V_i y.$$

This is GMRES.

D.0.2 Application to the solution of the D-bar equation

In our numerical solution of the D-bar equation, we wish to solve

$$m(z, s) = 1 + \frac{1}{4\pi^2} \int_{\mathbb{R}^2} \frac{\mathbf{t}(k)}{\bar{k}(s-k)} e_{-z}(k) \overline{m(z,k)} dk.$$

Denoting the Green's function of the $\bar{\partial}$ operator $G(k) = \frac{1}{\pi k}$, we can write this as

$$m(x,s) = 1 + G * \left(\frac{\mathbf{t}(k)}{\bar{k}} e_{-z}(k) \overline{m(z,k)} \right).$$

Noting that the Fourier transform of a convolution is the product of the Fourier transforms, we can write this as

$$m(x,s) = 1 + \mathcal{F}^{-1}\left(\mathcal{F}(G) \mathcal{F}\left(\frac{\mathbf{t}(k)}{\bar{k}} e_{-z}(k) \overline{m(z,k)} \right) \right). \quad (D.4)$$

Denote

$$T(x,k) = -\frac{\mathbf{t}(k)}{\bar{k}} e_{-z}(k)$$

and the operator $\mathcal{A}$ by

$$\mathcal{A}(T\phi) = \mathcal{F}^{-1}(\mathcal{F}(G)\mathcal{F}(T\bar{\phi})).$$

Then we have

$$[I + \mathcal{A}(T\cdot)]m = 1.$$

The system is then discretized as described above and is very suitable for solution by GMRES, since the action of $[I + \mathcal{A}(T\cdot)]$ is well-defined and is not so easily expressed as a simple preconditioned matrix system.

D.0.3 A simple example

Here we explain how to avoid the construction of the system matrix A in a very simple example. Let

$$A = \begin{bmatrix} 0 & 1 \\ 1 & 0 \end{bmatrix}, \qquad b = \begin{bmatrix} 5 \\ 3 \end{bmatrix}.$$

We can find the solution $x = [3\ 5]^T$ approximately using GMRES by writing in MATLAB

```
A=[[0 1];[1 0]];
b=[5;3];
x=gmres(A,b)
```

However, the matrix-free option is to define a routine swap as follows:

```
function result = swap(v)
result = zeros(size(v));
result(1) = v(2);
result(2) = v(1);
```

Note that swap just interchanges the two first components in the argument vector, just as multiplication by matrix A does to a vector in $\mathbb{R}^2$. Then we can solve the system by passing the function swap as an argument to GMRES: x=gmres(@swap,b).

For example, in the D-bar solver explained in Section 15.4, we implement the action of the appropriate real-linear operator using FFT instead of constructing the system matrix explicitly. The resulting matrix-free algorithm is much faster as it has lower computational complexity than the matrix-based algorithm.

Bibliography

[1] M. J. ABLOWITZ AND A. S. FOKAS, *Complex Variables: Introduction and Applications*, Cambridge University Press, Cambridge, UK, 2003. [204, 224]

[2] E. ABOUFADEL AND S. SCHLICKER, *Discovering Wavelets*, John Wiley & Sons, New York, 1999. [95]

[3] A. ABUBAKAR, T. M. HABASHY, M. LI, AND J. LIU, *Inversion algorithms for large-scale geophysical electromagnetic measurements*, Inverse Problems, 25 (2009), 123012. [162]

[4] R. A. ADAMS, *Sobolev Spaces*, Academic Press, New York, 1975. [206, 297, 303]

[5] G. ALESSANDRINI, *Stable determination of conductivity by boundary measurements*, Applicable Analysis, 27 (1988), pp. 153–172. [165, 227]

[6] M. ALLAIN AND J. IDIER, *Efficient binary reconstruction for non-destructive evaluation using gammagraphy*, Inverse Problems, 23 (2007), pp. 1371–1393. [113]

[7] M. ALTSCHULER, *Reconstruction of the global-scale three-dimensional solar corona*, in Image Reconstruction from Projections, Gabor Herman, ed., vol. 32 of Topics in Applied Physics, Springer, Berlin, Heidelberg, 1979, pp. 105–145. [113]

[8] A. ALÙ AND N. ENGHETA, *Achieving transparency with plasmonic and metamaterial coatings*, Physical Review E, 72 (2005), 016623. [183]

[9] W. AMES, *Numerical Methods for Partial Differential Equations*, Academic Press, San Diego, 1992. [18]

[10] H. AMMARI, O. KWON, J. K. SEO, AND E. J. WOO, *T-Scan electrical impedance imaging system for anomaly detection*, SIAM Journal on Applied Mathematics, 65 (2004), pp. 252–266. [161]

[11] S. R. ARRIDGE, *Optical tomography in medical imaging*, Inverse Problems, 15 (1999), pp. R41–R93. [155]

[12] S. R. ARRIDGE, O. DORN, J. P. KAIPIO, V. KOLEHMAINEN, M. SCHWEIGER, T. TARVAINEN, M. VAUHKONEN, AND A. ZACHAROPOULOS, *Reconstruction of subdomain boundaries of piecewise constant coefficients of the radiative transfer equation from optical tomography data*, Inverse Problems, 22 (2006), pp. 2175–2196. [155]

[13] M. ASSENHEIMER, O. LAVER-MOSKOVITZ, D. MALONEK, D. MANOR, U. NAHALIEL, R. NITZAN, AND A. SAAD, *The T-SCAN technology: Electrical impedance as a diagnostic tool for breast cancer detection*, Physiological Measurement, 22 (2001), pp. 1–8. [161]

[14] K. ASTALA, T. IWANIEC, AND G. MARTIN, *Elliptic Partial Differential Equations and Quasiconformal Mappings in the Plane*, vol. 48 of Princeton Mathematical Series, Princeton University Press, Princeton, NJ, 2009. [224, 273]

[15] K. ASTALA, M. LASSAS, AND L. PÄIVÄRINTA, *The borderlines of the invisibility and visibility for Calderón's inverse problem*, preprint, arXiv:1109.2749v1 (2012). [184, 200]

[16] K. ASTALA, J. L. MUELLER, L. PÄIVÄRINTA, A. PERÄMÄKI, AND S. SILTANEN, *Direct electrical impedance tomography for nonsmooth conductivities*, Inverse Problems and Imaging, 5 (2011), pp. 531–549. [145, 200, 266, 273, 274, 275, 276]

[17] K. ASTALA, J. L. MUELLER, L. PÄIVÄRINTA, AND S. SILTANEN, *Numerical computation of complex geometrical optics solutions to the conductivity equation*, Applied and Computational Harmonic Analysis, 29 (2010), pp. 391–403. [200, 216]

[18] K. ASTALA AND L. PÄIVÄRINTA, *A boundary integral equation for Calderón's inverse conductivity problem*, in Proceedings of the 7th International Conference on Harmonic Analysis, Collectanea Mathematica, 2006. [200, 266, 268]

[19] ———, *Calderón's inverse conductivity problem in the plane*, Annals of Mathematics, 163 (2006), pp. 265–299. [199, 200, 215, 216, 217, 249, 266, 267, 268, 273]

[20] K. ASTALA, L. PÄIVÄRINTA, AND M. LASSAS, *Calderón's inverse problem for anisotropic conductivity in the plane*, Communications in Partial Differential Equations, 30 (2005), pp. 207–224. [184]

[21] R. C. ASTER, C. H. THURBER, AND B. BORCHERS, *Parameter Estimation and Inverse Problems*, Elsevier Academic Press, Amsterdam, 2005. [140]

[22] S. BABAEIZADEH, D. H. BROOKS, AND D. ISAACSON, *A 3-D boundary element solution to the forward problem of electrical impedance tomography*, in Prodceedings of the 26th Annual International Conference of the IEEE Engineering in Medicine and Biology Society, Vol. 1, 2004, pp. 960–963. [185]

[23] S. BABAEIZADEH, D. H. BROOKS, D. ISAACSON, AND J. C. NEWELL, *Electrode boundary conditions and experimental validation for BEM-based EIT forward and inverse solutions*, IEEE Transactions on Medical Imaging, 25 (2006), pp. 1180–1188. [185]

[24] A. P. BAGSHAW, A. D. LISTON, R. H. BAYFORD, A. TIZZARD, A. P. GIBSON, A. T. TIDSWELL, M. K. SPARKES, H. DEHGHANI, C. D. BINNIE, AND D. S. HOLDER, *Electrical impedance tomography of human brain function using reconstruction algorithms based on the finite element method*, NeuroImage, 20 (2003), pp. 752–764. [161]

[25] J. A. BAKER AND J. Y. LO, *Breast tomosynthesis: State-of-the-art and review of the literature*, Academic Radiology, 18 (2011), pp. 1298–1310. [123]

[26] N. V. BANICHUK AND P. J. NEITTAANMÄKI, *Structural Optimization with Uncertainties*, vol. 162 of Solid Mechanics and Its Applications, Springer, Dordrecht, 2010. [140]

[27] D. C. BARBER AND B. H. BROWN, *Progress in electrical impedance tomography*, in Inverse Problems in Partial Differential Equations (Arcata, CA, 1989), SIAM, Philadelphia, PA, 1990, pp. 151–164. [160]

[28] J. A. BARCELÓ, T. BARCELÓ, AND A. RUIZ, *Stability of the inverse conductivity problem in the plane for less regular conductivities*, Journal of Differential Equations, 173 (2001), pp. 231–270. [200, 231, 250]

[29] T. BARCELÓ, D. FARACO, AND A. RUIZ, *Stability of Calderón inverse conductivity problem in the plane*, Journal de Mathématiques Pures et Appliqués, 88 (2007), pp. 522–556. [200]

[30] D. BAROUDI AND E. SOMERSALO, *Gas temperature mapping using impedance tomography*, Inverse Problems, 13 (1997), p. 1177. [3]

[31] J. BARZILAI AND J. M. BORWEIN, *Two-point step size gradient methods*, IMA Journal of Numerical Analysis, 8 (1988), pp. 141–148. [90, 91]

[32] X. L. BATTLE, G. S. CUNNINGHAM, AND K. M. HANSON, *3D tomographic reconstruction using geometrical models*, in Society of Photo-Optical Instrumentation Engineers (SPIE) Conference Series, K. M. Hanson, ed., vol. 3034 of Proceedings of SPIE, presented at the Society of Photo-Optical Instrumentation Engineers (SPIE) Conference, 1997, pp. 346–357. [113]

[33] X. L. BATTLE, K. M. HANSON, AND G. S. CUNNINGHAM, *Tomographic reconstruction using 3D deformable models*, Physics in Medicine and Biology, 43 (1998), pp. 983–990. [113]

[34] R. BEALS AND R. R. COIFMAN, *Scattering, transformations spectrales et équations d'évolution non linéaires*, in Goulaouic-Meyer-Schwartz Seminar, 1980–1981, École Polytechnique, Palaiseau, France, 1981, XXII,10. [230]

[35] ———, *Scattering, transformations spectrales et équations d'évolution non linéaire II*, in Goulaouic-Meyer-Schwartz Seminar, 1981–1982, École Polytechnique, Palaiseau, France, 1982, XXI, 9. [230]

[36] ———, *Multidimensional inverse scatterings and nonlinear partial differential equations*, in Pseudodifferential Operators and Applications (Notre Dame, IN, 1984), AMS, Providence, RI, 1985, pp. 45–70. [230]

[37] ———, *The D-bar approach to inverse scattering and nonlinear evolutions*, Physica D, 18 (1986), pp. 242–249. [230]

[38] ———, *Linear spectral problems, nonlinear equations and the $\bar{\partial}$-method*, Inverse Problems, 5 (1989), pp. 87–130. [224, 230]

[39] J. BEAR, *Dynamics of Fluids in Porous Media*, Elsevier, New York, 1972. [133]

[40] J. V. BECK, B. BLACKWELL, AND C. R. ST. CLAIR, *Inverse Heat Conduction: Ill-Posed Problems*, Wiley-Interscience, New York, 1985. [133]

[41] L. BEILINA AND M.V. KLIBANOV, *Approximate Global Convergence and Adaptivity for Coefficient Inverse Problems*, Springer, New York, 2012. [140]

[42] M. A. BENNETT AND R. A. WILLIAMS, *Monitoring the operation of an oil/water separator using impedance tomography*, Minerals Engineering, 17 (2004), pp. 605–614. [162]

[43] Y. BENVENISTE AND T. MILOH, *Neutral inhomogeneities in conduction phenomena*, Journal of the Mechanics and Physics of Solids, 47 (1999), pp. 1873–1892. [183]

[44] E. BERETTA AND E. FRANCINI, *Lipschitz stability for the electrical impedance tomography problem: The complex case*, Communications in Partial Differential Equations, 36 (2011), pp. 1723–1749. [200]

[45] J. BIAN, X. HAN, E. Y. SIDKY, G. CAO, J. LU, O. ZHOU, AND X. PAN, *Investigation of sparse data mouse imaging using micro-CT with a carbon-nanotube-based X-ray source*, Tsinghua Science & Technology, 15 (2010), pp. 74–78. [113]

[46] J. BIAN, J. H. SIEWERDSEN, X. HAN, E. Y. SIDKY, J. L. PRINCE, C. A. PELIZZARI, AND X. PAN, *Evaluation of sparse-view reconstruction from flat-panel-detector cone-beam CT*, Physics in Medicine and Biology, 55 (2010), pp. 6575–6599. [113]

[47] J. BIKOWSKI, K. KNUDSEN, AND J. L. MUELLER, *Direct numerical reconstruction of conductivities in three dimensions using scattering transforms*, Inverse Problems, 27 (2011), 015002. [180, 185, 200]

[48] J. BIKOWSKI AND J. MUELLER, *2D EIT reconstructions using Calderón's method*, Inverse Problems and Imaging, 2 (2008), pp. 43–61. [200, 263]

[49] Å. BJÖRCK, *Numerical Methods for Least Squares Problems*, SIAM, Philadelphia, PA, 1996. [82]

[50] R. BLUE, *Real-Time Three-Dimensional Electrical Impedance Tomography*, Ph.D. thesis, Rensselaer Polytechnic Institute, Troy, NY, 1997. [180]

[51] M. BOITI, J. P. LEON, M. MANNA, AND F. PEMPINELLI, *On a spectral transform of a KdV-like equation related to the Schrödinger operator in the plane*, Inverse Problems, 3 (1987), pp. 25–36. [205, 208]

[52] K. BOONE, A. M. LEWIS, AND D. S. HOLDER, *Imaging of cortical spreading depression by EIT: Implications for localization of epileptic foci*, Physiological Measurement, 15 (1994), pp. A189–A198. [161]

[53] L. BORCEA, *Electrical impedance tomography*, Inverse Problems, 18 (2002), pp. 99–136. [160]

[54] G. BOVERMAN, D. ISAACSON, T.-J. KAO, SAULNIER, G. J., AND J. C. NEWELL, *Methods for direct image reconstruction for EIT in two and three dimensions*, in Proceedings of the 2008 Electrical Impedance Tomography Conference, Dartmouth College, Hanover, NH, 2008. [180]

[55] G. BOVERMAN, T.-J. KAO, D. ISAACSON, AND G. J. SAULNIER, *An implementation of Calderón's method for 3-D limited view EIT*, IEEE Transactions on Medical Imaging, 1 (2008), pp. 1–10. [200]

[56] H. BREZIS, *Functional Analysis, Sobolev Spaces and Partial Differential Equations*, Springer, New York, 2011. [297]

[57] B. H. BROWN, *Electrical impedance tomography (EIT): A review*, Journal of Medical Engineering & Technology, 27 (2003), pp. 97–108. [161]

[58] B. H. BROWN, D. C. BARBER, A. H. MORICE, AND A. D. LEATHARD, *Cardiac and respiratory related electrical impedance changes in the human thorax*, IEEE Transactions on Biomedical Engineering, 41 (1994), pp. 729–734. [161]

[59] R. M. BROWN AND G. UHLMANN, *Uniqueness in the inverse conductivity problem for nonsmooth conductivities in two dimensions*, Communications in Partial Differential Equations, 22 (1997), pp. 1009–1027. [199, 200]

[60] M. BRÜHL AND M. HANKE, *Numerical implementation of two non-iterative methods for locating inclusions by impedance tomography*, Inverse Problems, 16 (2000), pp. 1029–1042. [277]

[61] J.-F. CAI, S. OSHER, AND Z. SHEN, *Linearized Bregman iterations for frame-based image deblurring*, SIAM Journal on Imaging Sciences, 2 (2009), pp. 226–252. [83]

[62] F. CAKONI AND D. COLTON, *Qualitative Methods in Inverse Scattering Theory: An Introduction*, Interaction of Mechanics and Mathematics, Springer, Berlin, 2006. [140]

[63] F. CAKONI, D. COLTON, AND P. MONK, *The Linear Sampling Method in Inverse Electromagnetic Scattering*, vol. 80 of CBMS-NSF Regional Conference Series in Applied Mathematics, SIAM, Philadelphia, PA, 2011. [140, 287]

[64] A.-P. CALDERÓN, *On an inverse boundary value problem*, in Seminar on Numerical Analysis and Its Applications to Continuum Physics (Rio de Janeiro, 1980), Soc. Brasil. Mat., Rio de Janeiro, 1980, pp. 65–73. [159, 180, 199, 200, 249, 259, 263, 265, 285, 286]

[65] D. CALVETTI, G. LANDI, L. REICHEL, AND F. SGALLARI, *Non-negativity and iterative methods for ill-posed problems*, Inverse Problems, 20 (2004), pp. 1747–1758. [6]

[66] D. CALVETTI AND E. SOMERSALO, *Priorconditioners for linear systems*, Inverse Problems, 21 (2005), pp. 1397–1418. [113]

[67] ———, *Introduction to Bayesian Scientific Computing: Ten Lectures on Subjective Computing*, Vol. 2, Springer, New York, 2007. [6]

[68] ———, *Microlocal sequential regularization in imaging*, Inverse Problems and Imaging, 1 (2007), pp. 1–11. [113]

[69] E. J. CANDÈS, J. ROMBERG, AND T. TAO, *Robust uncertainty principles: Exact signal reconstruction from highly incomplete frequency information*, IEEE Transactions on Information Theory, 52 (2006), pp. 489–509. [88, 113]

[70] A. CEDERLUND, M. KALKE, AND U. WELANDER, *Volumetric tomography—a new tomographic technique for panoramic units*, Dentomaxillofacial Radiology, 38 (2009), pp. 104–111. [129]

[71] K. CHADAN, D. COLTON, L. PÄIVÄRINTA, AND W. RUNDELL, *An Introduction to Inverse Scattering and Inverse Spectral Problems*, SIAM, Philadelphia, PA, 1997. [140]

[72] A. CHAMBOLLE, S. E. LEVINE, AND B. J. LUCIER, *An upwind finite-difference method for total variation–based image smoothing*, SIAM Journal on Imaging Sciences, 4 (2011), pp. 277–299. [83]

[73] A. CHAMBOLLE AND P.-L. LIONS, *Image recovery via total variation minimization and related problems*, Numerische Mathematik, 76 (1997), pp. 167–188. [83]

[74] T. F. CHAN AND K. CHEN, *On a nonlinear multigrid algorithm with primal relaxation for the image total variation minimisation*, Numerical Algorithms, 41 (2006), pp. 387–411. [83]

[75] T. F. CHAN, G. H. GOLUB, AND P. MULET, *A nonlinear primal-dual method for total variation-based image restoration*, SIAM Journal on Scientific Computing, 20 (1999), pp. 1964–1977. [83]

[76] T. F. CHAN AND J. SHEN, *Image processing and analysis: Variational, PDE, Wavelet, and Stochastic Methods*, SIAM, Philadelphia, PA, 2005. [83]

[77] C. CHAUX, P. L. COMBETTES, J.-C. PESQUET, AND V. R. WAJS, *A variational formulation for frame-based inverse problems*, Inverse Problems, 23 (2007), pp. 1495–1518. [83]

[78] M. CHENEY AND D. ISAACSON, *Distinguishability in impedance imaging*, IEEE Transactions on Biomedical Engineering, 39 (1992), pp. 852–860. [176, 178]

[79] ———, *Issues in electrical impedance imaging*, IEEE Computational Science and Engineering, 2 (1995), pp. 53–62. [173]

[80] M. CHENEY, D. ISAACSON, AND J. C. NEWELL, *Electrical impedance tomography*, SIAM Review, 41 (1999), pp. 85–101. [160]

[81] M. CHENEY, D. ISAACSON, J. C. NEWELL, S. SIMSKE, AND J. GOBLE, *NOSER: An algorithm for solving the inverse conductivity problem*, International Journal of Imaging Systems and Technology, 2 (1990), pp. 66–75. [172]

[82] K. S. CHENG, D. ISAACSON, J. C. NEWELL, AND D. G. GISSER, *Electrode models for electric current computed tomography*, IEEE Transactions on Biomedical Engineering, 36 (1989), pp. 918–924. [173, 191]

[83] V. A. CHEREPENIN, A. Y. KARPOV, A. V. KORJENEVSKY, V. N. KORNIENKO, Y. S. KULTIASOV, M. B. OCHAPKIN, O. V. TROCHANOVA, AND J. D. MEISTER, *Three-dimensional EIT imaging of breast tissues: System design and clinical testing*, IEEE Transactions on Medical Imaging, 21 (2002), pp. 662–667. [161]

[84] V. CHEREPENIN, A. KARPOV, A. KORJENEVSKY, V. KORNIENKO, A. MAZALETSKAYA, D. MAZOUROV, AND D. MEISTER, *A 3D electrical impedance tomography (EIT) system for breast cancer detection*, Physiological Measurement, 22 (2001), pp. 9–18. [161]

[85] K. CHOI, J. WANG, L. ZHU, T.-S. SUH, S. BOYD, AND L. XING, *Compressed sensing based cone-beam computed tomography reconstruction with a first-order method*, Medical Physics, 37 (2010), pp. 5113–5125. [113]

[86] C. K. CHUI, *Wavelets: A Mathematical Tool for Signal Analysis*, SIAM, Philadelphia, PA 1997. [95]

[87] A. CLOP, D. FARACO, AND A. RUIZ, *Integral stability of Calderón inverse conductivity problem in the plane*, Inverse Problems and Imaging, 4 (2010), pp. 49–91. [200]

[88] C. F. COLEMAN AND J. R. MCLAUGHLIN, *Solution of the inverse spectral problem for an impedance with integrable derivative. I, II*, Communications on Pure and Applied Mathematics, 46 (1993), pp. 145–184, 185–212. [152]

[89] D. COLTON AND A. KIRSCH, *An approximation problem in inverse scattering problems theory*, Applicable Analysis, 41 (1991), pp. 23–32. [151]

[90] ——, *A simple method for solving inverse scattering problems in the resonance region*, Inverse Problems, 12 (1996), pp. 383–393. [287]

[91] D. COLTON AND R. KRESS, *Inverse Acoustic and Electromagnetic Scattering Theory*, second ed., Springer, Berlin, 1998. [140, 148, 151, 288]

[92] D. COLTON AND P. MONK, *The inverse scattering problem for acoustic waves in an inhomogeneous medium*, Quarterly Journal of Mechanics and Applied Mathematics, 41 (1988), pp. 97–125. [151]

[93] ——, *A new method for solving the inverse scattering problem for acoustic waves in an inhomogeneous medium*, Inverse Problems, 5 (1989), pp. 1013–1026. [151]

[94] ——, *A new method for solving the inverse scattering problem for acoustic waves in an inhomogeneous medium II*, Inverse Problems, 6 (1990), pp. 935–947. [151]

[95] ——, *A comparison of two methods for solving the inverse scattering problem for acoustic waves in an inhomogeneous medium*, Journal on Computational and Applied Mathematics, 42 (1992), pp. 5–16. [151]

[96] ——, *The detection and monitoring of leukemia using electromagnetic waves: Mathematical theory*, Inverse Problems, 10 (1994), pp. 1235–1251. [151]

[97] ——, *The detection and monitoring of leukemia using electromagnetic waves: Numerical analysis*, Inverse Problems, 11 (1995), pp. 329–342. [151]

[98] ——, *A linear sampling method for the detection of leukemia using microwaves*, SIAM Journal on Applied Mathematics, 58 (1998), pp. 926–941. [287]

[99] P. L. COMBETTES AND V. R. WAJS, *Signal recovery by proximal forward-backward splitting*, Multiscale Modeling and Simulation, 4 (2005), pp. 1168–1200. [83]

[100] A. M. CORMACK, *Representation of a function by its line integrals, with some radiological applications* I, Journal of Applied Physics, 34 (1963), pp. 2722–2727. [114]

[101] H. CORNEAN, K. KNUDSEN, AND S. SILTANEN, *Towards a D-bar reconstruction method for three-dimensional EIT*, Journal of Inverse and Ill-Posed Problems, 14 (2006), pp. 111–134. [180]

[102] E. L. COSTA, C. N. CHAVES, S. GOMES, M. A. BERALDO, M. S. VOLPE, M. R. TUCCI, I. A. SCHETTINO, S. H. BOHM, C. R. CARVALHO, H. TANAKA, R. G. LIMA, AND M. B. AMATO, *Real-time detection of pneumothorax using electrical impedance tomography*, Critical Care Medicine, 36 (2008), pp. 1230–1238. [161]

[103] E. L. V. COSTA, R. G. LIMA, AND M. B. P. AMATO, *Electrical impedance tomography*, Current Opinion in Critical Care, 15 (2009), pp. 18–24. [161]

[104] J. DAHL, P. C. HANSEN, S. H. JENSEN, AND T. L. JENSEN, *Algorithms and software for total variation image reconstruction via first-order methods*, Numerical Algorithms, 53 (2010), pp. 67–92. [83]

[105] Y.-H. DAI AND R. FLETCHER, *Projected Barzilai-Borwein methods for large-scale box-constrained quadratic programming*, Numerische Mathematik, 100 (2005), pp. 21–47. [92]

[106] W. DAILY AND A. RAMIREZ, *Electrical resistance tomography during in-situ trichloroethylene remediation at the Savannah River Site*, Journal of Applied Geophysics, 33 (1995), pp. 239–249. [162]

[107] W. DAILY, A. RAMIREZ, AND R. JOHNSON, *Electrical impedance tomography of a perchloroethylene release*, Journal of Environmental & Engineering Geophysics, 2 (1998), pp. 189–201. [162]

[108] W. DAILY, A. RAMIREZ, D. LABRECQUE, AND J. NITAO, *Electrical resistivity tomography of vadose water movement*, Water Resources Research, 28 (1992), pp. 1429–1442. [162]

[109] R. A. DANFORTH AND D. E. CLARK, *Effective dose from radiation absorbed during a panoramic examination with a new generation machine*, Oral Surgery, Oral Medicine, Oral Pathology, Oral Radiology, and Endodontology, 89 (2000), pp. 236–243. [130]

[110] I. DAUBECHIES, *Orthonormal bases of compactly supported wavelets*, Communications on Pure and Applied Mathematics, 41 (1988), pp. 909–996. [96]

[111] I. DAUBECHIES, *Ten Lectures on Wavelets*, vol. 61 of CBMS-NSF Regional Conference Series in Applied Mathematics, SIAM, Philadelphia, PA 2006. [95, 96, 97, 108]

[112] I. DAUBECHIES, M. DEFRISE, AND C. DE MOL, *An iterative thresholding algorithm for linear inverse problems with a sparsity constraint*, Communications on Pure and Applied Mathematics, 57 (2004), pp. 1413–1457. [83, 99]

[113] M. E. DAVISON, *The ill-conditioned nature of the limited angle tomography problem*, SIAM Journal on Applied Mathematics, 43 (1983), pp. 428–448. [120]

[114] M. DEANGELO AND J. L. MUELLER, *2D D-bar reconstructions of human chest and tank data using an improved approximation to the scattering transform*, Physiological Measurement, 31 (2010), pp. 221–232. [254, 257, 262]

[115] A. H. DELANEY AND Y. BRESLER, *Globally convergent edge-preserving regularized reconstruction: An application to limited-angle tomography*, IEEE Transactions on Image Processing, 7 (1998), pp. 204–221. [113]

[116] F. DELBARY, P. C. HANSEN, AND K. KNUDSEN, *Electrical impedance tomography: 3D reconstructions using scattering transforms*, Applicable Analysis, 91 (2012), pp. 737–755. [180, 200]

[117] ———, *A direct numerical reconstruction algorithm for the 3D Calderón problem*, Journal of Physics: Conference Series, 290 (2011), 012003. [180]

[118] G. DE MARSILY, *Quantitative Hydrogeology*, Academic Press, New York, 1986. [146]

[119] J. DENNIS AND R. SCHNABEL, *Numerical Methods for Unconstrained Optimization and Nonlinear Equations*, Prentice-Hall, Englewood Cliff, NJ, 1983. [82]

[120] Z. DES PLANTES, *Eine neue Methode zur Differenzierung in der Röntgenographie (Planigraphies)*, Acta Radiologica [Old Series], 13 (1932), pp. 182–192. [123]

[121] F. DIEKMANN AND U. BICK, *Tomosynthesis and contrast-enhanced digital mammography: Recent advances in digital mammography*, European Radiology, 17 (2007), pp. 3086–3092. [123]

[122] S. DO, W. C. KARL, M. K. KALRA, T. J. BRADY, AND H. PIEN, *A variational approach for reconstructing low dose images in clinical helical CT*, in Proceedings of the 2010 IEEE International Symposium on Biomedical Imaging: From Nano to Macro, 2010, pp. 784–787. [113]

[123] J. T. DOBBINS III, *Tomosynthesis imaging: At a translational crossroads*, Medical Physics, 36 (2009), pp. 1956–1967. [123]

[124] J. T. DOBBINS III, H. P. MCADAMS, D. J. GODFREY, AND C. M. LI, *Digital tomosynthesis of the chest*, Journal of Thoracic Imaging, 23 (2008), pp. 86–92. [123]

[125] D. C. DOBSON AND F. SANTOSA, *Resolution and stability analysis of an inverse problem in electrical impedance tomography: Dependence on the input current patterns*, SIAM Journal on Applied Mathematics, 54 (1994), pp. 1542–1560. [179]

[126] ———, *Recovery of blocky images from noisy and blurred data*, SIAM Journal on Applied Mathematics, 56 (1996), pp. 1181–1198. [83]

[127] D. C. DOBSON AND C. R. VOGEL, *Convergence of an iterative method for total variation denoising*, SIAM Journal on Numerical Analysis, 34 (1997), pp. 1779–1791. [83]

[128] D. L. DONOHO, *Compressed sensing*, IEEE Transactions on Information Theory, 52 (2006), pp. 1289–1306. [88]

[129] X. DUAN, L. ZHANG, Y. XING, Z. CHEN, AND J. CHENG, *Few-view projection reconstruction with an iterative reconstruction-reprojection algorithm and TV constraint*, IEEE Transactions on Nuclear Science, 56 (2009), pp. 1377–1382. [113]

[130] P. EDIC, G. J. SAULNIER, J. C. NEWELL, AND D. ISAACSON, *A real-time electrical impedance tomograph*, IEEE Transactions on Biomedical Engineering, 42 (1995), pp. 849–859. [254]

[131] T. EIROLA, M. HUHTANEN, AND J. VON PFALER, *Solution methods for $\mathbb{R}$-linear problems in $\mathbb{C}^n$*, SIAM Journal on Matrix Analysis and Applications, 25 (2004), pp. 804–828. [237, 240]

[132] H. W. ENGL, M. HANKE, AND A. NEUBAUER, *Regularization of Inverse Problems*, Kluwer Academic Publishers, Dordrecht, 1996. [6, 48, 140, 232]

[133] C. L. EPSTEIN, *Introduction to the Mathematics of Medical Imaging*, second ed., SIAM, Philadelphia, PA, 2008. [6, 113]

[134] C. L. EPSTEIN AND J. SCHOTLAND, *The bad truth about Laplace's transform*, SIAM Review, 50 (2008), pp. 504–520. [133]

[135] E. ESSER, X. ZHANG, AND T. F. CHAN, *A general framework for a class of first order primal-dual algorithms for convex optimization in imaging science*, SIAM Journal on Imaging Sciences, 3 (2010), pp. 1015–1046. [83]

[136] L. C. EVANS, *Partial Differential Equations*, AMS, Providence, RI, 1998. [297]

[137] L. FABRIZI, R. YERWORTH, A. MCEWAN, O. GILAD, R. BAYFORD, AND D. S. HOLDER, *A method for removing artefacts from continuous EEG recordings during functional electrical impedance tomography for the detection of epileptic seizures*, Physiological Measurement, 31 (2010), pp. S57–S72. [161]

[138] L. D. FADDEEV, *Increasing solutions of the Schrödinger equation*, Soviet Physics Doklady, 10 (1966), pp. 1033–1035. [199]

[139] D. FANELLI AND O. ÖKTEM, *Electron tomography: A short overview with an emphasis on the absorption potential model for the forward problem*, Inverse Problems, 24 (2008), 013001. [3, 113]

[140] A. FARIDANI, *Introduction to the mathematics of computed tomography*, in Inside Out: Inverse Problems and Applications, vol. 47 of Mathematical Sciences Research Institute Publications, Cambridge University Press, Cambridge, UK, 2003, pp. 1–46. [113]

[141] H. FENG, W. C. KARL, AND D. A. CASTANON, *A curve evolution approach to object-based tomographic reconstruction*, IEEE Transactions on Image Processing, 12 (2003), pp. 44–57. [113]

[142] X. FENG AND A. PROHL, *Analysis of total variation flow and its finite element approximations*, M2AN Mathematical Modelling and Numerical Analysis, 37 (2003), pp. 533–556. [83]

[143] A. V. FIACCO AND G. P MCCORMICK, *Nonlinear Programming: Sequential Unconstrained Minimization Techniques*, John Wiley & Sons, New York, 1968. [100]

[144] A. S. FOKAS AND M. J. ABLOWITZ, *Method of solution for a class of multidimensional nonlinear evolution equations*, Physical Review Letters, 51 (1983), pp. 7–10. [230]

[145] ———, *The inverse scattering transform for multidimensional 2 + 1 problems*, Nonlinear Phenomena, Springer, Berlin, 1984, pp. 137–183. [230]

[146] M. FORNASIER, A. LANGER, AND C.-B. SCHÖNLIEB, *A convergent overlapping domain decomposition method for total variation minimization*, Numerische Mathematik, 116 (2010), pp. 645–685. [83]

[147] M. FORNASIER AND C.-B. SCHÖNLIEB, *Subspace correction methods for total variation and l_1-minimization*, SIAM Journal on Numerical Analysis, 47 (2009), pp. 3397–3428. [83]

[148] E. FRANCINI, *Recovering a complex coefficient in a planar domain from Dirichlet-to-Neumann map*, Inverse Problems, 16 (2000), pp. 107–119. [181, 200]

[149] D. FREIMARK, M. ARAD, R. SOKOLOVER, S. ZLOCHIVER, AND S. ABBOUD, *Monitoring lung fluid content in CHF patients under intravenous diuretics treatment using bio-impedance measurements*, Physiological Measurement, 28 (2007), pp. S269–S277. [161]

[150] I. FRERICHS, *Electrical impedance tomography (EIT) in applications related to lung and ventilation: a review of experimental and clinical activities*, Physiological Measurement, 21 (2000), pp. R1–R21. [161]

[151] I. FRERICHS, J. HINZ, P. HERRMANN, G. WEISSER, G. HAHN, T. DUDYKEVYCH, M. QUINTEL, AND G. HELLIGE, *Detection of local lung air content by electrical impedance tomography compared with electron beam CT*, Journal of Applied Physiology, 93 (2002), pp. 660–666. [161]

[152] I. FRERICHS, S. PULLETZ, G. ELKE, F. REIFFERSCHEID, D. SCHADLER, J. SCHOLZ, AND N. WEILER, *Assessment of changes in distribution of lung perfusion by electrical impedance tomography*, Respiration, 77 (2009), pp. 282–291. [161]

[153] I. FRERICHS, G. SCHMITZ, S. PULLETZ, D. SCHÄDLER, G. ZICK, J. SCHOLZ, AND N. WEILER, *Reproducibility of regional lung ventilation distribution determined by electrical impedance tomography during mechanical ventilation*, Physiological Measurement, 28 (2007), pp. S261–S267. [161]

[154] H. U. FREY, S. FREY, D. LARSON, T. NYGRÉN, AND J. SEMETER, *Tomographic methods for magnetospheric applications*, Science Closure and Enabling Technologies for Constellation Class Missions, University of California Printing Services, University of California, Berkeley, and NASA, Goddard Space Flight Center, 1998, pp. 72–77. [113]

[155] L. F. FUKS, M. CHENEY, D. ISAACSON, D. G. GISSER, AND J. C. NEWELL, *Detection and imaging of electric conductivity and permittivity at low frequency*, IEEE Transactions on Biomedical Engineering, 38 (1991), pp. 1106–1110. [160]

[156] I. M. GELFAND AND B. M. LEVITAN, *On identification of the differential expression via its spectral function*, Izv. Akad. Nauk SSSR, Sek. Matem, 15 (1951), pp. 309–360 (in Russian). [154]

[157] A. P. GIBSON, J. C. HEBDEN, AND S. R. ARRIDGE, *Recent advances in diffuse optical imaging*, Physics in Medicine and Biology, 50 (2005), pp. R1–R43. [155]

[158] D. GILBARG AND N. S. TRUDINGER, *Elliptic Partial Differential Equations of Second Order*, Springer, New York, 1977. [164]

[159] D. G. GISSER, D. ISAACSON, AND J. C. NEWELL, *Electric current computed tomography and eigenvalues*, SIAM Journal on Applied Mathematics, 50 (1990), pp. 1623–1634. [168, 178, 179, 185, 186]

[160] G. GLADWELL, *Inverse Problems in Vibration*, Martinus Hijhoff, Dordrecht, 1986. [152]

[161] G. GLADWELL, S. DODS, AND S. CHAUDHURI, *Nonuniform transmission-line synthesis using inverse eigenvalue analysis*, IEEE Transactions on Circuits and Systems, 35 (1988), pp. 659–665. [152]

[162] J. C. GOBLE, M. CHENEY, AND D. ISAACSON, *Electrical impedance tomography in three dimensions*, The Applied Computational Electromagnetics Society Journal, 7 (1992), pp. 128–147. [180]

[163] D. GOLDFARB AND W. YIN, *Second-order cone programming methods for total variation-based image restoration*, SIAM Journal on Scientific Computing, 27 (2005), pp. 622–645. [83]

[164] T. GOLDSTEIN AND S. OSHER, *The split Bregman method for L1-regularized problems*, SIAM Journal on Imaging Sciences, 2 (2009), pp. 323–343. [83]

[165] G. H. GOLUB, M. HEATH, AND G. WAHBA, *Generalized cross-validation as a method for choosing a good ridge parameter*, Technometrics, 21 (1979), pp. 215–223. [72]

[166] T. GOMI AND H. HIRANO, *Clinical potential of digital linear tomosynthesis imaging of total joint arthroplasty*, Journal of Digital Imaging, 21 (2008), pp. 312–322. [123]

[167] C. GORDON, D. L. WEBB, AND S. WOLPERT, *One cannot hear the shape of a drum*, Bulletin of the American Mathematical Society, 27 (1992), pp. 134–138. [152]

[168] J. GOTTLIEB AND P. DUCHATEAU, EDS., *Parameter Identification and Inverse Problems in Hydrology, Geology, and Ecology*, Kluwer Academic, Dordrecht, 1996. [148]

[169] D. G. GRANT, *Tomosynthesis: A three-dimensional radiographic imaging technique*, IEEE Transactions on Biomedical Engineering, 19 (1972), pp. 20–28. [123]

[170] M. GRASMAIR, M. HALTMEIER, AND O. SCHERZER, *Sparse regularization with l^q penalty term*, Inverse Problems, 24 (2008), 055020. [108]

[171] ——, *Necessary and sufficient conditions for linear convergence of l^1-regularization*, Communications on Pure and Applied Mathematics, 64 (2011), pp. 161–182. [108]

[172] A. GREENLEAF, Y. KURYLEV, M. LASSAS, AND G. UHLMANN, *Full-wave invisibility of active devices at all frequencies*, Communications in Mathematical Physics, 275 (2007), pp. 749–789. [184]

[173] ——, *Cloaking devices, electromagnetic wormholes, and transformation optics*, SIAM Review, 51 (2009), pp. 3–33. [184]

[174] A. GREENLEAF, Y. KURYLEV, M. LASSAS, AND G. UHLMANN, *Invisibility and inverse problems*, Bulletin of the American Mathematical Society, 46 (2009), pp. 55–79. [184]

[175] A. GREENLEAF, M. LASSAS, AND G. UHLMANN, *Anisotropic conductivities that cannot be detected by EIT*, Physiological Measurement, 24 (2003), pp. 413–419. [183, 184]

[176] ——, *The Calderón problem for conormal potentials I: Global uniqueness and reconstruction*, Communications on Pure and Applied Mathematics, 56 (2003), pp. 328–352. [184]

[177] ——, *On nonuniqueness for Calderón's inverse problem*, Mathematical Research Letters, 10 (2003), pp. 685–694. [183, 184]

[178] A. GREENLEAF AND G. UHLMANN, *Nonlocal inversion formulas for the X-ray transform*, Duke Mathematical Journal, 58 (1989), pp. 205–240. [120, 121]

[179] P. G. GRINEVICH AND R. G. NOVIKOV, *Transparent potentials at fixed energy in dimension two. Fixed-energy dispersion relations for the fast decaying potentials*, Communications in Mathematical Physics, 174 (1995), pp. 409–446. [205]

[180] R. A. J. GROENHUIS, R. L. WEBBER, AND U. E. RUTTIMANN, *Computerized tomosynthesis of dental tissues*, Oral Surgery, Oral Medicine, Oral Pathology, 56 (1983), pp. 206–214. [123]

[181] R. B. GUENTHER AND J. W. LEE, *Partial Differential Equations of Mathematical Physics and Integral Equations*, Dover, Mineola, NY, 1988. [14]

[182] G. HAHN, I. SIPINKOVA, F. BAISCH, AND G. HELLIGE, *Changes in the thoracic impedance distribution under different ventilatory conditions*, Physiological Measurement, 16 (1995), pp. A161–A173. [161]

[183] E. T. HALE, W. YIN, AND Y. ZHANG, *Fixed-point continuation for l_1-minimization: Methodology and convergence*, SIAM Journal on Optimization, 19 (2008), pp. 1107–1130. [83]

[184] ———, *Fixed-point continuation applied to compressed sensing: Implementation and numerical experiments*, Journal of Computational Mathematics, 28 (2010), pp. 170–194. [83]

[185] D. HALLIKAINEN, *History of panoramic radiography*, Acta Radiologica, 37 (1996), pp. 441–445. [126]

[186] K. HÄMÄLÄINEN, A. KALLONEN, V. KOLEHMAINEN, M. LASSAS, E. NIEMI, K. NIINIMÄKI, AND S. SILTANEN, *Total variation regularized tomography with automatic parameter choice*, unpublished manuscript (2012). [119]

[187] K. HÄMÄLÄINEN, A. KALLONEN, V. KOLEHMAINEN, M. LASSAS, K. NIINIMÄKI, AND S. SILTANEN, *Sparse tomography*, submitted manuscript (2012). [117]

[188] U. HÄMARIK, R. PALM, AND T. RAUS, *A family of rules for parameter choice in Tikhonov regularization of ill-posed problems with inexact noise level*, Journal of Computational and Applied Mathematics, 236 (2012), pp. 2146–2157. [72]

[189] S. J. HAMILTON, C. N. L. HERRERA, J. L. MUELLER, AND A. VON HERRMANN, *A direct D-bar reconstruction algorithm for recovering a complex conductivity in 2-D*, preprint, Arxiv:2012.1785v1 (2012). [181, 200, 259, 263]

[190] M. HANKE, *Limitations of the l-curve method in ill-posed problems*, BIT, 36 (1996), pp. 287–301. [76]

[191] ———, *Iterative regularization techniques in image reconstruction*, in Surveys on Solution Methods for Inverse Problems, Springer, Vienna, 2000, pp. 35–52. [6]

[192] M. HANKE AND M. BRÜHL, *Recent progress in electrical impedance tomography*, Inverse Problems, 19 (2003), pp. S65–S90. [160]

[193] M. HANKE, J. NAGY, AND R. PLEMMONS, *Preconditioned iterative regularization for ill-posed problems*, in Numerical Linear Algebra (Kent, OH, 1992), de Gruyter, Berlin, 1993, pp. 141–163. [6]

[194] P. C. HANSEN, *Rank-Deficient and Discrete Ill-Posed Problems: Numerical Aspects of Linear Inversion*, SIAM, Philadelphia, PA, 1998. [6, 44, 45, 75, 76]

[195] ———, *Discrete Inverse Problems: Insight and Algorithms*, vol. 7 of Fundamentals of Algorithms, SIAM, Philadelphia, PA, 2010. [6, 83]

[196] P. C. HANSEN, J. G. NAGY, AND D. P. O'LEARY, *Deblurring Images: Matrices, Spectra, and Filtering*, vol. 3 of Fundamentals of Algorithms, SIAM, Philadelphia, PA, 2006. [6]

[197] K. M. HANSON, G. S. CUNNINGHAM, JR., G. R. JENNINGS, AND D. R. WOLF, *Tomographic reconstruction based on flexible geometric models*, in Proceedings of the IEEE International Conference on Image Processing, Vol. 2, 1994, pp. 145–147. [113]

[198] K. M. HANSON, G. S. CUNNINGHAM, AND R. J. MCKEE, *Uncertainties in tomographic reconstructions based on deformable models*, in Medical Imaging: Image Processing, vol. 3034 of Proceedings of SPIE, K. M. Hanson, ed., 1997, pp. 276–286. [113]

[199] B. HARRACH AND J. K. SEO, *Detecting inclusions in electrical impedance tomography without reference measurements*, SIAM Journal on Applied Mathematics, 69 (2009), pp. 1662–1681. [182]

[200] A. HASANOV AND J. L. MUELLER, *A numerical method for backward parabolic problems with non-selfadjoint elliptic operators*, Applied Numerical Mathematics, 37 (2001), pp. 55–78. [134, 135]

[201] J. HASLINGER AND P. NEITTAANMÄKI, *Finite Element Approximation for Optimal Shape, Material and Topology Design*, second ed., John Wiley & Sons, Chichester, UK, 1996. [140]

[202] A. HAUPTMANN, K. HÄMÄLÄINEN, A. KALLONEN, E. NIEMI, AND S. SILTANEN, *Total variation regularization for X-ray tomography* (2012). [83]

[203] J. C. HEBDEN, A. GIBSON, T. AUSTIN, R. M. YUSOF, N. EVERDELL, D. T. DELPY, S. R. ARRIDGE, J. H. MEEK, AND J. S. WYATT, *Imaging changes in blood volume and oxygenation in the newborn infant brain using three-dimensional optical tomography*, Physics in Medicine and Biology, 49 (2004), pp. 1117–1130. [155]

[204] L. M. HEIKKINEN, J. KOURUNEN, T. SAVOLAINEN, P. J. VAUHKONEN, J. P. KAIPIO, AND M. VAUHKONEN, *Real time three-dimensional electrical impedance tomography applied in multiphase flow imaging*, Measurement Science and Technology, 17 (2006), pp. 2083–2087. [162]

[205] J. HEINO AND E. SOMERSALO, *Estimation of optical absorption in anisotropic background*, Inverse Problems, 18 (2002), pp. 559–573. [155]

[206] G. T. HERMAN AND R. DAVIDI, *Image reconstruction from a small number of projections*, Inverse Problems, 24 (2008), 045011. [113]

[207] G. T. HERMAN AND A. KUBA, *Discrete Tomography: Foundations, Algorithms, and Applications*, Birkhäuser, Boston, MA, 1999. [113]

[208] ———, *Advances in Discrete Tomography and Its Applications*, Applied and Numerical Harmonic Analysis, Birkhäuser Boston, MA, 2007. [113]

[209] E. M. C. HILLMAN, J. C. HEBDEN, M. SCHWEIGER, H. DEHGHANI, F. E. W. SCHMIDT, D. T. DELPY, AND S. R. ARRIDGE, *Time resolved optical tomography of the human forearm*, Physics in Medicine and Biology, 46 (2001), pp. 1117–1130. [155]

[210] M. HINTERMÜLLER AND K. KUNISCH, *Total bounded variation regularization as a bilaterally constrained optimization problem*, SIAM Journal on Applied Mathematics, 64 (2004), pp. 1311–1333. [83]

[211] S.-E. HJELT, *Pragmatic Inversion of Geophysical Data*, no. 39 in Lecture Notes in Earth Sciences, Springer, New York, 1992. [140]

[212] H. HOCHSTADT, *On determination of the density of a vibrating string from spectral data*, Journal of Mathematical Analysis and Applications, 55 (1976), pp. 673–685. [154]

[213] B. HOFMANN, *Regularization for Applied Inverse and Ill-Posed Problems*, Teubner, Stuttgart, 1986. [45]

[214] D. HOLDER, *Clinical and Physiological Applications of Electrical Impedance Tomography*, UCL Press, London, 1993. [161]

[215] D. S. HOLDER, ED., *Electrical Impedance Tomography; Methods, History, and Applications*, IOP Publishing Ltd., Bristol, UK, 2005. [140, 161]

[216] P. HUA, J. WOO, J. G. WEBSTER, AND W. TOMPKINS, *Finite element modeling of electrode-skin contact impedance in electrical impedance tomography*, IEEE Transactions on Biomedical Engineering, 40 (1993), pp. 335–343. [192]

[217] M. HUHTANEN AND A. PERÄMÄKI, *Numerical solution of the R-linear Beltrami equation*, Mathematics of Computation, 81 (2012), pp. 387–397. [216, 217, 218, 273]

[218] N. HYVÖNEN, M. KALKE, M. LASSAS, H. SETÄLÄ, AND S. SILTANEN, *Three-dimensional dental X-ray imaging by combination of panoramic and projection data*, Inverse Problems and Imaging, 4 (2010), pp. 257–271. [113, 129, 130]

[219] T. IDE, H. ISOZAKI, S. NAKATA, AND S. SILTANEN, *Local detection of three-dimensional inclusions in electrical impedance tomography*, Inverse Problems, 26 (2010), 035001. [145, 205, 277]

[220] T. IDE, H. ISOZAKI, S. NAKATA, S. SILTANEN, AND G. UHLMANN, *Probing for electrical inclusions with complex spherical waves*, Communications on Pure and Applied Mathematics, 60 (2007), pp. 1415–1442. [145, 189, 205, 277]

[221] M. IKEHATA, *Reconstruction of the support function for inclusion from boundary measurements*, Journal of Inverse and Ill-Posed Problems, 8 (2000), pp. 367–378. [277, 278]

[222] ——, *A regularized extraction formula in the enclosure method*, Inverse problems, 18 (2002), pp. 435–440. [224, 278]

[223] ——, *Mittag-Leffler's function and extracting from Cauchy data*, in Inverse Problems and Spectral Theory: Proceedings of the Workshop on Spectral Theory of Differential Operators and Inverse Problems, vol. 348 of Contemporary Mathematics, AMS, Providence, RI, 2004, pp. 41–52. [145]

[224] M. IKEHATA, E. NIEMI, AND S. SILTANEN, *Inverse obstacle scattering with limited-aperture data*, Inverse Problems and Imaging, 6 (2012), pp. 77–94. [277, 287]

[225] M. IKEHATA AND T. OHE, *A numerical method for finding the convex hull of inclusions using the enclosure method*, in Electromagnetic Nondestructive Evaluation (VI), IOS Press, Amsterdam, 2002, pp. 21–28. [277]

[226] ——, *The enclosure method for an inverse crack problem and the Mittag-Leffler function*, Inverse Problems, 24 (2008), 015006. [205, 277]

[227] M. IKEHATA AND S. SILTANEN, *Numerical method for finding the convex hull of an inclusion in conductivity from boundary measurements*, Inverse Problems, 16 (2000), pp. 1043–1052. [145, 205, 277, 278]

[228] ——, *Electrical impedance tomography and Mittag-Leffler's function*, Inverse Problems, 20 (2004), pp. 1325–1348. [224, 277]

[229] ——, *Numerical solution of the Cauchy problem for the stationary Schrödinger equation using Faddeev's Green function*, SIAM Journal on Applied Mathematics, 64 (2004), pp. 1907–1932. [205, 208]

[230] D. ISAACSON, *Distinguishability of conductivities by electric current computed tomography*, IEEE Transactions on Medical Imaging, 5 (1986), pp. 91–95. [173, 177, 179]

[231] D. ISAACSON AND M. CHENEY, *Current problems in impedance imaging*, in Inverse Problems in Partial Differential Equations, D. Colton, R. Ewing, and W. Rundell, eds., SIAM, Philadelphia, PA, 1990, pp. 139–148. [162, 178]

[232] ——, *Effects of measurement precision and finite numbers of electrodes on linear impedance imaging algorithms*, SIAM Journal on Applied Mathematics, 51 (1991), pp. 1705–1731. [179, 236]

[233] D. ISAACSON, J. L. MUELLER, J. C. NEWELL, AND S. SILTANEN, *Imaging cardiac activity by the D-bar method for electrical impedance tomography*, Physiological Measurement, 27 (2006), pp. S43–S50. [145, 161, 200, 250, 254, 256, 257, 259]

[234] ——, *Reconstructions of chest phantoms by the D-bar method for electrical impedance tomography*, IEEE Transactions on Medical Imaging, 23 (2004), pp. 821–828. [145, 200, 249, 254]

[235] E. L. ISAACSON AND E. TRUBOWITZ, *The inverse Sturm-Liouville problem* I, IEEE Transactions on Medical Imaging, 5 (1986), pp. 91–95. [154]

[236] V. ISAKOV, *Inverse Problems for Partial Differential Equations*, Springer, New York, 1998. [6, 140]

[237] T. JENSEN, J. JØRGENSEN, P. HANSEN, AND S. JENSEN, *Implementation of an optimal first-order method for strongly convex total variation regularization*, BIT Numerical Mathematics, (2011), pp. 1–28. [83, 113]

[238] X. JIA, B. DONG, Y. LOU, AND S. B. JIANG, *GPU-based iterative cone-beam CT reconstruction using tight frame regularization*, Physics in Medicine and Biology, 56 (2011), p. 3787. [113]

[239] X. JIA, Y. LOU, R. LI, W. Y. SONG, AND S. B. JIANG, *GPU-based fast cone beam CT reconstruction from undersampled and noisy projection data via total variation*, Medical Physics, 37 (2010), pp. 1757–1760. [113]

[240] F. JOHN, *Numerical solution of the equation of heat conduction for preceding times*, Annali di Matematica Pura ed Applicata, Series IV, 40 (1955), pp. 129–142. [36]

[241] J. JORDANA, M. GASULLA, AND R. PALLÀS-ARENY, *Electrical resistance tomography to detect leaks from buried pipes*, Measurement Science and Technology, 12 (2001), pp. 1061–1068. [162]

[242] J. JOSSINET, *Variability of impeditivity in normal and pathological breast tissue*, Medical and Biological Engineering and Computing, 34 (1996), pp. 346–350. [161]

[243] ———, *The impeditivity of freshly excised human breast tissue*, Physiological Measurement, 19 (1998), pp. 61–75. [161]

[244] M. KAC, *Can one hear the shape of a drum?*, The American Mathematical Monthly, 73 (1966), pp. 1–23. [151, 152]

[245] J. P. KAIPIO, A. SEPPÄNEN, E. SOMERSALO, AND H. HAARIO, *Statistical inversion approach for optimizing current patterns in EIT*, in Proceedings of 3rd World Congress on Industrial Process Tomography, 2003, pp. 683–688. [179]

[246] ———, *Posterior covariance related optimal current patterns in electrical impedance tomography*, Inverse Problems, 20 (2004), pp. 919–936. [179]

[247] J. KAIPIO AND E. SOMERSALO, *Statistical and Computational Inverse Problems*, vol. 160 of Applied Mathematical Sciences, Springer, New York, 2005. [6, 113, 140]

[248] G. KAISER, *A Friendly Guide to Wavelets*, Birkhäuser, Boston, MA, 1994. [95]

[249] A. C. KAK AND M. SLANEY, *Principles of Computerized Tomographic Imaging*, IEEE Press, New York, 1988. [6, 25, 113]

[250] T. KAKO AND K. TOUDA, *Numerical approximation of Dirichlet-to-Neumann mapping and its application to voice generation problem*, in Domain Decomposition Methods in Science and Engineering, vol. 40 of Lecture Notes in Computational Science and Engineering, Springer, Berlin, 2005, pp. 51–65. [146]

[251] ———, *Numerical method for voice generation problem based on finite element method*, Journal of Computational Acoustics, 14 (2006), pp. 45–56. [146]

[252] M. KALKE, *Method for Limited Angle Tomography*, U.S. patent 7 853 056 B2, 2010. [129]

[253] H. KALLASJOKI, K. PALOMÄKI, C. MAGI, P. ALKU, AND M. KURIMO, *Noise robust LVCSR feature extraction based on stabilized weighted linear prediction*, in Proceedings of the 13th International Conference on Speech and Computer, St. Petersburg, Russia, 2009. [146]

[254] B. KALTENBACHER, A. NEUBAUER, AND O. SCHERZER, *Iterative Regularization Methods for Nonlinear Ill-Posed Problems*, vol. 6 of Radon Series on Computational and Applied Mathematics, de Gruyter, Berlin, 2008. [6, 140, 145]

[255] T.-J. KAO, G. BOVERMAN, B. S. KIM, D. ISAACSON, G. J. SAULNIER, J. C. NEWELL, M. H. CHOI, R. H. MOORE, AND D. B. KOPANS, *Regional admittivity spectra with tomosynthesis images for breast cancer detection: Preliminary patient study*, IEEE Transactions on Medical Imaging, 27 (2008), pp. 1762–1768. [161, 182]

[256] T.-J. KAO, G. J. SAULNIER, H. XIA, C. TAMMA, J. C. NEWELL, AND D. ISAACSON, *A compensated radiolucent electrode array for combined EIT and mammography*, Physiological Measurement, 28 (2007), p. S291. [161]

[257] K. KARHUNEN, A. SEPPÄNEN, A. LEHIKOINEN, P. J. M. MONTEIRO, AND J. P. KAIPIO, *Electrical resistance tomography imaging of concrete*, Cement and Concrete Research, 40 (2010), pp. 137–145. [162]

[258] A. KATSEVICH, *A general scheme for constructing inversion algorithms for cone beam CT*, International Journal of Mathematics and Mathematical Sciences, (2003), pp. 1305–1321. [113]

[259] A. KATSEVICH AND M. KAPRALOV, *Filtered backprojection inversion of the cone beam transform for a general class of curves*, SIAM Journal on Applied Mathematics, 68 (2007), pp. 334–353. [113]

[260] A. KATCHALOV, Y. KURYLEV, AND M. LASSAS, *Inverse Boundary Spectral Problems*, vol. 123 of Chapman & Hall/CRC Monographs and Surveys in Pure and Applied Mathematics, Chapman & Hall/CRC, Boca Raton, FL, 2001. [140, 152]

[261] A. KATSEVICH AND G. LAURITSCH, *Filtered backprojection algorithms for spiral cone beam CT*, in Sampling, Wavelets, and Tomography, Applied Numerical and Harmonic Analysis, Birkhäuser, Boston, MA, 2004, pp. 255–287. [113]

[262] F. KEINERT, *Wavelets and Multiwavelets*, Chapman & Hall, Boca Raton, FL, 2004. [95]

[263] C. T. KELLEY, *Iterative Methods for Optimization*, SIAM, Philadelphia, PA, 1999. [82]

[264] A. KEMNA, A. BINLEY, A. RAMIREZ, AND W. DAILY, *Complex resistivity tomography for environmental applications*, Chemical Engineering Journal, 77 (2000), pp. 11–18. [162]

[265] A. KEMNA, B. KULESSA, AND H. VEREECKEN, *Imaging and characterisation of subsurface solute transport using electrical resistivity tomography (ERT) and equivalent transport models*, Journal of Hydrology, 267 (2002), pp. 125–146. [162]

[266] M. KERKER, *Invisible bodies*, Journal of the Optical Society of America, 65 (1975), pp. 376–379. [183]

[267] T. E. KERNER, K. D. PAULSEN, A. HARTOV, S. K. SOHO, AND S. P. POPLACK, *Electrical impedance spectroscopy of the breast: clinical imaging results in 26 subjects*, IEEE Transactions on Medical Imaging, 21 (2002), pp. 638–645. [161, 182]

[268] B. S. KIM, D. ISAACSON, H. XIA, T. J. KAO, J. C. NEWELL, AND G. J. SAULNIER, *A method for analyzing electrical impedance spectroscopy data from breast cancer patients*, Physiological Measurement, 28 (2007), pp. S237–S246. [161, 182]

[269] A. KIRSCH, *An Introduction to the Mathematical Theory of Inverse Problems*, second ed., Springer, New York, 2011. [6, 48, 140, 232, 287]

[270] A. KIRSCH AND N. GRINBERG, *The Factorization Method for Inverse Problems*, Oxford University Press, Oxford, 2008. [140, 145, 287]

[271] E. KLANN, R. RAMLAU, AND L. REICHEL, *Wavelet-based multilevel methods for linear ill-posed problems*, BIT, 51 (2011), pp. 669–694. [113]

[272] E. KLANN, R. RAMLAU, AND W. RING, *A Mumford-Shah level-set approach for the inversion and segmentation of SPECT/CT data*, Inverse Problems and Imaging, 5 (2011), pp. 137–166. [113]

[273] D. KLOTZ, K. P. SEILER, H. MOSER, AND F. NEUMAIER, *Dispersivity and velocity relationship from laboratory and field experiments*, Journal of Hydrology, 45 (1980), pp. 169–184. [147]

[274] K. KNUDSEN, *A new direct method for reconstructing isotropic conductivities in the plane*, Physiological Measurement, 24 (2003), pp. 391–403. [199, 200]

[275] K. KNUDSEN, M. LASSAS, J. L. MUELLER, AND S. SILTANEN, *D-bar method for electrical impedance tomography with discontinuous conductivities*, SIAM Journal on Applied Mathematics, 67 (2007), pp. 893–913. [250, 251, 252, 253, 264]

[276] ——, *Reconstructions of piecewise constant conductivities by the D-bar method for electrical impedance tomography*, in Journal of Physics: Conference Series, 124 (2008), 012029. [200, 257]

[277] ——, *Regularized D-bar method for the inverse conductivity problem*, Inverse Problems and Imaging, 3 (2009), pp. 599–624. [145, 200, 223, 225, 232, 233]

[278] K. KNUDSEN AND J. L. MUELLER, *The Born approximation and Calderón's method for reconstructions of conductivities in 3-D*, Discrete and Continuous Dynamical Systems, 2011 (2011), pp. 884–853. [180]

[279] K. KNUDSEN, J. L. MUELLER, AND S. SILTANEN, *Numerical solution method for the dbar-equation in the plane*, Journal of Computational Physics, 198 (2004), pp. 500–517. [200, 237]

[280] K. KNUDSEN AND A. TAMASAN, *Reconstruction of less regular conductivities in the plane*, Communications in Partial Differential Equations, 29 (2004), pp. 361–381. [200, 231]

[281] R. V. KOHN, H. SHEN, M. S. VOGELIUS, AND M. I. WEINSTEIN, *Cloaking via change of variables in electric impedance tomography*, Inverse Problems, 24 (2008), 015016. [184]

[282] R. V. KOHN AND M. VOGELIUS, *Identification of an unknown conductivity by means of measurements at the boundary*, in Inverse Problems (New York, 1983), vol. 14 of SIAM-AMS Proceedings, AMS, Providence, RI, 1984, pp. 113–123. [184]

[283] V. KOLEHMAINEN, M. LASSAS, K. NIINIMÄKI, AND S. SILTANEN, *Sparsity-promoting Bayesian inversion*, Inverse Problems, 28 (2012), 025005. [83, 89, 99, 100, 107, 108, 109]

[284] V. KOLEHMAINEN, M. LASSAS, AND P. OLA, *The inverse conductivity problem with an imperfectly known boundary*, SIAM Journal on Applied Mathematics, 66 (2005), pp. 365–383. [181]

[285] ——, *Electrical impedance tomography problem with inaccurately known boundary and contact impedances*, IEEE Transactions on Medical Imaging, 27 (2008), pp. 1404–1414. [181]

[286] ——, *Calderón's inverse problem with an imperfectly known boundary and reconstruction up to a conformal deformation*, SIAM Journal on Mathematical Analysis, 42 (2010), pp. 1371–1381. [181]

[287] V. KOLEHMAINEN, M. LASSAS, P. OLA, AND S. SILTANEN, *Recovering boundary shape and conductivity in electrical impedance tomography*, Inverse Problems and Imaging, to appear (2012). [181]

[288] V. KOLEHMAINEN, M. LASSAS, AND S. SILTANEN, *Limited data X-ray tomography using nonlinear evolution equations*, SIAM Journal on Scientific Computing, 30 (2008), pp. 1413–1429. [113]

[289] V. KOLEHMAINEN, S. SILTANEN, S. JÄRVENPÄÄ, J. P. KAIPIO, P. KOISTINEN, M. LASSAS, J. PIRTTILÄ, AND E. SOMERSALO, *Statistical inversion for medical X-ray tomography with few radiographs: II. Application to dental radiology*, Physics in Medicine and Biology, 48 (2003), pp. 1465–1490. [83, 113, 126]

[290] V. KOLEHMAINEN, A. VANNE, S. SILTANEN, S. JÄRVENPÄÄ, J. P. KAIPIO, M. LASSAS, AND M. KALKE, *Parallelized Bayesian inversion for three-dimensional dental X-ray imaging*, IEEE Transactions on Medical Imaging, 25 (2006), pp. 218–228. [113, 129]

[291] ———, *Bayesian inversion method for 3D dental X-ray imaging*, Elektrotechnik und Informationstechnik, 124 (2007), pp. 248–253. [129]

[292] R. KRESS, *On the numerical solution of a hypersingular integral equation in scattering theory*, Journal of Computational and Applied Mathematics, 61 (1995), pp. 345–360. [289]

[293] T. KUBO, P.-J. PAUL LIN, W. STILLER, M. TAKAHASHI, H.-U. KAUCZOR, Y. OHNO, AND H. HATABU, *Radiation dose reduction in chest CT: A review*, American Journal of Roentgenology, 190 (2008), pp. 335–343. [113]

[294] P. W. A. KUNST, A. V. NOORDEGRAAF, E. RAAIJMAKERS, J. BAKKER, A. B. GROENEVELD, P. E. POSTMUS, AND P. M. J. M. DE VRIES, *Electrical impedance tomography in the assessment of extravascular lung water in noncardiogenic acute respiratory failure*, Chest, 116 (1999), pp. 1695–1702. [161]

[295] J. KYBIC, T. BLU, AND M. A. UNSER, *Variational approach to tomographic reconstruction*, in Society of Photo-Optical Instrumentation Engineers (SPIE) Conference Series, M. Sonka and K. M. Hanson, eds., vol. 4322 of Society of Photo-Optical Instrumentation Engineers (SPIE) Conference Series, 2001, pp. 30–39. [113]

[296] E. KYRÖLÄ, J. TAMMINEN, G. W. LEPPELMEIER, V. SOFIEVA, S. HASSINEN, J. L. BERTAUX, A. HAUCHECORNE, F. DALAUDIER, C. COT, O. KORABLEV, O. FANTON D'ANDON, G. BARROT, A. MANGIN, B. THÉODORE, M. GUIRLET, F. ETANCHAUD, P. SNOEIJ, R. KOOPMAN, L. SAAVEDRA, R. FRAISSE, D. FUSSEN, AND F. VANHELLEMONT, *Gomos on envisat: An overview*, Advances in Space Research, 33 (2004), pp. 1020–1028. [3, 113]

[297] H. LANGET, C. RIDDELL, Y. TROUSSET, A. TENENHAUS, E. LAHALLE, G. FLEURY, AND N. PARAGIOS, *Compressed sensing based 3D tomographic reconstruction for rotational angiography*, in Medical Image Computing and Computer-Assisted Intervention MICCAI 2011, G. Fichtinger, A. Martel, and T. Peters, eds., vol. 6891 of Lecture Notes in Computer Science, Springer, Berlin, Heidelberg, 2011, pp. 97–104. [113]

[298] M. L. LAPIDUS, *Can one hear the shape of a fractal drum? Partial resolution of the Weyl-Berry conjecture*, in Geometric Analysis and Computer Graphics (Berkeley, CA, 1988), vol. 17 of Mathematical Sciences Research Institute Publications, Springer, New York, 1991, pp. 119–126. [152]

[299] M. LASSAS, M. MATAICH, S. SILTANEN, AND E. SOMERSALO, *Wind velocity observation with a CW doppler radar*, IEEE Transactions on Geoscience and Remote Sensing, 40 (2002), pp. 2427–2437. [3]

[300] M. LASSAS, J. L. MUELLER, AND S. SILTANEN, *Mapping properties of the nonlinear Fourier transform in dimension two*, Communications in Partial Differential Equations, 32 (2007), pp. 591–610. [205]

[301] M. LASSAS, J. L MUELLER, S. SILTANEN, AND A. STAHEL, *The Novikov-Veselov Equation and the Inverse Scattering Method, Part I: Analysis*, ArXiv e-prints, (2011). [205]

[302] ———, *The Novikov-Veselov Equation and the Inverse Scattering Method, Part II: Computation*, (2011). [205]

[303] M. LASSAS, E. SAKSMAN, AND S. SILTANEN, *Discretization-invariant Bayesian inversion and Besov space priors*, Inverse Problems and Imaging, 3 (2009), pp. 87–122. [108]

[304] M. LASSAS AND S. SILTANEN, *Can one use total variation prior for edge-preserving Bayesian inversion?*, Inverse Problems, 20 (2004), pp. 1537–1564. [83, 106, 107]

[305] M. LASSAS, M. TAYLOR, AND G. UHLMANN, *The Dirichlet-to-Neumann map for complete Riemannian manifolds with boundary*, Communications in Analysis and Geometry, 11 (2003), pp. 207–222. [183, 184]

[306] M. LASSAS AND G. UHLMANN, *On determining a Riemannian manifold from the Dirichlet-to-Neumann map*, Annales Scientifiques de l'École Normale Supérieure (4), 34 (2001), pp. 771–787. [184]

[307] R. LATTES AND J.-L. LIONS, *The Method of Quasi-Reversibility. Applications to Partial Differential Equations*, Elsevier, New York, 1969. [133]

[308] A. LECHLEITER, *A regularization technique for the factorization method*, Inverse problems, 22 (2006), pp. 1605–1625. [224]

[309] ———, *Factorization Methods for Photonics and Rough Surfaces*, KIT Scientific Publishing, 2008. [140]

[310] J. LEE AND G. UHLMANN, *Determining anisotropic real-analytic conductivities by boundary measurements*, Communications on Pure and Applied Mathematics, 42 (1989), pp. 1097–1112. [184]

[311] U. LEONHARDT, *Optical conformal mapping*, Science, 312 (2006), pp. 1777–1780. [183, 184]

[312] Y. LI AND F. SANTOSA, *A computational algorithm for minimizing total variation in image restoration*, IEEE Transactions on Image Processing, 5 (1996), pp. 987–995. [83]

[313] H. Y. LIAO, *A gradually unmasking method for limited data tomography*, in Proceedings of the 4th IEEE International Symposium on Biomedical Imaging: From Nano to Macro, 2007, pp. 820–823. [113]

[314] H. Y. LIAO AND G. SAPIRO, *Sparse representations for limited data tomography*, in Proceedings of the 5th IEEE International Symposium on Biomedical Imaging: From Nano to Macro, 2008, pp. 1375–1378. [113]

[315] L. H. LIEU AND L. A. VESE, *Image restoration and decomposition via bounded total variation and negative Hilbert-Sobolev spaces*, Applied Mathematics and Optimization, 58 (2008), pp. 167–193. [83]

[316] Y. T. LIN, A. ORTEGA, AND A. G. DIMAKIS, *Sparse recovery for discrete tomography*, in Proceedings of the 17th IEEE International Conference on Image Processing (ICIP), 2010, pp. 4181–4184. [113]

[317] A. LINNINGER, S. BASATI, R. DAWE, AND R. PENN, *An impedance sensor to monitor and control cerebral ventricular volume*, Medical Engineering & Physics, 31 (2009), pp. 838–845. [161]

[318] L. LIU, *Stability Estimates for the Two-Dimensional Inverse Conductivity Problem*, Ph.D. thesis, University of Rochester, 1997. [200, 250]

[319] B. F. LOGAN, *The uncertainty principle in reconstructing functions from projections*, Duke Mathematical Journal, 42 (1975), pp. 661–706. [114]

[320] A. K. LOUIS, *Nonuniqueness in inverse Radon problems: The frequency distribution of the ghosts*, Mathematische Zeitschrift, 185 (1984), pp. 429–440. [114]

[321] ———, *Orthogonal function series expansions and the null space of the Radon transform*, SIAM Journal of Mathematical Analysis, 15 (1984), pp. 621–633. [114]

[322] ———, *Incomplete data problems in X-ray computerized tomography I. Singular value decomposition of the limited angle transform*, Numerische Mathematik, 48 (1986), pp. 251–262. [120]

[323] B. D. LOWE AND W. RUNDELL, *An inverse problem for a Sturm-Liouville operator*, Journal of Mathematical Analysis and Applications, 181 (1994), pp. 188–199. [154]

[324] J. B. LUDLOW, L. E. DAVIES-LUDLOW, S. L. BROOKS, AND W. B. HOWERTON, *Dosimetry of 3 CBCT devices for oral and maxillofacial radiology: CB Mercuray, NewTom 3G and i-CAT*, Dentomaxillofacial Radiology, 35 (2006), pp. 219–226. [130]

[325] D. G. LUENBERGER, *Linear and Nonlinear Programming*, second ed., Kluwer Academic, Boston, 2003. [82]

[326] P. MAASS, *The X-ray transform: Singular value decomposition and resolution*, Inverse Problems, 3 (1987), pp. 729–741. [114]

[327] A. MALICH, T. FRITSCH, R. ANDERSON, T. BOEHM, M. G. FREESMEYER, M. FLECK, AND W. A. KAISER, *Electrical impedance scanning for classifying suspicious breast lesions: First results*, European Radiology, 10 (2000), pp. 1555–1561. [161]

[328] C. H. McCollough, A. N. Primak, N. Braun, J. Kofler, L. Yu, and J. Christner, *Strategies for reducing radiation dose in CT*, Radiologic Clinics of North America, 47 (2009), pp. 27–40. [113]

[329] J. R. McLaughlin, *Inverse spectral theory using nodal points as data—a uniqueness result*, Journal of Differential Equations, 73 (1988), pp. 354–362. [152, 154]

[330] S. Mehrotra, *On the implementation of a primal-dual interior point method*, SIAM Journal on Optimization, 2 (1992), pp. 575–601. [100]

[331] W. Menke, *Geophysical Data Analysis: Discrete Inverse Theory*, International Geophysics Series, Academic Press, San Diego, CA, 1989. [6, 140]

[332] P. Metherall, D. C. Barber, and R. H. Smallwood, *Three dimensional electrical impedance tomography*, in Proceedings of the IX International Conference on Electrical Bio-Impedance, Heidelberg, Germany, 1995. [180]

[333] P. Metherall, D. C. Barber, R. H. Smallwood, and B. H. Brown, *Three-dimensional electrical impedance tomography*, Nature, 380 (1996), pp. 509–512. [180]

[334] P. Metherall, R. H. Smallwood, and D. C. Barber, *Three dimensional electrical impedance tomography of the human thorax*, in Proceedings of the 18th Annual International Conference of the IEEE Engineering in Medicine and Biology Society, 1996. [180]

[335] Y. Meyer, *Wavelets and Operators*, Vol. 1, Cambridge University Press, Cambridge, UK, 1995. [95, 98]

[336] G. W. Milton and N.-A. P. Nicorovici, *On the cloaking effects associated with anomalous localized resonance*, Proceedings of the Royal Society A: Mathematical, Physical and Engineering Science, 462 (2006), pp. 3027–3059. [183]

[337] A. Mohammad-Djafari and K. Sauer, *Shape reconstruction in X-ray tomography from a small number of projections using deformable models*, in The 17th International Workshop on Maximum Entropy and Bayesian Methods (MaxEnt97), Boise, Idaho, USA, August 1997. [113]

[338] A. Mohammad-Djafari and C. Soussen, *Reconstruction of compact homogeneous 3D objects from their projections*, in Discrete Tomography—Foundations, Algorithms and Applications, Birkhäuser, Boston, MA, 1999, pp. 317–342. [113]

[339] P. Monk, *Finite Element Methods for Maxwell's Equations*, Oxford University Press, New York, 2003. [162]

[340] V. A. Morozov, *Methods for Solving Incorrectly Posed Problems*, Springer, New York, 1984. Translated from the Russian by A. B. Aries; translation edited by Z. Nashed. [6, 140]

[341] J. P. Morucci, M. Granie, M. Lei, M. Chabert, and P. M. Marsili, *3D reconstruction in electrical impedance imaging using a direct sensitivity matrix approach*, Physiological Measurement, 16 (1995), pp. A123–A128. [180]

[342] J. L. MUELLER, D. ISAACSON, AND J. C. NEWELL, *A reconstruction algorithm for electrical impedance tomography data collected on rectangular electrode arrays*, IEEE Transactions on Biomedical Engineering, 49 (1999), pp. 1379–1386. [161]

[343] ———, *Reconstruction of conductivity changes due to ventilation and perfusion from EIT data collected on a rectangular electrode array*, Physiological Measurement, 22 (2001), pp. 97–106. [161]

[344] J. L. MUELLER AND S. SILTANEN, *Direct reconstructions of conductivities from boundary measurements*, SIAM Journal on Scientific Computing, 24 (2003), pp. 1232–1266. [185, 200, 210, 225, 251, 257]

[345] E. K. MURPHY, *2-D D-bar Conductivity Reconstructions on Non-circular Domains*, Ph.D. thesis, Colorado State University, Fort Collins, CO, 2007. [181, 250]

[346] E. K. MURPHY AND J. L. MUELLER, *Effect of domain-shape modeling and measurement errors on the 2-d D-bar method for electrical impedance tomography*, IEEE Transactions on Medical Imaging, 28 (2009), pp. 1576–1584. [181, 200, 250, 251, 253]

[347] E. K. MURPHY, J. L. MUELLER, AND J. C. NEWELL, *Reconstructions of conductive and insulating targets using the D-bar method on an elliptical domain*, Physiological Measurement, 28 (2007), pp. S101–S144. [181, 200, 250]

[348] A. I. NACHMAN, *Reconstructions from boundary measurements*, Annals of Mathematics, 128 (1988), pp. 531–576. [150, 180, 200]

[349] ———, *Global Uniqueness for a Two-Dimensional Inverse Boundary Value Problem*, University of Rochester, Deptartment of Mathematics Preprint Series, 1993. [207]

[350] ———, *Global uniqueness for a two-dimensional inverse boundary value problem*, Annals of Mathematics, 143 (1996), pp. 71–96. [199, 200, 205, 206, 207, 224, 225, 226, 229, 230, 231, 237]

[351] G. NAKAMURA, P. RONKANEN, S. SILTANEN, AND K. TANUMA, *Recovering conductivity at the boundary in three-dimensional electrical impedance tomography*, Inverse Problems and Imaging, 5 (2011), pp. 485–510. [180]

[352] G. NAKAMURA, S. SILTANEN, K. TANUMA, AND S. WANG, *Numerical recovery of conductivity at the boundary from the localized Dirichlet-to-Neumann map*, Computing, 75 (2005), pp. 197–213. [180]

[353] F. NATTERER, *The Mathematics of Computerized Tomography*, vol. 32 of SIAM Classics in Applied Mathematics, SIAM, Philadelphia, PA, 2001. [6, 25, 113, 119]

[354] F. NATTERER AND F. WÜBBELING, *Mathematical Methods in Image Reconstruction*, vol. 5 of Monographs on Mathematical Modeling and Computation, SIAM, Philadelphia, PA, 2001. [6, 25, 113, 140]

[355] P. NEITTAANMÄKI, M. RUDNICKI, AND A. SAVINI, *Inverse Problems and Optimal Design in Electricity and Magnetism*, Monographs in Electrical and Electronic Engineering, Clarendon Press, 1996. [140]

[356] Y. NESTEROV, *Barrier subgradient method*, Mathematical Programming, 127 (2011), pp. 31–56. [83]

[357] J. C. NEWELL, R. S. BLUE, D. ISAACSON, G. J. SAULNIER, AND A. S. ROSS, *Phasic three-dimensional impedance imaging of cardiac activity*, Physiological Measurement, 23 (2002), pp. 203–209. [161]

[358] R. NEWTON, *Inverse Schrödinger Scattering in Three Dimensions*, Springer, Berlin, 1989. [140]

[359] K. NIINIMÄKI, S. SILTANEN, AND V. KOLEHMAINEN, *Bayesian multiresolution method for local tomography in dental X-ray imaging*, Physics in Medicine and Biology, 52 (2007), pp. 6663–6678. [113]

[360] L. T. NIKLASON, B. T. CHRISTIAN, L. E. NIKLASON, D. B. KOPANS, D. E. CASTLEBERRY, B. H. OPSAHL-ONG, C. E. LANDBERG, P. J. SLANETZ, A. A. GIARDINO, R. MOORE, D. ALBAGLI, M. C. DEJULE, P. F. FITZGERALD, D. F. FOBARE, B. W. GIAMBATTISTA, R. F. KWASNICK, J. LIU, S. J. LUBOWSKI, G. E. POSSIN, J. F. RICHOTTE, C. Y. WEI, AND R. F. WIRTH, *Digital tomosynthesis in breast imaging*, Radiology, 205 (1997), pp. 399–406. [113]

[361] A. NISSINEN, L. M. HEIKKINEN, AND J. P. KAIPIO, *The Bayesian approximation error approach for electrical impedance tomography—experimental results*, Measurement Science and Technology, 19 (2008), 015501. [181]

[362] A. NISSINEN, L. M. HEIKKINEN, V. KOLEHMAINEN, AND J. P. KAIPIO, *Compensation of errors due to discretization, domain truncation and unknown contact impedances in electrical impedance tomography*, Measurement Science and Technology, 20 (2009), 105504. [181]

[363] A. NISSINEN, V. KOLEHMAINEN, AND J. P. KAIPIO, *Compensation of errors due to incorrect model geometry in electrical impedance tomography*, Journal of Physics: Conference Series, 224 (2010), 012050. [181]

[364] ——, *Compensation of modelling errors due to unknown domain boundary in electrical impedance tomography*, IEEE Transaction on Medical Imaging, 30 (2011), pp. 231–242. [181]

[365] J. NOCEDAL AND S. J. WRIGHT, *Numerical Optimization*, second ed., Springer Series in Operations Research, Springer, New York, 2006. [100]

[366] R. G. NOVIKOV, *A multidimensional inverse spectral problem for the equation* $-\delta\psi + (v(x) - eu(x))\psi = 0$, Functional Analysis and Its Applications, 22 (1988), pp. 263–272. [150, 180, 200]

[367] S. P. NOVIKOV AND A. P. VESELOV, *Two-dimensional Schrödinger operator: Inverse scattering transform and evolutional equations*, Physica D, 18 (1986), pp. 267–273. [205]

[368] H. NUMATA, *Consideration of the parabolic radiography of the dental arch*, J. Shimazu Stud., 10 (1933), pp. 13–21. [125]

[369] S. OSHER, M. BURGER, D. GOLDFARB, J. XU, AND W. YIN, *An iterative regularization method for total variation-based image restoration*, Multiscale Modeling and Simulation, 4 (2005), pp. 460–489. [83]

[370] S. OSHER AND R. FEDKIW, *Level Set Methods and Dynamic Implicit Surfaces*, vol. 153 of Applied Mathematical Sciences, Springer, New York, 2003. [6, 83]

[371] S. OSHER, Y. MAO, B. DONG, AND W. YIN, *Fast linearized Bregman iteration for compressive sensing and sparse denoising*, Communications in Mathematical Sciences, 8 (2010), pp. 93–111. [83]

[372] S. OSHER AND J. A. SETHIAN, *Fronts propagating with curvature-dependent speed: Algorithms based on Hamilton-Jacobi formulations*, Journal of Computational Physics, 79 (1988), pp. 12–49. [83]

[373] Y. V. PAATERO, *Suunnittelemastani uudesta hampaiden röntgenkuvaustekniikasta*, Suomen Hammaslääkäriseuran Toimituksia, 86 (1946), p. 37. [125]

[374] ———, *A new tomographical method for radiographing curved outer surfaces*, Acta Radiologica, 32 (1949), pp. 177–184. [125]

[375] X. PAN, E. Y. SIDKY, AND M. VANNIER, *Why do commercial CT scanners still employ traditional, filtered back-projection for image reconstruction?*, Inverse Problems, 25 (2009), 123009. [113]

[376] J. B. PENDRY, D. SCHURIG, AND D. R. SMITH, *Controlling electromagnetic fields*, Science, 312 (2006), pp. 1780–1782. [183, 184]

[377] M. PERSSON, D. BONE, AND H. ELMQVIST, *Total variation norm for three-dimensional iterative reconstruction in limited view angle tomography*, Physics in Medicine and Biology, 46 (2001), pp. 853–866. [113]

[378] T. M. T. PHAM, M. YUILL, C. DAKIN, AND A. SCHIBLER, *Regional ventilation distribution in the first 6 months of life*, European Respiratory Journal, 37 (2011), pp. 919–924. [161]

[379] C. POPA AND R. ZDUNEK, *Kaczmarz extended algorithm for tomographic image reconstruction from limited-data*, Mathematics and Computers in Simulation, 65 (2004), pp. 579–598. [113]

[380] J. PÖSCHEL AND E. TRUBOWITZ, *Inverse Spectral Theory*, Academic Press, New York, 1987. [154]

[381] R. POTTHAST, *Point Sources and Multipoles in Inverse Scattering Theory*, CRC Press, Boca Raton, FL, 2001. [140, 287]

[382] S. PRÖSSDORF AND J. SARANEN, *A fully discrete approximation method for the exterior Neumann problem of the Helmholtz equation*, Zeitschrift für Analysis und ihre Anwendungen, 13 (1994), pp. 683–695. [288]

[383] E. T. QUINTO, *Singularities of the X-ray transform and limited data tomography in $\mathbb{R}^2$ and $\mathbb{R}^3$*, SIAM Journal on Mathematical Analysis, 24 (1993), pp. 1215–1225. [120, 121]

[384] ——, *Local algorithms in exterior tomography*, Journal of Computational and Applied Mathematics, 199 (2007), pp. 141–148. [113]

[385] ——, *Exterior and limited-angle tomography in non-destructive evaluation*, Inverse Problems, 14 (1998), pp. 339–353. [113]

[386] T. RAITIO, A. SUNI, J. YAMAGISHI, H. PULAKKA, J. NURMINEN, M. VAINIO, AND P. ALKU, *HMM-based speech synthesis utilizing glottal inverse filtering*, IEEE Transactions on Audio, Speech, and Language Processing, 19 (2011), pp. 153–165. [146]

[387] A. RAMIREZ, W. DAILY, D. J. LABREQUE, E. OWEN, AND D. CHESNUT, *Monitoring an underground steam injection process using electrical resistance tomography*, Water Resources Research, 29 (1993), pp. 73–88. [162]

[388] A. RAMIREZ, W. DAILY, D. J. LABREQUE, AND D. ROELANT, *Detection of leaks in underground storage tanks using electrical resistance method*, Journal of Environmental and Engineering Geophysics, 1 (1996), pp. 189–203. [162]

[389] R. RAMLAU AND W. RING, *A Mumford-Shah level-set approach for the inversion and segmentation of X-ray tomography data*, Journal of Computational Physics, 221 (2007), pp. 539–557. [113]

[390] A. G. RAMM, *Recovery of the potential from fixed energy scattering data*, Inverse Problems, 4 (1988), pp. 877–886. [150]

[391] ——, *Multidimensional Inverse Scattering Problems*, Longman-Wiley, New York, 1992. [150]

[392] M. RANTALA, S. VÄNSKÄ, S. JÄRVENPÄÄ, M. KALKE, M. LASSAS, J. MOBERG, AND S. SILTANEN, *Wavelet-based reconstruction for limited-angle X-ray tomography*, IEEE Transactions on Medical Imaging, 25 (2006), pp. 210–217. [113, 127]

[393] M. RAYDAN, *The Barzilai and Borwein gradient method for the large scale unconstrained minimization problem*, SIAM Journal on Optimization, 7 (1997), pp. 26–33. [92]

[394] M. REED AND B. SIMON, *Methods of Modern Mathematical Physics. Vol. I: Functional Analysis*, Academic Press, New York, 1980. [297]

[395] C. B. REID, M. M. BETCKE, D. CHANA, AND R. D. SPELLER, *The development of a pseudo-3D imaging system (tomosynthesis) for security screening of passenger baggage*, Nuclear Instruments and Methods in Physics Research Section A: Accelerators, Spectrometers, Detectors and Associated Equipment, 652 (2011), pp. 108–111. [113]

[396] M. RENARDY AND R. C. ROGERS, *An Introduction to Partial Differential Equations*, vol. 13 of Texts in Applied Mathematics, Springer, New York, 1993. [40, 41, 291, 294]

[397] M. C. ROBINI, A. LACHAL, AND I. E. MAGNIN, *A stochastic continuation approach to piecewise constant reconstruction*, IEEE Transactions on Image Processing, 16 (2007), pp. 2576–2589. [113]

[398] H. L. ROYDEN, *Real Analysis*, third ed., Macmillan, New York, 1988. [83]

[399] L. I. RUDIN, S. OSHER, AND E. FATEMI, *Nonlinear total variation based noise removal algorithms*, Physica D: Nonlinear Phenomena, 60 (1992), pp. 259–268. [83]

[400] W. RUDIN, *Functional Analysis*, Tata McGraw-Hill, New York, 1974. [297]

[401] A. R. RUUSKANEN, A. SEPPÄNEN, S. DUNCAN, E. SOMERSALO, AND J. P. KAIPIO, *Using process tomography as a sensor for optimal control*, Applied Numerical Mathematics, 56 (2006), pp. 37–54. [162]

[402] E. SAKSMAN, T. NYGRÉN, AND M. MARKKANEN, *Ionospheric structures invisible in satellite radiotomography*, Radio Science, 32 (1997), pp. 605–616. [114]

[403] J. C. SANTAMARINA AND D. FRATTA, *Discrete Signals and Inverse Problems*, John Wiley & Sons, New York, 2005. [6, 140]

[404] J. SARANEN AND G. VAINIKKO, *Periodic Integral and Pseudodifferential Equations with Numerical Approximation*, Springer, 2002. [210, 288]

[405] O. SCHERZER, M. GRASMAIR, H. GROSSAUER, M. HALTMEIER, AND F. LENZEN, *Variational Methods in Imaging*, vol. 167 of Applied Mathematical Sciences, Springer, New York, 2009. [6, 83, 140]

[406] D. SCHURIG, J. J. MOCK, B. J. JUSTICE, S. A. CUMMER, J. B. PENDRY, A. F. STARR, AND D. R. SMITH, *Metamaterial electromagnetic cloak at microwave frequencies*, Science, 314 (2006), pp. 977–980. [183]

[407] T. SCHUSTER, *The Method of Approximate Inverse: Theory and Applications*, vol. 1906 of Lecture Notes in Mathematics, Springer, Berlin, 2007. [6]

[408] A. SEPPÄNEN, M. VAUHKONEN, P. J. VAUHKONEN, E. SOMERSALO, AND J. P. KAIPIO, *State estimation with fluid dynamical evolution models in process tomography—an application to impedance tomography*, Inverse Problems, 17 (2001), pp. 467–483. [162]

[409] K. SHUNG, M. B. SMITH, AND B. M. W. TSUI, *Principles of Medical Imaging*, Academic Press, San Diego, CA, 1992. [151]

[410] E. Y. SIDKY, C. M. KAO, AND X. PAN, *Effect of the data constraint on few-view, fan-beam CT image reconstruction by TV minimization*, in Nuclear Science Symposium Conference Record, vol. 4, IEEE, 2006, pp. 2296–2298. [113]

[411] E. Y. SIDKY, C.-M. KAO, AND P. XIAOCHUAN, *Accurate image reconstruction from few-views and limited-angle data in divergent-beam CT*, Journal of X-Ray Science and Technology, 14 (2006), pp. 119–139. [113]

[412] E. Y. SIDKY AND X. PAN, *Image reconstruction in circular cone-beam computed tomography by constrained, total-variation minimization*, Physics in Medicine and Biology, 53 (2008), p. 4777. [113]

[413] S. SILTANEN, *Electrical Impedance Tomography and Faddeev Green's Functions*, vol. 121 of Annales Academiae Scientiarum Fennicae Mathematica Dissertationes, Dissertation, Helsinki University of Technology, Espoo, 1999. [205, 208, 213]

[414] S. SILTANEN, V. KOLEHMAINEN, S. JÄRVENPÄÄ, J. P. KAIPIO, P. KOISTINEN, M. LASSAS, J. PIRTTILÄ, AND E. SOMERSALO, *Statistical inversion for medical X-ray tomography with few radiographs: I. General theory*, Physics in Medicine and Biology, 48 (2003), pp. 1437–1463. [24, 113, 126]

[415] S. SILTANEN, J. MUELLER, AND D. ISAACSON, *An implementation of the reconstruction algorithm of A. Nachman for the 2-D inverse conductivity problem*, Inverse Problems, 16 (2000), pp. 681–699. [145, 185, 186, 200, 208, 225, 226, 237, 242, 249, 257]

[416] S. SILTANEN AND J. P. TAMMINEN, *Reconstructing conductivities with boundary corrected D-bar method*, Journal of Inverse and Ill-Posed Problems, (2012). [200]

[417] V. SINGH, L. MUKHERJEE, P. M. DINU, J. XU, AND K. R. HOFFMANN, *Limited view CT reconstruction and segmentation via constrained metric labeling*, Computer Vision and Image Understanding, 112 (2008), pp. 67–80. [113]

[418] H. J. SMIT, A. V. NOORDEGRAAF, J. T. MARCUS, A. BOONSTRA, P. M. VRIES, AND P. E. POSTMUS, *Determinants of pulmonary perfusion measured by electrical impedance tomography*, European Journal of Applied Physiology, 92 (2004), pp. 45–49. [161]

[419] K. T. SMITH AND F. KEINERT, *Mathematical foundations of computed tomography*, Applied Optics, 24 (1985), pp. 3950–3957. [113]

[420] K. T. SMITH, D. C. SOLMON, AND S. L. WAGNER, *Practical and mathematical aspects of the problem of reconstructing objects from radiographs*, Bulletin of the American Mathematical Society, 83 (1977), pp. 1227–1270. [114, 119]

[421] F. SMITHIES, *The eigenvalues and singular values of integral equations*, Proceedings of the London Mathematical Society, (1937), pp. 255–279. [45]

[422] E. SOMERSALO, M. CHENEY, AND D. ISAACSON, *Existence and uniqueness for electrode models for electric current computed tomography*, SIAM Journal on Applied Mathematics, 52 (1992), pp. 1023–1040. [172, 173, 185, 192]

[423] M. SONHDI, *A survey of the vocal tract inverse problem: Theory, computations and experiments*, in Medical Imaging: Image Processing. Proceedings of SPIE, F. Santosa, Y.-H. Pao, W. Symes, and C. Holland, eds., SIAM, 1984, pp. 1–19. [152]

[424] C. SOUSSEN AND J. IDIER, *Reconstruction of three-dimensional localized objects from limited angle X-ray projections: An approach based on sparsity and multigrid image representation*, Journal of Electronic Imaging, 17 (2008), 033011. [113]

[425] C. SOUSSEN AND A. MOHAMMAD-DJAFARI, *Polygonal and polyhedral contour reconstruction in computed tomography*, IEEE Transactions on Image Processing, 13 (2004), pp. 1507–1523. [113]

[426] S. STEFANESCO, C. SCHLUMBERGER, AND M. SCHLUMBERGER, *Sur la distribution électrique potentielle autour d'une prise de terre ponctuelle dans un terrain à couches horizontales, homogènes et isotropes*, Journal de Physique et Le Radium, 7 (1930), pp. 132–140. [162]

[427] W. G. STRANG AND G. J. FIX, *Analysis of the Finite Element Method*, Prentice-Hall Series in Automatic Computation, Prentice-Hall, Englewood Cliffs, NJ, 1973. [185]

[428] N. SUN, *Mathematical Modeling of Groundwater Pollution*, Springer, New York, 1996. [147]

[429] Z. SUN AND G. UHLMANN, *Anisotropic inverse problems in two dimensions*, Inverse Problems, 19 (2003), pp. 1001–1010. [184]

[430] J. SYLVESTER, *An anisotropic inverse boundary value problem*, Communications on Pure and Applied Mathematics, 43 (1990), pp. 201–232. [184]

[431] J. SYLVESTER AND G. UHLMANN, *A global uniqueness theorem for an inverse boundary value problem*, Annals of Mathematics, 125 (1987), pp. 153–169. [200]

[432] K. C. TAM AND V. PEREZ-MENDEZ, *Tomographical imaging with limited-angle input*, Journal of the Optical Society of America, 71 (1981), pp. 582–592. [120]

[433] J. TANG, B. E. NETT, AND G.-H. CHEN, *Performance comparison between total variation (TV)-based compressed sensing and statistical iterative reconstruction algorithms*, Physics in Medicine and Biology, 54 (2009), pp. 5781–5804. [113]

[434] A. TARANTOLA, *Inverse Problem Theory and Methods for Model Parameter Estimation*, SIAM, Philadelphia, PA, 2005. [6, 140]

[435] T. TARVAINEN, M. VAUHKONEN, V. KOLEHMAINEN, AND J. P. KAIPIO, *Finite element model for the coupled radiative transfer equation and diffusion approximation*, International Journal for Numerical Methods in Engineering, 65 (2006), pp. 383–405. [155]

[436] Z. TIAN, X. JIA, K. YUAN, T. PAN, AND S. B. JIANG, *Low-dose CT reconstruction via edge-preserving total variation regularization*, Physics in Medicine and Biology, 56 (2011), pp. 5949–5967. [113]

[437] R. TIBSHIRANI, *Regression shrinkage and selection via the lasso*, Journal of the Royal Statistical Society: Series B, 58 (1996), pp. 267–288. [83]

[438] T. TIDSWELL, A. GIBSON, R. H. BAYFORD, AND D. S. HOLDER, *Three-dimensional electrical impedance tomography of human brain activity*, NeuroImage, 13 (2001), pp. 283–294. [161]

[439] A. N. TIKHONOV AND V. Y. ARSENIN, *Solutions of Ill-Posed Problems*, V. H. Winston & Sons, Washington, D.C., John Wiley & Sons, New York, 1977. Translated from the Russian; preface by translation editor Fritz John, Scripta Series in Mathematics. [6, 140]

[440] A. TINGBERG, *X-ray tomosynthesis: A review of its use for breast and chest imaging*, Radiation Protection Dosimetry, 139 (2010), pp. 100–107. [123]

[441] H. TRIEBEL, *Function Spaces and Wavelets on Domains*, vol. 7 of Tracts in Mathematics, European Mathematical Society, 2008. [95, 99, 108]

[442] T. TSAI, *The Schrödinger operator in the plane*, Inverse Problems, 9 (1993), pp. 763–787. [205]

[443] ———, *The associated evolution equations of the Schrödinger operator in the plane*, Inverse Problems, 10 (1994), pp. 1419–1432. [205]

[444] V. TSAPAKI, J. E. ALDRICH, R. SHARMA, M. A. STANISZEWSKA, A. KRISANACHINDA, M. REHANI, A. HUFTON, C. TRIANTOPOULOU, P. N. MANIATIS, J. PAPAILIOU, AND M. PROKOP, *Dose reduction in CT while maintaining diagnostic confidence: Diagnostic reference levels at routine head, chest, and abdominal CT—IAEA-coordinated research project*, Radiology, 240 (2006), pp. 828–834. [113]

[445] G. UHLMANN, ED., *Inside Out: Inverse Problems and Applications*, vol. 47 of Mathematical Sciences Research Institute Publications, Cambridge University Press, Cambridge, UK, 2003. [140]

[446] G. UHLMANN, *Electrical impedance tomography and Calderón's problem*, Inverse Problems, 25 (2009), 123011. [199]

[447] G. UHLMANN AND J.-N. WANG, *Reconstructing discontinuities using complex geometrical optics solutions*, SIAM Journal on Applied Mathematics, 68 (2008), pp. 1026–1044. [145, 205, 277]

[448] G. VAINIKKO, *Multidimensional Weakly Singular Integral Equations*, vol. 1549 of Lecture Notes in Mathematics, Springer, New York, 1993. [43]

[449] ———, *Fast solvers of the Lippmann-Schwinger equation*, in Direct and Inverse Problems of Mathematical Physics (Newark, DE, 1997), vol. 5 of International Society for Analysis, Applications and Computation, Kluwer Academic, Dordrecht, 2000, pp. 423–440. [210, 211, 237]

[450] M. T. VAN GENUCHTEN AND J. C. PARKER, *Boundary conditions for displacement experiments through short laboratory soil columns*, Soil Sciences Society of America Journal, 48 (1984), pp. 703–708. [148]

[451] R. J. VANDERBEI, *Linear programming: Foundations and Extensions*, third ed., vol. 114 of International Series in Operations Research & Management Science, Springer, New York, 2008. [100]

[452] S. VÄNSKÄ, M. LASSAS, AND S. SILTANEN, *Statistical X-ray tomography using empirical Besov priors*, International Journal of Tomography & Statistics, 11 (2009), pp. 3–32. [113]

[453] P. J. VAUHKONEN, *Image Reconstruction in Three-Dimensional Electrical Impedance Tomography*, Ph.D. thesis, University of Kuopio, 2005. [180]

[454] P. J. VAUHKONEN, M. VAUHKONEN, T. SAVOLAINEN, AND J. P. KAIPIO, *Static three-dimensional electrical impedance tomography*, Annals of the New York Academy of Sciences, 873 (1999), pp. 472–481. [180]

[455] ———, *Three-dimensional electrical impedance tomography based on the complete electrode model*, IEEE Transactions on Biomedical Engineering, 46 (1999), pp. 1150–1160. [180, 192]

[456] I. N. VEKUA, *Generalized Analytic Functions*, Pergamon Press, London, 1962. [230]

[457] L. A. VESE AND S. J. OSHER, *Image denoising and decomposition with total variation minimization and oscillatory functions*, Journal of Mathematical Imaging and Vision, 20 (2004), pp. 7–18. Special issue on mathematics and image analysis. [83]

[458] V. G. VESELAGO, *The electrodynamics of substances with simultaneously negative values of ϵ and μ*, Soviet Physics Uspekhi, 10 (1968), p. 509. [184]

[459] A. P. VESELOV AND S. P. NOVIKOV, *Finite-zone, two-dimensional, potential Schrödinger operators, explicit formulas and evolution equations*, Soviet Mathematics Doklady, 30 (1984), pp. 558–591. [205]

[460] J. A. VICTORINO, J. B. BORGES, V. N. OKAMOTO, G. F. J. MATOS, M. R. TUCCI, M. P. R. CARAMEZ, H. TANAKA, F. S. SIPMANN, D. C. B. SANTOS, C. S. V. BARBAS, ET AL., *Imbalances in regional lung ventilation: A validation study on electrical impedance tomography*, American Journal of Respiratory and Critical Care Medicine, 169 (2004), pp. 791–800. [161]

[461] C. R. VOGEL, *Computational Methods for Inverse Problems*, no. 23 in Frontiers in Applied Mathematics, SIAM, Philadelphia, PA, 2002. [6, 83]

[462] ———, *Non-convergence of the L-curve regularization parameter selection method*, Inverse Problems, 12 (1996), pp. 535–547. [76]

[463] C. R. VOGEL AND M. E. OMAN, *Fast, robust total variation-based reconstruction of noisy, blurred images*, IEEE Transactions on Image Processing, 7 (1998), pp. 813–824. [83]

[464] A. V. NOORDEGRAAF, TH. J. C. FAES, A. JANSE, J. T. MARCUS, J. G. F. BRONZWAER, P. E. POSTMUS, AND P. M. DE VRIES, *Noninvasive assessment of right ventricular diastolic function by electrical impedance tomography*, Chest, 111 (1997), pp. 1222–1228. [161]

[465] A. V. Noordegraaf, Th. J. C. Faes, A. Janse, J. T. Marcus, R. M. Heethaar, P. E. Postmus, and P. M. de Vries, *Improvement of cardiac imaging in electrical impedance tomography by means of a new electrode configuration*, Physiological Measurement, 17 (1996), pp. 179–188. [161]

[466] A. V. Noordegraaf, P. W. Kunst, A. Janse, J. T. Marcus, P. E. Postmus, T. J. Faes, and P. M. de Vries, *Pulmonary perfusion measured by means of electrical impedance tomography*, Physiological Measurement, 19 (1998), pp. 263–273. [161]

[467] Y. Wang and S. Ma, *Projected Barzilai-Borwein method for large-scale nonnegative image restoration*, Inverse Problems in Science and Engineering, 15 (2007), pp. 559–583. [92]

[468] Y. Wang, A. G. Yagola, and C. Yang, eds., *Optimization and Regularization for Computational Inverse Problems and Applications*, Springer, New York, 2011. [6]

[469] Y. Wang, J. Yang, W. Yin, and Y. Zhang, *A new alternating minimization algorithm for total variation image reconstruction*, SIAM Journal on Imaging Sciences, 1 (2008), pp. 248–272. [83]

[470] A. J. Ward and J. B. Pendry, *Refraction and geometry in Maxwell's equations*, Journal of Modern Optics, 43 (1996), pp. 773–793. [183]

[471] R. L. Webber, R. A. Horton, T. E. Underhill, J. B. Ludlow, and D. A. Tyndall, *Comparison of film, direct digital, and tuned-aperture computed tomography images to identify the location of crestal defects around endosseous titanium implants*, Oral Surgery, Oral Medicine, Oral Pathology, Oral Radiology, and Endodontology, 81 (1996), pp. 480–490. [124]

[472] R. L. Webber, R. A. Horton, D. A. Tyndall, and J. B. Ludlow, *Tuned aperture computed tomography (TACT). Theory and application for three-dimensional dento-alveolar imaging*, Dentomaxillofacial Radiology, 26 (1997), pp. 53–62. [124]

[473] R. L. Webber and J. K. Messura, *An in vivo comparison of diagnostic information obtained from tuned-aperture computed tomography and conventional dental radiographic imaging modalities*, Oral Surgery, Oral Medicine, Oral Pathology, Oral Radiology, and Endodontology, 88 (1999), pp. 239–247. [124]

[474] R. L. Webber, H. R. Underhill, and R. I. Freimanis, *A controlled evaluation of tuned-aperture computed tomography applied to digital spot mammography*, Journal of Digital Imaging, 13 (2000), pp. 90–97. [124]

[475] R. Weder, *A rigorous analysis of high-order electromagnetic invisibility cloaks*, Journal of Physics A, 41 (2008), 065207. [184]

[476] A. Wexler, *Electrical impedance imaging in two and three dimensions*, Clinical Physics and Physiological Measurement, Supplement A, 9 (1988), pp. 29–33. [180]

[477] S. C. WHITE, *1992 assessment of radiation risk from dental radiography*, Dentomaxillofacial Radiology, 21 (1992), pp. 118–26. [130]

[478] S. J. WRIGHT, *Primal-Dual Interior-Point Methods*, SIAM, Philadelphia, PA, 1997. [100]

[479] Q. WU AND F. FRICKE, *Determination of blocking locations and cross-sectional area in a duct by eigenfrequency shifts*, Journal of the Acoustical Society, 87 (1990), pp. 67–75. [152]

[480] T. WU, R. H. MOORE, E. A. RAFFERTY, AND D. B. KOPANS, *A comparison of reconstruction algorithms for breast tomosynthesis*, Medical Physics, 31 (2004), pp. 2636–2647. [113]

[481] C. G. XIE, S. M. HUANG, B. S. HOYLE, AND M. S. BECK, *Tomographic imaging of industrial process equipment–development of system model and image reconstruction algorithm for capacitive tomography*, Sensors & Their Applications, 5 (1991), pp. 203–208. [162]

[482] C. G. XIE, A. PLASKOWSKI, AND M. S. BECK, *8-electrode capacitance system for two-component flow identification*, IEEE Proceedings A, 136 (1989), pp. 173–190. [162]

[483] T. YATES, J. C. HEBDEN, A. GIBSON, N. EVERDELL, S. R. ARRIDGE, AND M. DOUEK, *Optical tomography of the breast using a multi-channel time-resolved imager*, Physics in Medicine and Biology, 50 (2005), pp. 2503–2517. [155]

[484] W. YIN, S. OSHER, D. GOLDFARB, AND J. DARBON, *Bregman iterative algorithms for l_1-minimization with applications to compressed sensing*, SIAM Journal on Imaging Sciences, 1 (2008), pp. 143–168. [83]

[485] S. YOON, A. R. PINEDA, AND R. FAHRIG, *Simultaneous segmentation and reconstruction: A level set method approach for limited view computed tomography*, Medical Physics, 37 (2010), pp. 2329–2340. [113]

[486] T. YORK, *Status of electrical tomography in industrial applications*, Journal of Electronic Imaging, 10 (2001), pp. 608–619. [162]

[487] T. A. YORK, J. L. DAVIDSON, L. MAZURKIEWICH, R. MANN, AND B. D. GRIEVE, *Towards process tomography for monitoring pressure filtration*, IEEE Sensors Journal, 5 (2005), pp. 139–152. [162]

[488] K. YOSHIDA, *Functional Analysis*, Springer, New York, 1966. [297]

[489] D. F. YU AND J. A. FESSLER, *Edge-preserving tomographic reconstruction with nonlocal regularization*, IEEE Transactions on Medical Imaging, 21 (2002), pp. 159–173. [113]

[490] H. YU AND G. WANG, *Compressed sensing based interior tomography*, Physics in Medicine and Biology, 54 (2009), p. 2791. [113]

[491] ——, *A soft-threshold filtering approach for reconstruction from a limited number of projections*, Physics in Medicine and Biology, 55 (2010), pp. 3905–3916. [113]

[492] L. YU, X. LIU, S. LENG, J. M. KOFLER, J. C. RAMIREZ-GIRALDO, M. QU, J. CHRISTNER, J. G. FLETCHER, AND C. H. MCCOLLOUGH, *Radiation dose reduction in computed tomography: Techniques and future perspective*, Imaging in Medicine, 1 (2009), pp. 65–84. [113]

[493] X. ZHANG, M. BURGER, AND S. OSHER, *A unified primal-dual algorithm framework based on Bregman iteration*, Journal of Scientific Computing, 46 (2011), pp. 20–46. [83]

[494] Y. ZHANG, H.-P. CHAN, B. SAHINER, J. WEI, M. M. GOODSITT, L. M. HADJI-ISKI, J. GE, AND C. ZHOU, *A comparative study of limited-angle cone-beam reconstruction methods for breast tomosynthesis*, Medical Physics, 33 (2006), pp. 3781–3795. [113]

[495] T. ZHOU, *Reconstructing electromagnetic obstacles by the enclosure method*, Inverse Problems and Imaging, 4 (2010), pp. 547–569. [277]

[496] C. M. ZIEGLER, M. FRANETZKI, T. DENIG, J. MÜHLING, AND S. HASSFELD, *Digital tomosynthesis–experiences with a new imaging device for the dental field*, Clinical Oral Investigations, 7 (2003), pp. 41–45. [123]

[497] E. ZIMMERMANN, A. KEMNA, J. BERWIX, W. GLAAS, H. M. MÜNCH, AND J. A. HUISMAN, *A high-accuracy impedance spectrometer for measuring sediments with low polarizability*, Measurement Science and Technology, 19 (2008), 105603. [162]

[498] Y. ZOU AND Z. GUO, *A review of electrical impedance techniques for breast cancer detection*, Medical Engineering & Physics, 25 (2003), pp. 79–90. [161]

Index

acoustic scattering, 148
Alessandrini's identity, 227
anisotropic conductivity, 183

Beltrami equation, 216, 266
Beurling transform, 217–219
blurring, 8, 13
Born approximation, 151, 255, 260, 261, 283, 284
boundary data
    Dirichlet, 164, 166, 188, 229, 277
    Neumann, 164, 170, 189
    Robin, 173, 192
boundary integral equation, 228, 234, 257, 266, 268, 271, 274
breast cancer, 161, 181
Brownian motion, 14

calculus of $\bar{\partial}$ operator, 204
Calderón, 159, 180, 182, 201, 262, 264, 285
camera
    digital, 132
Cauchy integral formula
    generalized, 204
central slice theorem, 25
circulant matrix, 11
cloaking, 183, 184
complex-linear, 237, 240
condition number, 51
conductivity equation, 147, 164, 188, 200, 206, 216, 229
conjugate direction method, 79
conjugate gradient method, 79, 81
contact impedance, 172, 192
convection-diffusion equation, 134
convolution, 7, 51, 146, 207, 211, 212, 218, 230, 238, 240, 263
    matrix, 11, 58
    operator, 207
current pattern, 173
    adjacent, 174, 175
    Fourier, 173
    optimal, 160, 179
    pairwise, 174, 178, 254
    skip 3, 176
    trigonometric, 173–175, 254
    Walsh, 174
cutoff
    frequency, 223, 232, 274
    function, 238, 264, 265

data
    noise-free, 144
deconvolution
    blind, 146
degree of ill-posedness, 45, 53
density, 151–153
diffusion, 14
    coefficient, 16
diffusion equation, 156
dispersion coefficient, 147
drum
    determining the shape of, 151

ECG, 161
EEG, 161
eigenfunction
    Dirichlet-to-Neumann (DN) map, 186, 284
    Neumann-to-Dirichlet (ND) map, 178, 179
eigenvalue
    Dirichlet-to-Neumann (DN) map, 186, 284
    Neumann-to-Dirichlet (ND) map, 178, 179

problem, 152, 153
electrode model
　complete, 173, 191
　continuum, 163, 170, 185
　gap, 170
　shunt, 172
exceptional points, 207
existence
　lack of, 142

far field pattern, 150, 287
fast Fourier transform (FFT), 11, 212, 213, 219, 240, 241
flowchart
　idealized D-bar reconstruction method, 225
　regularized D-bar reconstruction method, 233
Fourier symbol, 208
Fréchet differential, 201
Fredholm equation
　first kind, 37, 41, 42, 44
　second kind, 207, 230, 289
Fredholm operator, 207
fundamental solution, 17, 205–207, 209, 287

Galerkin method, 38, 46
GMRES, 213, 217, 218, 240, 241, 271, 285, 289, 307
Green's function, 205
　Faddeev, 208, 210, 228, 231, 235, 259
　for the Laplacian, 207, 209, 210, 234

heat equation
　backward, 133

indicator function, 277, 282
inverse crime, 7, 12, 13, 19, 20, 32, 131, 145

kernel
　Hilbert–Schmidt, 41
　weakly singular, 42

L-curve method, 63, 72, 73, 89, 105, 106
Laplace

　equation, generalized, 163, 166
　transform, 133
least-squares, 32, 53, 68, 84, 134
linear operators, 40, 234
　compact, 40, 41, 43
　injective and bounded, 48
　unbounded, 40
linearized problem, 259
　EIT, 201, 285
Lippmann–Schwinger equation, 150, 151, 206, 210, 228, 230, 237, 238
low-pass filter, 223, 255, 265, 272
　nonlinear regularization, 245
low-pass transport matrix, 266, 272

minimization
　least-squares, 134
minimization problem, 66, 70, 81, 82, 84, 89, 90, 99, 100, 104, 119, 135, 144, 145
　nonlinear, 144, 223
　quadratic, 81, 99
minimum
　global, 145
　local, 145
mollifier, 262
Morozov discrepancy principle, 72

Neumann problem, 170, 189
nonuniqueness, 36, 50, 55, 140

optical tomography, 154
optimization, 144

panoramic (dental) imaging, 123, 124, 126
parabolic problem
　backward, 134, 146
parallelization, 145
perfusion, 254, 256, 257, 262
permittivity, 159, 163, 181, 184, 263
photograph, 7, 117, 119, 132
Picard condition, 45
point spread function, 8, 13, 132
　discrete, 10
power, 165
power method, 179
preconditioning, 82
pseudoinverse, 55

pulmonary imaging, 161, 181

quadratic
    form, 200, 201
    functional, 67, 91
    problem, 79, 81
    programming, 83, 86, 89, 99, 116, 119

reaction-diffusion equation, 146
real-linear, 237, 238, 240, 270
refractive index, 149
regularization, 47
    admissible parameter, 47, 143, 232, 233
    direct, 145
    iterative, 144
    nonlinear, 231
    nonlinear low-pass filter, 244
    parameter, 63, 223, 283
    strategy, 47, 140, 143–145, 232
    Tikhonov, xi, xii, 6, 63, 83, 84, 104, 106, 111, 113, 114, 116, 120, 121, 129, 144
regularization method
    nonlinear, 278
regularized inversion, 47, 53, 143, 231
relative error, 56

S-curve method, 89, 119
scattered field, 149
Schrödinger equation, 206, 224, 283
shunting, 172
signal-to-noise ratio, 177
singular
    functions, 44
    values, 44, 49, 51, 53, 58, 65
    vectors, 56, 58, 60, 61, 65
singular values
    generalized, 75
sinogram, 3, 31, 32, 111, 114
solute transport, 146, 147
Sommerfeld radiation condition, 149
sound speed, 148, 151
sound-hard object, 286
sound-hard, object, 149
sound-soft, object, 149
space
    Banach, 40
    Besov, xii, 95, 98, 99, 108, 114, 117, 120
    Hilbert, 40
    Sobolev, 164, 176, 187, 189, 206, 297, 299, 301, 302
sparsity, xi, 72, 88, 95, 99, 101, 108, 117, 120
stability, 36, 250
    conditional, 48, 141, 142
    criterion, 18
    final time, 134
    numerical, 55
stacked form, 68
Sturm–Liouville problem, 153

tangential derivative map, 270
transformation optics, 183
transport equation, 156
transport matrix, 268

ultrasound, 148

variational form, 192
vocal tract, 146, 152

wave equation, 148, 151
weak form
    of Dirichlet-to-Neumann (DN) map, 165
wedge product, 204
well-posed, 35, 40